AF595360

Advances in Applied Marine Sciences and Engineering

Advances in Applied Marine Sciences and Engineering

Editors

Enjin Zhao
Lin Mu
Hao Qin

MDPI • Basel • Beijing • Wuhan • Barcelona • Belgrade • Manchester • Tokyo • Cluj • Tianjin

Editors
Enjin Zhao
College of Marine Science
and Technology,
China University
of Geosciences,
Wuhan, China

Lin Mu
College of Life Sciences
and Oceanography,
Shenzhen University,
Shenzhen, China

Hao Qin
College of Marine Science
and Technology,
China University
of Geosciences,
Wuhan, China

Editorial Office
MDPI
St. Alban-Anlage 66
4052 Basel, Switzerland

This is a reprint of articles from the Special Issue published online in the open access journal *Applied Sciences* (ISSN 2076-3417) (available at: https://www.mdpi.com/journal/applsci/special_issues/Signal_Image_Processing).

For citation purposes, cite each article independently as indicated on the article page online and as indicated below:

LastName, A.A.; LastName, B.B.; LastName, C.C. Article Title. *Journal Name* **Year**, *Volume Number*, Page Range.

ISBN 978-3-0365-8208-5 (Hbk)
ISBN 978-3-0365-8209-2 (PDF)

Contents

About the Editors

Enjin Zhao

Enjin Zhao received the B.S. and Ph.D. degrees from the College of Engineering, Ocean University of China, Qingdao, China, in 2012 and 2017, respectively. He is currently with the College of Marine Science and Technology, China University of Geosciences, Wuhan, China. His research interests include marine environment, ocean engineering and marine technology.

Lin Mu

Lin Mu received the B.S., M.S., and Ph.D. degrees in physical oceanography from the Ocean University of China, Qingdao, China, in 2000, 2002, and 2007, respectively. He is currently a Professor of physical oceanography with Shenzhen University and the Shenzhen Research Institute, China University of Geosciences, Guangdong, China. His research interests include physical oceanography: prevention and mitigation of marine disasters, maritime search and rescue, and emergency response management of offshore oil spills.

Hao Qin

Hao Qin received his Ph.D. degree from Shanghai Jiao Tong University. He is currently working as an associate professor at the College of Marine Science and Technology, China University of Geosciences, Wuhan, China. He has authored or coauthored over 30 research papers. He has deep understanding and experience in the fields of ocean engineering, coastal engineering and marine disaster prevention.

Preface to "Advances in Applied Marine Sciences and Engineering"

In recent years, marine development has become the strategy of maritime countries around the world. Deeper understanding of the development characteristics and trends of marine science and engineering will help us better develop marine resources and protect the marine environment. As such, applied marine sciences and engineering have become an important research field in the society and attracted the worldwide attention. Basic theoretical research and applied research on marine hydrology, meteorology, physics and geology are focused. In addition, new viewpoints, new developments, new theories, new methods, new applications and new experiences in the investigation, research and management in the fields of marine environment and big data, ocean and coastal engineering, marine/coastal disaster and risk assessment, river dynamics, marine ecological environment, marine instruments and equipment, etc. are also included.

The aim of this book is to present new insights to the readers in the field of advances in applied marine sciences and engineering and to bring together the state of the art in marine technology. The main attention was paid to such key aspects as solid–fluid interaction, image processing, meteorological tide, sediment transport and freak wave. An overview of the different research studies in this area was carried out by researchers from different countries and with different professional backgrounds. The reader can find examples of the use of ecofriendly solutions and new materials in marine technology and descriptions of interesting case studies regarding advanced monitoring systems. The book also includes a description of the various mathematical tools used to develop new solutions in applied marine sciences and engineering. The effectiveness of their use to simulate the behavior of real engineering structures is examined. The presented research results indicate growing awareness of the importance of ocean engineering in the professional community, both for researchers and engineers. There is no turning back from the possibility/necessity of seeking new solutions in line with this spirit. This, in turn, creates new challenges that must be met during the design and construction of new marine engineering structures. Such challenges will be easier to address if the knowledge about the currently used solutions in this area is more complete and congregated. That is what this book intends to achieve.

Enjin Zhao, Lin Mu, and Hao Qin

Editors

Editorial

Special Issue on Advances in Applied Marine Sciences and Engineering

Enjin Zhao [1,2], Hao Qin [1,2,*] and Lin Mu [3]

[1] College of Marine Science and Technology, China University of Geosciences, Wuhan 430074, China; zhaoej@cug.edu.cn
[2] Shenzhen Research Institute, China University of Geosciences, Shenzhen 518057, China
[3] College of Life Sciences and Oceanography, Shenzhen University, Shenzhen 518060, China
[*] Correspondence: qinhao@cug.edu.cn

1. Introduction

Marine Science and technology are the basis for human beings to understand, develop and protect the ocean. Applied marine sciences and engineering have attracted worldwide attention, and include basic theoretical research and applied research on marine hydrology, meteorology, physics and geology [1,2]. The articles published in this Special Issue represent the latest research of various branches and interdisciplinary subjects of marine science, promote the sustainable development, utilization and protection of the ocean, and provide an academic channel of communication for researchers studying basic marine science and applied marine science. Moreover, this Special Issue reports academic papers and research reports on physical oceanography, marine oil spills, maritime search and rescue, marine environments and big data, marine geology, ocean and coastal engineering, marine/coastal disaster and risk assessment, the marine ecological environment, river dynamics, marine chemistry, marine water literature and remote sensing, as well as the latest research trends of related disciplines. A total of 15 papers have been collected around the pioneering applications and advances of marine technology in areas such as the marine environment, marine geology and ocean engineering. The papers presented in this collection demonstrate the innovative utilization and application of marine science data, and contribute to the expansion of knowledge and the refinement of methodologies in these domains.

2. Advances in Applied Marine Sciences and Engineering

This Special Issue presents a comprehensive exploration of the latest advancements in applied marine sciences and engineering, focusing on their application in ocean engineering, marine technology, the marine environment and marine geology. In the field of ocean engineering, Liu et al. [3] experimentally investigated the flow-induced rotation of two mechanically tandem-coupled cylinders. Cao and Wen [4] combined the large eddy simulated turbulence model and Lighthill's acoustic analogy theory and extracted the transient flow field data as the excitation conditions for acoustic calculations. Zhang et al. [5] examined the structural responses of ship bow structures under the influence of green-water loads caused by freak waves with an unusually high amplitude and concentrated energy. In the field of marine technology, Carrillo-Aguilar et al. [6] suggested algorithms to identify the dry fins of 37 shark species involved in the shark fin trade from 14 countries, which demonstrated a high sensitivity and specificity of image processing. The first methodology used a non-linear composite filter using Fourier transform for each species. The second methodology was a neural network. Wang et al. [7] proposed a novel swarm-intelligence-based approach for a sparse sensor array design to reduce PSLL under spacing constrains. Xiong et al. [8] proposed a new algorithm for a higher-order vector finite element method based on two new types of second-order edge elements to solve the electromagnetic-field diffusion problem in a 3D anisotropic medium. Based on the

Citation: Zhao, E.; Qin, H.; Mu, L. Special Issue on Advances in Applied Marine Sciences and Engineering. *Appl. Sci.* **2023**, *13*, 7875. https://doi.org/10.3390/app13137875

Received: 29 June 2023
Accepted: 1 July 2023
Published: 5 July 2023

data-driven principle, Li et al. [9] employed active disturbance rejection control with a tracking differentiator (LADRC-TD) algorithm for AUH depths and heading control. Kim and Oh [10] proposed a deep learning-based stacked autoencoder model for the first time to analyze ship operating modes. In the field of the marine environment, Soldani et al. [11] presented statistical analysis aimed at correctly estimating the hydrobarometric transfer factor in harbors, which can play a fundamental role in optimizing the management of port waters. The research results provided a reference for the noise-reduction design of underwater vehicles, thus improving their concealment in combat. Wu et al. [12] established a hydrodynamic model by FVCOM (Finite Volume Coastal Ocean Model) to research the influence of breakwater and sediment transport in Shantou Offshore Area. Fu et al. [13] adopted a unified, unstructured wave- and sediment-transport model and a topographic evolution model to carry out a high-spatial-temporal-resolution numerical simulation of the offshore dynamic environment under the disturbance of an artificial island. In the field of marine geology, Wang et al. [14] carried out a multi-proxy analysis that included micropaleontology and grain size to obtain information on the sedimentary environment, sea level change and climate change to study the sediment evolution of the Hanjiang River Delta during the late quaternary. Jiang et al. [15] analyzed the characteristics and causes of coastal water chemistry in Qionghai City, China. Fu et al. [16] studied the effects of shading on the growth and carbon storage of *Enhalus acoroides*. Yuan et al. [17] obtained and systematically analyzed soil and sediment samples from the surroundings of a lagoon in the Shamei Inland Sea, Qionghai City, Hainan Province and investigated the spatial distribution of nitrogen forms.

3. Future Applications

Although this Special Issue includes 15 papers on advances in applied marine sciences and engineering regarding ocean engineering, marine technology, the marine environment and marine geology, there is still a future expectation for more investigations on marine science and technology with artificial intelligence and machine learning algorithms. In addition to this, the ocean, a cradle of life and a bounty of resources is strategically important to connect the world and promote high-quality development. Ocean engineering is an essential subject and a key support for the sustainable exploration and utilization of the ocean resources. We hope that more and more people will pay attention to the development of ocean engineering and technology.

Funding: This research received no external funding.

Acknowledgments: We would like to express our sincere appreciation to all the authors and the editorial team of this Special Issue for their exceptional contributions to the publication of *Applied Sciences*. Through your dedicated efforts and extensive expertise, this Special Issue has become an invaluable resource, offering valuable insights and information for research in the field of remote sensing applications in archaeology, geography, and the earth sciences. Finally, I offer my special thanks to the section managing editor of this Special Issue from MDPI branch office.

Conflicts of Interest: The authors declare no conflict of interest.

References

1. Sun, Q.L.; Wang, Q.; Shi, F.Y.; Alves, T.; Gao, S.; Xie, X.N.; Wu, S.G.; Li, J.B. Runup of landslide-generated tsunamis controlled by paleogeography and sea-level change. *Commun. Earth Environ.* **2022**, *3*, 244. [CrossRef]
2. Zhao, E.J.; Dong, Y.K.; Tang, Y.Z.; Sun, J.K. Numerical investigation of hydrodynamic characteristics and local scour mechanism around submarine pipelines under joint effect of solitary waves and currents. *Ocean Eng.* **2021**, *222*, 108553. [CrossRef]
3. Liu, F.; Feng, W.; Yan, X.; Ran, D.; Shao, N.; Wang, X.; Yang, D. Experimental Investigation on Flow-Induced Rotation of Two Mechanically Tandem-Coupled Cylinders. *Appl. Sci.* **2022**, *12*, 10604. [CrossRef]
4. Cao, H.; Wen, L. High-Precision Numerical Research on Flow and Structure Noise of Underwater Vehicle. *Appl. Sci.* **2022**, *12*, 12723. [CrossRef]
5. Zhang, C.; Zhang, W.; Qin, H.; Han, Y.; Zhao, E.; Mu, L.; Zhang, H. Numerical Study on the Green-Water Loads and Structural Responses of Ship Bow Structures Caused by Freak Waves. *Appl. Sci.* **2023**, *13*, 6791. [CrossRef]

6. Carrillo-Aguilar, L.; Guerra-Rosas, E.; Álvarez-Borrego, J.; Echavarría-Heras, H.; Hernández-Muñóz, S. Identification of Shark Species Based on Their Dry Dorsal Fins through Image Processing. *Appl. Sci.* **2022**, *12*, 11646. [CrossRef]
7. Wang, L.; Zhao, H.; Wang, Q. Underwater Sparse Acoustic Sensor Array Design under Spacing Constraints Based on a Global Enhancement Whale Optimization Algorithm. *Appl. Sci.* **2022**, *12*, 11825. [CrossRef]
8. Xiong, B.; Lu, Y.; Chen, H.; Cheng, Z.; Liu, H.; Han, Y. A New 3D Marine Controlled-Source Electromagnetic Modeling Algorithm Based on Two New Types of Quadratic Edge Elements. *Appl. Sci.* **2023**, *13*, 2821. [CrossRef]
9. Li, H.; An, X.; Feng, R.; Chen, Y. Motion Control of Autonomous Underwater Helicopter Based on Linear Active Disturbance Rejection Control with Tracking Differentiator. *Appl. Sci.* **2023**, *13*, 3836. [CrossRef]
10. Kim, J.; Oh, J. A Pilot Study of Stacked Autoencoders for Ship Mode Classification. *Appl. Sci.* **2023**, *13*, 5491. [CrossRef]
11. Soldani, M.; Faggioni, O. Observing Meteorological Tides: Fifteen Years of Statistics in the Port of La Spezia (Italy). *Appl. Sci.* **2022**, *12*, 12202. [CrossRef]
12. Wu, Y.; Zhu, K.; Qin, H.; Wang, Y.; Sun, Z.; Jiang, R.; Wang, W.; Yi, J.; Wang, H.; Zhao, E. Numerical Investigation on the Influence of Breakwater and the Sediment Transport in Shantou Offshore Area. *Appl. Sci.* **2023**, *13*, 3011. [CrossRef]
13. Fu, G.; Li, J.; Yuan, K.; Song, Y.; Fu, M.; Wang, H.; Wan, X. Wind-Wave-Current Coupled Modeling of the Effect of Artificial Island on the Coastal Environment. *Appl. Sci.* **2023**, *13*, 7171. [CrossRef]
14. Wang, Y.; Zhou, L.; Wan, X.; Liu, X.; Wang, W.; Yi, J. Study on Sedimentary Evolution of the Hanjiang River Delta during the Late Quaternary. *Appl. Sci.* **2023**, *13*, 4579. [CrossRef]
15. Jiang, J.; Fu, G.; Feng, Y.; Gu, X.; Jiang, P.; Shen, C.; Chen, Z. Characteristics and Causes of Coastal Water Chemistry in Qionghai City, China. *Appl. Sci.* **2023**, *13*, 5579. [CrossRef]
16. Fu, M.; Song, Y.; Wang, Y.; Fu, G.; Zhang, X. Effects of Shading on the Growth and Carbon Storage of *Enhalus acoroides*. *Appl. Sci.* **2023**, *13*, 6035. [CrossRef]
17. Yuan, K.; Song, Y.; Fu, G.; Lin, B.; Fu, K.; Wang, Z. Spatial Distribution and Main Controlling Factors of Nitrogen in the Soils and Sediments of a Coastal Lagoon Area (Shameineihai, Hainan). *Appl. Sci.* **2023**, *13*, 7409. [CrossRef]

applied sciences

Article

Experimental Investigation on Flow-Induced Rotation of Two Mechanically Tandem-Coupled Cylinders

Fang Liu [1,2], Weipeng Feng [1,2], Xiang Yan [1,2,*], Danjie Ran [1,2], Nan Shao [3], Xiaoqun Wang [3] and Defeng Yang [1,2]

[1] State Key Laboratory of Hydraulic Engineering Simulation and Safety, Tianjin University, 135 Yaguan Road, Jinnan District, Tianjin 300350, China
[2] School of Civil Engineering, Tianjin University, 135 Yaguan Road, Jinnan District, Tianjin 300350, China
[3] School of Water Conservancy and Hydropower, Hebei University of Engineering, 178 Zhonghuanan Road, Handan District, Handan 056038, China
[*] Correspondence: xiangyan@tju.edu.cn; Tel.: +86-137-5229-0971

Abstract: The flow-induced rotational motion of tandem double cylinders has rarely been studied in existing papers. In order to further study the flow-induced rotation (FIR) of two mechanically tandem-coupled cylinders, an FIR device was designed in this paper, and the theoretical basis of this system was established. On this basis, a series of variable spacing ratio (L/D) tests were carried out in a recirculating water tunnel. The range of L/D was $4.0 \leq L/D \leq 9.0$. The main experimental conclusions can be summarized as follows: (1) When L/D = 4.0 and 4.5, the rotational response was similar to vortex-induced vibration (VIV), which is different from typical VIV, in that the rotational oscillation would appear to be a re-growth region when velocitycontinued to increase after the oscillation entered the lower branch of VIV. Additionally, the oscillation was at a low level and the maximum arc length ratio (A^*) was less than 0.55 in these two cases; (2) For L/D = 5.0, 5.5 and 6.0, the rotational responses all showed typical VIV. When the oscillation reached a high level, the maximum A^* was more than 0.85 for each case; (3) When L/D = 7.0, 8.0 and 9.0, the rotational responses still presented typical VIV. The oscillation was at a medium level, and the maximum A^* was between 0.53 and 0.72, but these three cases had a wider synchronization interval than the other cases, and the range showed an increasing trend with the growth of L/D.

Keywords: flow-induced rotation; mechanically tandem-coupled cylinders; spacing ratio

Citation: Liu, F.; Feng, W.; Yan, X.; Ran, D.; Shao, N.; Wang, X.; Yang, D. Experimental Investigation on Flow-Induced Rotation of Two Mechanically Tandem-Coupled Cylinders. *Appl. Sci.* **2022**, *12*, 10604. https://doi.org/10.3390/app122010604

Academic Editors: Lin Mu, Enjin Zhao and Hao Qin

Received: 5 September 2022
Accepted: 17 October 2022
Published: 20 October 2022

1. Introduction

Flow-induced motion (FIM) is a complicated non-linear vibration phenomenon caused by the interaction between moving fluid and elastic bodies [1], which is widely used in practical engineering, such as marine risers [2–5], bridge latches [6,7] and high-rise buildings [8]. The FIM phenomenon could cause great damage to engineering. Therefore, much research [9,10] has focused on how to predict and suppress FIM to ensure the safety of engineering structures in the early stages. However, with the development and utilization of ocean energy and in-depth research on FIM, more scholars are focusing on exploiting ocean energy by means of the FIM phenomenon. A series of new ocean current and tidal current power generation devices have been proposed [11–13]. Among them, a typical one is the Vortex Induced Vibration for Aquatic Clean Energy (VIVACE), a new low-speed current power generation device, proposed by Bernitsas from the University of Michigan [14–17]. At the same time, scholars have also carried out a large number of studies on oscillation characteristics and energy conversion of FIM for multi-oscillators, most of which have concentrated on circular cylinders [18–20].

Previously, scholars mainly studied the effect of spacing on the wake and oscillation characteristics of the upstream and downstream cylinders through experimental methods. King and Johns [21] found that when two identical cylinders were partially, or completely, immersed in fluid, with a spacing ratio of $0.25 < L/D < 6$, a complex interaction would occur between the wake vortex and the deflection of the cylinders. Zdravkovich [22–24] systematically studied the changes of wake vortex and flow force of tandem cylinders through the experimental method, and summarized the evolution process of the wake vortex of the upstream and downstream cylinders with the change of spacing. On this basis, Igarashi [25] analyzed the effect of spacing ratio and Reynolds number on vibration and wake of upstream and downstream cylinders, through experiments for a Reynolds number of $8.7 \times 10^3 \leq Re \leq 5.2 \times 10^4$ and the spacing ratio of $1.03 \leq L/D \leq 5.0$, and defined the wake path with different spacing ratios. Bokaian and Geoola [26,27] studied the vibration characteristics of upstream and downstream cylinders with different sizes through a series of physical experiments. It was shown that the oscillation wake of the upstream cylinder would be affected by the amplitude variation of the downstream cylinder. Yao et al. [28] studied the force of the upstream and downstream cylinders through the experimental method, under the condition that the Reynolds number range was $4.5 \times 10^4 < Re < 5.8 \times 10^5$ and the spacing ratio was $2.5 \leq L/D \leq 5.02$, and proved that the mutual disturbance between the two cylinders could disturb the lift and drag coefficients of the system. Mahir and Rockwell [29,30] conducted an experimental analysis on the wake of tandem double cylinders and their research showed that the wake of the double cylinder system was more complex and varied than that of a single cylinder under different phase conditions. Brika and Laneville [31] conducted an experimental study on the vibration of tandem cylinders, which showed that the resonance region of the tandem cylinders was larger than that of an isolated cylinder. Additionally, the resonance region decreased with increase of L/D.

Subsequently, with the continuous progress of CFD technology, and the diversification of test equipment, scholars have further analyzed the flow field and vibration characteristics of tandem double cylinders. Meneghini et al. [32] observed the flow field characteristics of two cylinders in series and calculated the vorticity lines and the force on the cylinders. Lin et al. [33] used particle image velocimetry (PIV) to measure the flow field of two tandem cylinders. Based on previous research results, Zhou and Yiu [34] concluded that there were three wake interference regions: extended-body regime, reattachment regime and co-shedding regime. Prasanth and Mittal [35,36] simulated the two cylinders in tandem arrangement at low Reynolds number and large spacing ratio ($L/D = 5.5$). The numerical results showed that the locking interval of the downstream cylinder was significantly larger than that of an isolated cylinder, due to the influence of the upstream cylinder, while the downstream cylinder had almost no influence on the upstream cylinder. Assi et al. [37] experimentally studied the vibration of tandem double cylinders with the upstream cylinder fixed, through a flume test. The results showed that wake galloping could be observed in the downstream cylinder, and the amplitude increased by nearly 50%, compared with that of a single cylinder, when the spacing ratio was $3.0 < L/D < 5.6$. Moreover, Assi et al. [38] proposed the concept of wake-induced vibration (WIV) in their further research. In order to develop a clearer understanding of the WIV excitation mechanism, the response of the downstream cylinder in a steady shear flow was compared with typical VIV responses of a single cylinder. The test results showed that the WIV phenomenon of the downstream cylinder was excited by the interaction between the unstable vortices from the wake of the upstream cylinder and the downstream cylinder. Bao et al. [39] performed a numerical simulation on the VIV of the tandem double cylinders with two degrees of freedom. The results showed that the double resonance phenomenon occurred in all frequency ratios, and the vibration of the upstream cylinder was similar to that of an isolated cylinder under various working conditions, while the dynamic response of the downstream cylinder was more sensitive to the incoming direction than to the lateral direction, due to the change of the frequency ratio.

In later papers, scholars paid more attention to analyzing the internal mechanism of vibration characteristics. Ji et al. [40] conducted a comprehensive analysis on the mechanism of the large-amplitude response of the downstream cylinder in tandem double cylinders at $Re = 100$, and found that the shedding vortices of the upstream cylinder induced a low pressure area, which, in turn, stimulated the downstream cylinder to generate a large amplitude response. Chen et al. [41] numerically investigated the three tandem cylinders in the range of L/D = 1.2 to 5.0, and analyzed the influence of three key factors on the galloping phenomenon, including balance position offset, vortex shedding and timing between cylinders. Qin et al. [42] experimentally studied tandem double cylinders. It was found that there were different vibration mechanisms, such as locking, VIV and galloping, under various frequency ratio conditions. Furthermore, differing from previous research on the FIM of cylinders, Arionfard and Nishi [43,44] identified four mechanisms of vibration with two mechanically coupled circular cylinders that were free to rotate around a pivot, and they proposed the concept of gap switching-induced vibration (GSIV). Zhu et al. [45] numerically investigated the flow-induced rotation of a circular cylinder with a detached splitter plate at Re = 100, and found the occurrence of bifurcation at $0 \leq G/D \leq 0.5$, which disappeared at $0.55 \leq G/D \leq 2.0$. Chen and Wu [46] undertook extensive simulations to investigate the VIV of a piggyback circular cylinder system at Re = 100. It was found that the motion trajectories would take the form of a classical pattern of 8 when the system was in tandem arrangement. Furthermore, many scholars also conducted in-depth research [47–49] on the FIM power generation of tandem double cylinders. The results showed that the power generation efficiency of tandem double cylinders was much higher than that of an isolated cylinder, and revealed the vibration enhancement conditions of tandem double cylinders.

In conclusion, due to the influence of upstream cylinder wake vortices on the downstream cylinder, the vibration intensity and power generation potential of tandem double cylinders are both higher than that of a single cylinder, and research on multi-cylinder tandem power generation is bound to become an important development direction of FIM. However, power generation from freely vibrating multi-cylinders needs numerous power take-off (PTO) devices. If the tandem double cylinders do not vibrate freely in the cross-flow direction, but are free to vibrate rotationally and cooperatively, the vibration energy of the two cylinders can be collected at the same time. In the mode of double cylinder rotational oscillation, the single-unit capacity of the supporting power generation equipment could be increased and the amount of equipment could be greatly reduced, so this power generation method would be more economica,l both in terms of dispatching and costs. Nevertheless, in the case of rotational vibration of two tandem-coupled cylinders, whether the oscillation level is increased or decreased, compared with free vibration, is rarely mentioned in the existing research. Therefore, it is necessary to thoroughly investigate the flow-induced rotation (FIR) of two tandem-coupled cylinders and to summarize the oscillation characteristics. This paper systematically explores the FIR of two mechanically tandem-coupled cylinders at different spacing ratios, from both theoretical and experimental aspects. The aim was to further guide and promote the design and practical engineering application of marine power generation FIM devices. For the framework of this study, the remaining parts proceed as follows: The principle of the FIR system is introduced and the mathematical expression is deduced in Section 2, followed by a detailed description of the method and content of the experiment in Section 3. The rotational oscillation characteristics of the FIR system at various spacing ratios are presented and analyzed in Section 4. Finally, the main conclusions are summarized in Section 5.

2. Theoretical Basis

2.1. Principle of Flow-Induced Rotation

Figure 1 is a simplified schematic diagram of the flow-induced rotation (FIR) system. As shown in Figure 1, the double cylinders are mechanically coupled and arranged in series. When the water flows through the FIR system, the fluid force, F_{fluid}, forces the tandem-coupled double cylinders to rotate around its geometric central axis, O. The motion of the cylinders drives the upper rotary shaft with the pivot point, P, to rotate synchronously through the transmission device, composed of the transmission belt and the synchronous wheel. The upper rotary shaft, P, is connected to a hollow vertical rod through a three-way shaft sleeve to ensure that they also rotate synchronously. Two tension springs are connected to this hollow vertical rod by using a spring connector to generate the restoring force. F_k. Note that the spring connector can slide freely on the vertical rod.

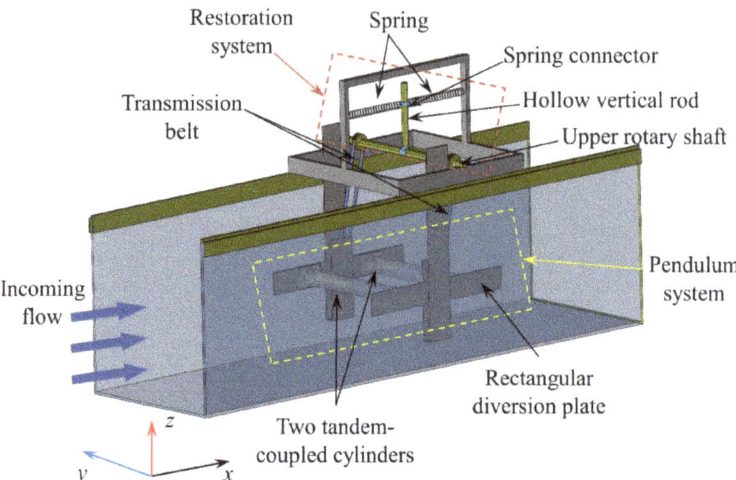

Figure 1. Schematic diagram of the flow-induced rotation.

In order to facilitate the establishment of a mathematical model of the FIR system, the original system was divided into three subsystems: pendulum system, transmission system and restoration system.

2.2. Dynamic Model of Flow-Induced Rotation (FIR) System

2.2.1. Pendulum System

The pendulum system is the rotation part of the FIR system, including double circular cylinders in tandem arrangement and two rectangular diversion plates, as shown in Figure 1. The function of this diversion plate is to mechanically couple the cylinders and facilitate changing the spacing between the two cylinders. In the process of motion, the force of the pendulum system is shown in Figure 2, including the fluid force, $F_{fluid,up}$, on the upstream cylinder, the fluid force, $F_{fluid,\,down}$, on the downstream cylinder, the total torque, $2M_1$, provided by the left and right transmission belts, the torques generated by viscous traction acting on each cylinder surface, $M_{VT,up}$ and $M_{VT,down}$, and the mechanical damping force (this force is not shown in Figure 2). According to this, the kinetic equation of the pendulum system is as follows:

$$J_1\ddot{\theta}_1 + c_1\dot{\theta}_1 = F_{fluid,up} \cdot \frac{L}{2} + F_{fluid,down} \cdot \frac{L}{2} - 2M_1 + M_{VT,up} + M_{VT,down} \quad (1)$$

where $\ddot{\theta}_1$, $\dot{\theta}_1$ and θ_1 are the angular acceleration, angular velocity and angular displacement of the pendulum system, respectively, and c_1 is the damping coefficient of the pendulum system. J_1 is the moment of inertia of the pendulum system to the rotary shaft O. L is the center distance of the upstream and downstream cylinders. M_1 is the torque provided by a single transmission belt.

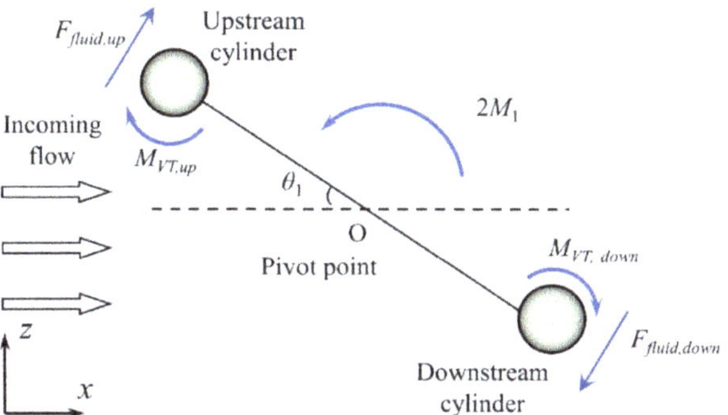

Figure 2. Force analysis of pendulum system.

According to previous research, the F_{fluid} acting on the structure includes two parts: the inertial force, $F_{inviscid}$, and the viscous force, $F_{viscous}$ [50]. Obviously, the $F_{inviscid}$ on the upstream and downstream cylinders is the same and the torques of $F_{inviscid}$ to shaft O is calculated as follows:

$$M_o(F_{inviscid}) = -J_a\ddot{\theta}_1 \tag{2}$$

where J_a is the additional moment of inertia of the pendulum system to the rotary shaft O. However, the torques of $F_{viscous}$ on the upstream and downstream cylinders to shaft O are calculated as follows:

$$M_O(F_{viscous,up}) = F_{viscous,up} \cdot \frac{L}{2} = \frac{1}{2}c_{tangent,up}\rho(Ucos\theta_1)^2 DH \cdot \frac{L}{2} = \frac{1}{4}c_{tangent,up}\rho U^2 DHLcos^2\theta_1 \tag{3}$$

$$M_O(F_{viscous,down}) = F_{viscous,down} \cdot \frac{L}{2} = \frac{1}{2}c_{tangent,down}\rho(Ucos\theta_1)^2 DH \cdot \frac{L}{2} = \frac{1}{4}c_{tangent,down}\rho U^2 DHLcos^2\theta_1 \tag{4}$$

where $c_{tangent,up}$ and $c_{tangent,down}$ is the instantaneous lift coefficient of the upstream and downstream cylinders, respectively, and P is the fluid density. U is the incoming flow velocity. D is the diameter of the circular cylinder. H is the length of the cylinder. Additionally, it is clear that $M_{VT,up}$ and $M_{VT,down}$, acting on the upstream and downstream cylinders, are equal, namely, $M_{VT,up} = M_{VT,down} = M_{VT}$. The shear stress acting on the surface of each cylinder is named τ_0, and the expression of τ_0 is:

$$\tau_0 = \frac{1}{2}C_f\rho\left(\frac{D}{2}\dot{\theta}_1\right)^2 = \frac{1}{8}C_f\rho D^2\left(\dot{\theta}_1\right)^2 \tag{5}$$

where C_f is the friction resistance coefficient of the boundary layer on the cylinder surface, $C_f = f(Re)$. By calculating the definite integral of the cylinder surface, the expression of M_{VT} can be obtained as:

$$M_{VT} = \int_0^{\pi D} \tau_0 \frac{D}{2} H dr \tag{6}$$

By substituting Equation (5) into Equation (6), then the expression of M_{VT} can be further expressed as:

$$M_{VT} = \frac{\pi}{16}C_f\rho HD^4\left(\dot{\theta}_1\right)^2 \tag{7}$$

Therefore, Equation (1) can be further expressed as:

$$(J_1 + 2J_a)\ddot{\theta}_1 + c_1\dot{\theta}_1 = \frac{1}{4}\left(c_{tangent,up} + c_{tangent,down}\right)\rho U^2 DHLcos^2\theta_1 - 2M_1 + \frac{\pi}{8}C_f\rho HD^4\left(\dot{\theta}_1\right)^2 \tag{8}$$

2.2.2. Restoration System

The restoration system is the part that provides restoring force. The restoring mechanism of the apparatus consists of two tension springs, the spring connector [43,44], the hollow vertical rod and the upper rotary shaft, P. The force of the restoration system after rotation through the angle θ_2 includes the total restoring force, $2F_k$, of the spring, the total torque, $2M_2$, provided by the left and right transmission belts and the mechanical damping force, as shown in Figure 3. According to this, the kinetic equation of restoration system is as follows:

$$J_2\ddot{\theta}_2 + c_2\dot{\theta}_2 = 2M_2 - 2F_k \cos\theta_2 \times \frac{B}{\cos\theta_2} \tag{9}$$

where $\ddot{\theta}_2$, $\dot{\theta}_2$ and θ_2 are the angular acceleration, angular velocity and angular displacement of the restoration system, respectively. J_2 is the moment of inertia of restoration system with respect to upper shaft, P, and c_2 is the mechanical damping coefficient of the restoration system. B is the distance between the upper rotating shaft and the spring connector when the restoration system is in the initial position (θ_2 = 0).

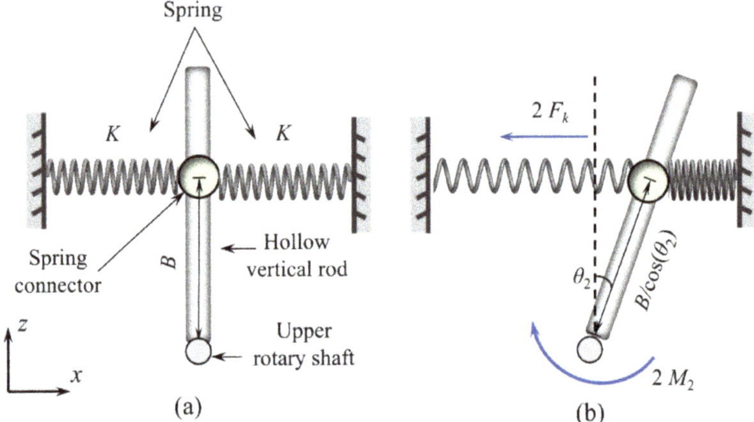

Figure 3. Force analysis of restoration system: (**a**) Equilibrium position; (**b**) Force after rotation angle θ_2.

The restoring force, F_k, produced by a single spring is expressed as:

$$F_k = K\Delta = KB\tan\theta_2 \tag{10}$$

where K is the stiffness of a single spring. Therefore, the final dynamic equation of the restoration system is:

$$J_2\ddot{\theta}_2 + c_2\dot{\theta}_2 + 2KB^2\tan\theta_2 = 2M_2 \tag{11}$$

2.2.3. Transmission System

The transmission system is the transfer motion part of the FIR system, which includes the left and right two transmission belts and the matching driving wheels. In the process of motion, the force of a single transmission belt and the matching driving wheels are shown in Figure 4a, including the torque, M'_1, provided by the pendulum system and the torque, M'_2, provided by the restoration system.

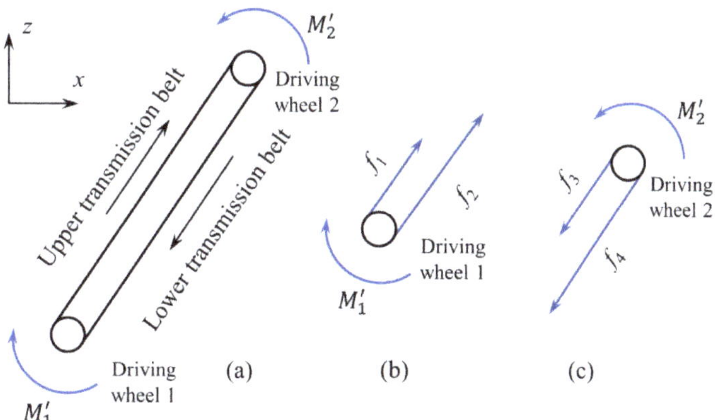

Figure 4. Force analysis of transmission system: (**a**) The force of a single transmission belt and the matching driving wheels; (**b**) The force of the driving wheel 1; (**c**) The force of the driving wheel 2.

The force of wheel 1 is shown in Figure 4b, including the torque, M'_1, provided by pendulum system, the tension, f_1, provided by the upper transmission belt and the tension, f_2, provided by the lower transmission belt. Then, the dynamic equation of wheel 1 is as follows:

$$J_{wheel,\,1}\ddot{\theta}_1 = M'_1 - (f_2 - f_1)\,R_1 \tag{12}$$

where $J_{wheel,1}$ is the moment of inertia of the driving wheel 1 around its own central axis. R_1 is the radius of wheel 1.

In the same way, the force of wheel 2 is shown in Figure 4c, including the torque, M'_2, provided by restoration system, the tension, f_3, provided by the upper transmission belt and the tension, f_4, provided by the lower transmission belt. Then, the dynamic equation of wheel 2 is as follows:

$$J_{wheel,2}\ddot{\theta}_2 = M'_2 - (f_4 - f_3)\,R_2 \tag{13}$$

where $J_{wheel,2}$ is the moment of inertia of the driving wheel 2 around its own central axis. R_2 is the radius of wheel 2. Additionally, the driving wheels 1 and 2 have consistent specifications ($R_1 = R_2$) and small mass, so their moments of inertia can be ignored. Clearly, the equations $f_1 = f_3$ and $f_2 = f_4$ are true, hence, it is concluded that $M'_1 = M'_2$ is established. In other words, the following equation is also true:

$$M_1 = M_2 \tag{14}$$

2.2.4. Total Kinetic Equation of FIR System

Since the geometric dimensions of the driving wheels 1 and 2 are exactly the same, the relation of speed ratio can be obtained as:

$$\frac{\theta_1}{\theta_2} = \frac{\dot{\theta}_1}{\dot{\theta}_2} = \frac{\ddot{\theta}_1}{\ddot{\theta}_2} = 1 \tag{15}$$

Both θ_1 and θ_2 are denoted by θ. Combined with Equation (14), the kinetic equation of the pendulum system is further defined as:

$$(J_1 + 2J_a)\ddot{\theta} + c_1\dot{\theta} = \frac{1}{4}\left(c_{tangent,up} + c_{tangent,down}\right)\rho U^2 DHLcos^2\theta - 2M_2 + \frac{\pi}{8}C_f\rho HD^4\left(\dot{\theta}\right)^2 \tag{16}$$

Based on the equation of the restoration system, the kinetic equation of the FIR system can be simplified as:

$$(J_1 + 2J_a + J_2)\ddot{\theta} + (c_1 + c_2)\dot{\theta} + 2KB^2 \tan\theta = \frac{1}{4}\left(c_{\text{tangent},up} + c_{\text{tangent},down}\right)\rho U^2 DHL\cos^2\theta + \frac{\pi}{8}C_f \rho HD^4 \left(\dot{\theta}\right)^2 \quad (17)$$

Definition $c_{total} = c_1 + c_2$ and $J_{osc} = J_1 + J_2$, hence, the kinetic equation of the FIR system is further written as:

$$(J_{osc} + 2J_a)\ddot{\theta} + c_{total}\dot{\theta} + 2KB^2 \tan\theta = \frac{1}{4}\left(c_{\text{tangent},up} + c_{\text{tangent},down}\right)\rho U^2 DHL\cos^2\theta + \frac{\pi}{8}C_f \rho HD^4 \left(\dot{\theta}\right)^2 \quad (18)$$

Considering the large Reynolds number ($2.0 \times 10^4 \leq Re \leq 9.7 \times 10^4$) and small angular displacement ($\theta \leq 30°$) in this experiment, it can be preliminarily judged that the order of magnitude of the term containing $\left(\dot{\theta}\right)^2$ in Equation (18) is small, so it is not, temporarily, considered in this paper, and the more explicit impact of this term on the FIR system is further analyzed in subsequent research. Therefore, Equation (18) can be further simplified as:

$$(J_{osc} + 2J_a)\ddot{\theta} + c_{total}\dot{\theta} + 2KB^2 \tan\theta = \frac{1}{4}\left(c_{\text{tangent},up} + c_{\text{tangent},down}\right)\rho U^2 DHL\cos^2\theta \quad (19)$$

The dimensionless parameter damping ratio (ζ_{total}) and the natural frequency ($f_{n,air}$) of the system in air are introduced, and the following equations are established:

$$\zeta_{total} = \frac{c_{total}}{2\sqrt{2KB^2 J_{osc}}} \quad (20)$$

$$f_{n,air} = \frac{1}{2\pi}\sqrt{\frac{2KB^2}{J_{osc}}} \quad (21)$$

Substituting Equations (20) and (21) into Equation (19) gives the final kinetic equation of the FIR system:

$$(J_{osc} + 2J_a)\ddot{\theta} + 2\zeta_{total}\sqrt{2KB^2 J_{osc}}\dot{\theta} + 4\pi^2 f_{n,air}^2 J_{osc} \tan\theta = \frac{1}{4}\left(c_{\text{tangent},up} + c_{\text{tangent},down}\right)\rho U^2 DHL\cos^2\theta \quad (22)$$

3. Experimental Methods

3.1. Physical Model

3.1.1. Recirculating Water Tunnel

All experiments were carried out in the recirculating water tunnel at Tianjin University. The test area of the tunnel had a width of 1.0 m, and a water depth of 1.34 m. The water in the tunnel was driven by a variable frequency power pump with a velocity range of 0.0–1.8 m/s, as shown in Figure 5. The differences in the flow velocity and turbulence in the vertical direction were small, which illustrated that the incoming flow in the test area of the circular cylinder was uniform [51]. Additionally, a rectifying grid was set in front of the test area to allow water flow to enter the test area smoothly. The test velocities were set at $0.229\ \text{m/s} \leq U \leq 1.1\ \text{m/s}$, with the corresponding Reynolds numbers of $20{,}105 \leq Re \leq 96{,}575$. It should be noted that Re is defined as $Re = UD/\nu$, where U is the characteristic velocity of the flow field, which is the test velocity in this paper; D is the characteristic length of the flow field, which is the diameter of the circular cylinder in this paper; ν is the kinematic viscosity.

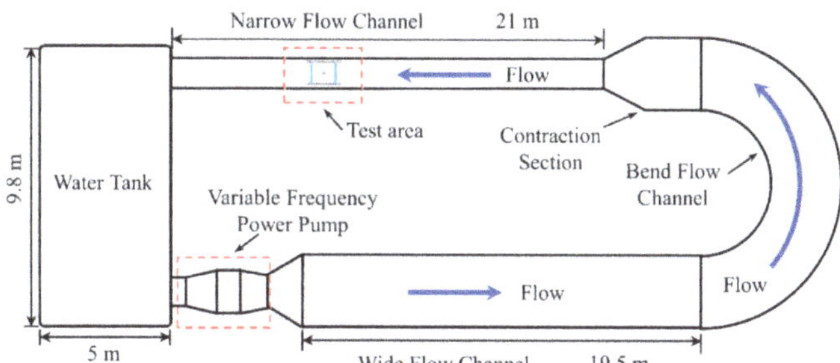

Figure 5. Plan view schematic of the recirculating water tunnel.

The relevant parameters of the FIR experiment are shown in Table 1.

Table 1. Test parameters.

Parameters	Symbol	Value
Full incoming velocity range	Y [m/s]	$0.0 \leq Y \leq 1.8$
Water depth	h [m]	1.34
Reynolds numbers	Re (UD/ν)	$20{,}105 \leq Re \leq 96{,}575$
Test velocity range	U [m/s]	$0.229 \leq U \leq 1.1$
Reduced velocity	U_r $(U/f_n\,D)$	$2.314 \leq U_r \leq 14.513$
Kinematic viscosity	ν [m^2/s]	1.139×10^{-6}
Water density	ρ [kg/m^3]	1000
Total stiffness of system	K [N/m]	1960
Distance between the upper shaft P and the spring connector when $\theta_2 = 0$	B [m]	0.21
Spacing ratio	L/D	4.0, 4.5, 5.0, 5.5, 6.0, 7.0, 8.0, 9.0

3.1.2. Flow-Induced Rotation Device

The arrangement and details of the experimental physical model are shown in Figure 6. Additionally, the ratio of moment of inertia J^* and other necessary parameters under each case are shown in Table 2. Here, the ratio of the moment of inertia J^* is:

$$J^* = J_{osc}/J_d \tag{23}$$

where J_{osc} is the total moment of inertia of the oscillation system, and J_d is the total moment of inertia of the displaced mass of two tandem-coupled cylinders with respect to axis O.

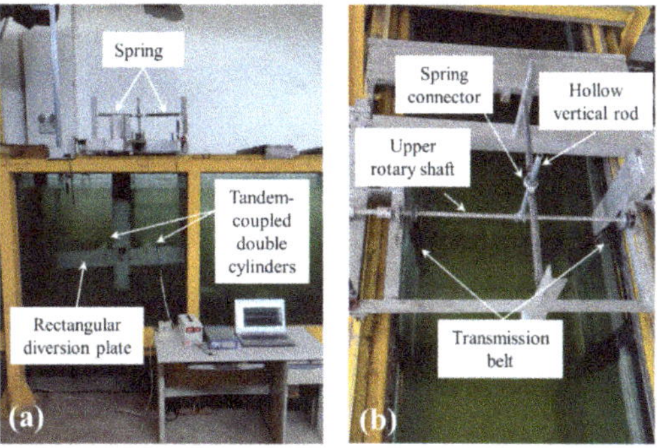

Figure 6. Physical models of FIR system: (**a**) Front view; (**b**) Vertical view.

Table 2. The parametric details of the experiments in this study.

L/D	J_{osc} (kg·m^2)	J_d (kg·m^2)	J^*
4.0	1.443	0.583	2.475
4.5	1.602	0.733	2.186
5.0	1.779	0.901	1.974
5.5	1.974	1.087	1.816
6.0	2.188	1.290	1.696
7.0	2.673	1.750	1.527
8.0	3.231	2.280	1.417
9.0	3.865	2.881	1.342

3.1.3. Test Circular Cylinder

The specifications of the upstream and downstream test circular cylinders were consistent. The parameters of the cylindrical oscillator are listed in Table 3. It was made of polymethy1 methacrylate, and the interior was hollow to facilitate counterweight. Both ends of the oscillator were set with end plates of equal thickness to reduce the influence of boundary conditions in the test, as shown in Figure 7.

Table 3. The parameters of the circular cylinder.

Item	**Symbol**	**Upstream Cylinder**	**Downstream Cylinder**
Diameter	D [m]	0.1	0.1
Height	H [m]	0.9	0.9
Thickness	d [m]	0.01	0.01

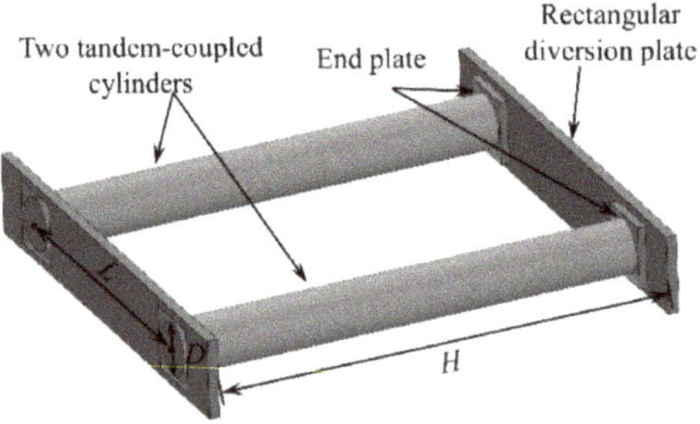

Figure 7. Test circular cylinders.

3.2. Data Acquisition and Processing System

In the experiment, the rotational angular displacement θ was measured by a magnetic-sensitive angular displacement sensor. The measurement method involved connecting the rotating shaft of the sensor with the upper rotary shaft, P, by using shaft-coupling to ensure that they rotated synchronously, as shown in Figure 8b. The sensor could output a voltage signal proportional to the rotational angle, and the instantaneous voltage signal was captured by the data acquisition system, shown in Figure 8c. Finally, MATLAB software was used to process the recorded data. The sampling frequency was set to 50 Hz. Additionally, the oscillation frequency (f_{osc}) of the FIR system was calculated by the Fast Fourier Transform (FFT) method under continuous oscillation for 60 s [18].

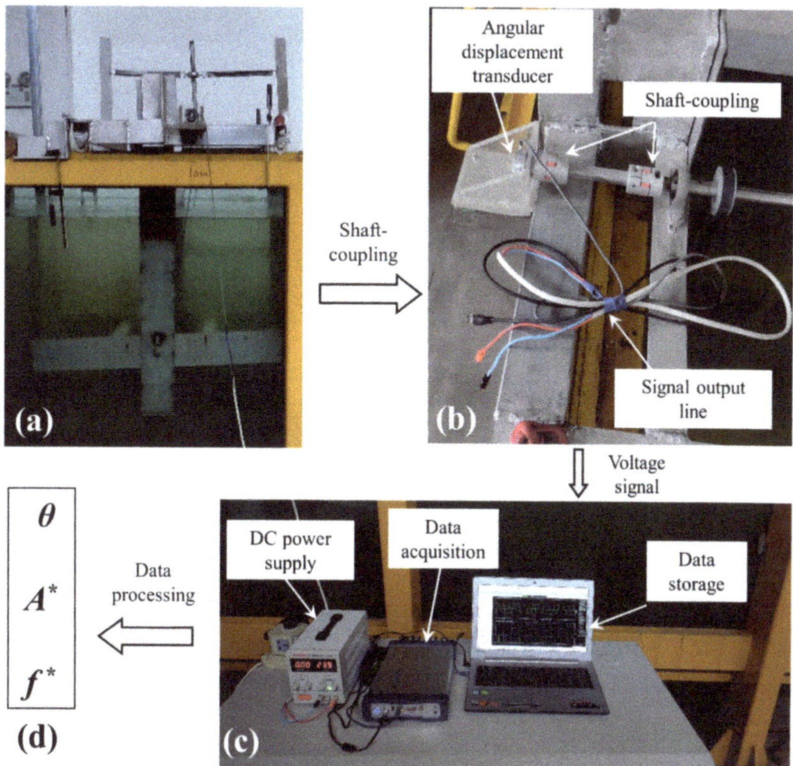

Figure 8. Investigation process of flow-induced rotation: (**a**) The FIR system; (**b**) Angular displacement transducer; (**c**) Data acquisition and processing system; (**d**) Experimental data.

3.3. Free Decay Tests

Free decay tests, with different spacing ratios, were conducted to obtain the damping ratio (ζ_{total}) and natural frequency ($f_{n,air}$) of the FIR system. The specific free decay test procedure was as follows:

First, the pendulum system was artificially deflected to a certain rotational angle under the condition of no water in the test water tunnel, and then released. Meanwhile, the test data was recorded with the help of a data acquisition system, shown in Figure 8c.

Four tests were performed at each L/D, and the average value of the four tests ($\zeta_{total,ave}$ and $f_{n,air,ave}$) was used as the result. Figure 9a–d corresponded to the free decay curves with L/D = 4.0, 6.0, 7.0 and 9.0, respectively. The free decay test results of all cases were shown in Tables 4 and 5. Here, the damping ratio ζ_{total} expression was defined as [18]:

$$\zeta_{total} = \frac{\ln \eta}{2\pi} = \frac{1}{2\pi} \ln\left(\frac{A_i}{A_{i+1}}\right) \tag{24}$$

where A_i is the ith peak of the free decay curve, and A_{i+1} is the i+1th peak of the free decay curve. Additionally, the relative error of natural frequency between various cases was defined as:

$$\delta = \frac{Case_i - Case_{i+1}}{Case_{i+1}} \times 100\% (i = 1, 2, \cdots\cdots, 7) \tag{25}$$

where $Case_i$ is the natural frequency of the ith case, and $Case_{i+1}$ is the natural frequency of the i+1th case. Moreover, the relative error of the damping ratio between various cases is defined as:

$$\varepsilon = \frac{Condition_j - Condition_{j+1}}{Condition_{j+1}} \times 100\%(j = 1, 2, \cdots\cdots, 7) \tag{26}$$

where $Condition_j$ is the damping ratio of the ith working condition, and $Condition_{j+1}$ is the damping ratio of the j+1th working condition.

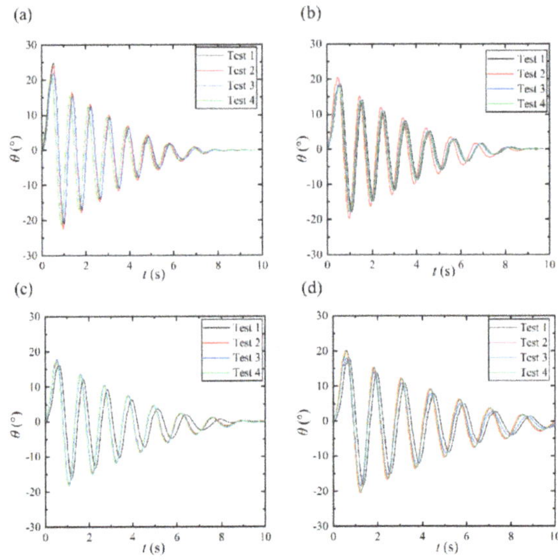

Figure 9. Free decay curves: (**a**) L/D = 4.0; (**b**) L/D = 6.0; (**c**) L/D = 7.0; (**d**) L/D = 9.0.

Table 4. Free decay test results of $f_{n,air}$ at various spacing ratios.

L/D	Test 1 $f_{n,air,1}$ (Hz)	Test 2 $f_{n,air,2}$ (Hz)	Test 3 $f_{n,air,3}$ (Hz)	Test 4 $f_{n,air,4}$ (Hz)	Average $f_{n,air,ave}$ (Hz)	Relative Error δ
4.0	1.146	1.147	1.147	1.146	1.1465	—
4.5	1.102	1.092	1.090	1.085	1.0923	0.0496
5.0	1.048	1.049	1.000	1.048	1.0363	0.0540
5.5	1.020	1.013	1.011	1.013	1.0142	0.0218
6.0	0.953	0.953	0.953	0.953	0.9530	0.0642
7.0	0.903	0.855	0.856	0.902	0.8790	0.0842
8.0	0.806	0.807	0.805	0.806	0.8060	0.0906
9.0	0.758	0.756	0.756	0.758	0.7570	0.0647

Table 5. Free decay test results of ζ_{total} at various spacing ratios.

L/D	Test 1 $\zeta_{total,1}$	Test 2 $\zeta_{total,2}$	Test 3 $\zeta_{total,3}$	Test 4 $\zeta_{total,4}$	Average $\zeta_{total,ave}$	Relative Error ε
4.0	0.060	0.055	0.052	0.052	0.0548	—
4.5	0.054	0.051	0.050	0.048	0.0508	0.0788
5.0	0.056	0.044	0.051	0.045	0.0490	0.0357
5.5	0.050	0.045	0.040	0.045	0.0450	0.0889
6.0	0.043	0.052	0.042	0.040	0.0443	0.0169
7.0	0.046	0.042	0.034	0.037	0.0398	0.1132
8.0	0.042	0.044	0.035	0.035	0.0390	0.0192
9.0	0.044	0.038	0.034	0.040	0.0390	0

As can be seen from Table 4, the $f_{n,air}$ would slightly decrease with an increase of L/D, but the $f_{n,air}$ of all cases was in the range of 0.757–1.147 Hz. Furthermore, the ζ_{total} of all work conditions was controlled within the interval of 0.039 to 0.054, as seen in Table 5.

3.4. Validation of Mathematical Model

Figure 10 compares the experimentally measured results and calculated results of the natural frequency of the FIR system in air at different spacing ratios. The calculated result was obtained by substituting the J_{osc} value in Table 2 into Equation (21). According to an analysis of Figure 10, there was little difference between the measured and the theoretical values of the natural frequency at each spacing ratio, and the relative errors (δ) were all less than 10%. The maximum δ was 6.86% in the case of L/D = 4.0. When L/D = 9.0, the δ was at its minimum, only 0.62%. Further analysis of Figure 10 shows that the calculated value of $f_{n,air}$ was larger than the measured value in each working condition, and the reason for this was that the masses of the transmission belt and driving wheel were omitted in the calculation process. The comparison results demonstrated that the present mathematical model for solving the natural frequency was reasonable, and could provide the theoretical basis for the following flow-induced rotation experiments.

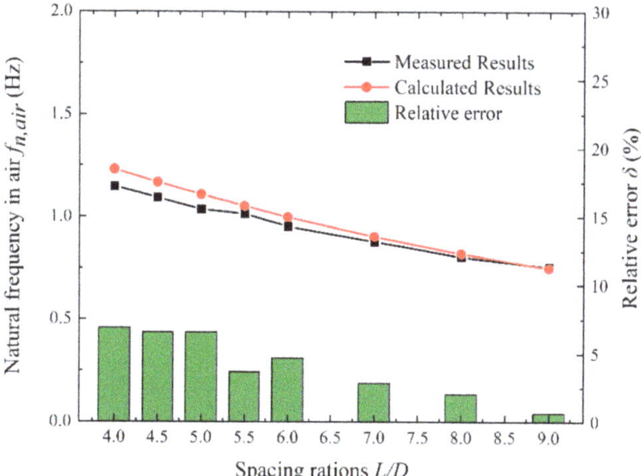

Figure 10. Comparison between experimental and calculated values of system natural frequency.

4. Results and Discussion

The results for the flow-induced rotation characteristics of two mechanically coupled cylinders in tandem arrangement with various spacing ratios are presented and analyzed in this section.

4.1. L/D = 4.0 and 4.5

4.1.1. Rotation Response

The rotational oscillation response curves with L/D = 4.0 and 4.5 were areshown in Figure 11. Herein, the rotation arc length ratio A^* (dimensionless parameter) was examined to characterize the oscillation level, and the calculation equation of A^* was as follows:

$$A^* = \frac{A}{D} = \frac{\frac{\pi\theta_{max}}{180} \times \frac{L}{2}}{D} = \frac{\pi L}{360D}\theta_{max} \tag{27}$$

where A is the average value of the peak of arc length in 60s of continuous oscillation. θ_{max} is the average value of the peak of θ in 60s of continuous oscillation. L is the center distance of the upstream and downstream cylinders. D is the diameter of the circular cylinder.

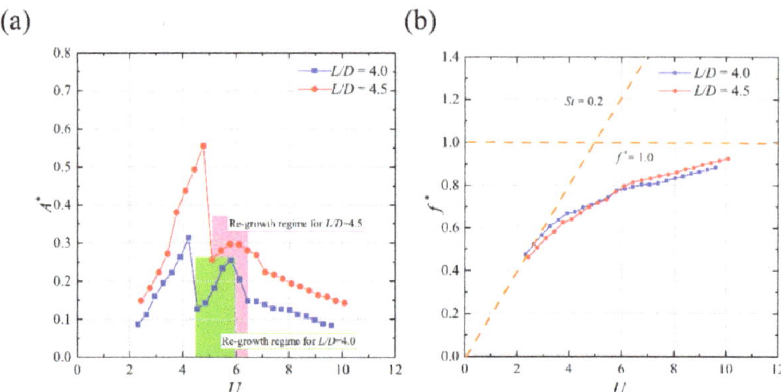

Figure 11. Complete rotation response of FIR system at L/D = 4.0 and 4.5: (**a**) Arc length response; (**b**) Frequency response.

Additionally, the frequency ratio f^* and reduced flow velocity U_r were defined as $f^* = f_{osc}/f_{n,air}$ and $U_r = U/f_{n,air}\,D$, respectively.

(1) L/D = 4.0: in the range of 2.314 < U_r < 4.531, the A^* first increased and, then, decreased sharply with increasing U_r, and the f^* showed a continuous increase with the rise of U_r. The results showed that the rotational oscillation conformed to the response law of VIV and the maximum A^* was 0.32 in this velocity range (in the complete A^*–U_r curve, the first peak was named A^*_{1st}, and the second peak was named A^*_{2nd}, which meant $A^*_{1st,4.0}$ = 0.32). Nevertheless, as 4.531 < U_r < 5.798, the A^* would present an increasing trend again, which meant that there was a re-growth region in the arc length responses. The reason might have been that with the increase of incoming flow velocity, the vortex shedding rate of the upstream cylinder was accelerated, which further promoted the rotational oscillation of the downstream cylinder, and finally led to the enhancement of the oscillation level of the two tandem-coupled cylinders. Note that the peak of A^* could only reach 0.26 in this re-growth region ($A^*_{2nd,4.0}$ = 0.26). The results indicated that, although the arc length grew again at 4.531 < U_r < 5.798, it was actually still in the lower branch of VIV, which meant that the oscillation was still suppressed. Continuing to increase U_r until the maximum of the experiment (U_r = 9.599), resulted in a steady decrease in the A^*. It showed that when U_r > 5.798, the rotational oscillation level of the FIR system decreased once again due to the suppression effect strengthening. The reason might be that, as the velocity continued to increase, the vortex shedding rate of the upstream cylinder became faster, resulting in the lower efficiency of reattachment to the downstream cylinder, and, finally, the overall oscillation level of the FIR system gradually decreasing.

(2) L/D = 4.5: for 2.430 < U_r < 5.093, the A^* also showed the response law of firstincreasing and, then, decreasing. Meanwhile, the f^* rose continuously. It was shown that the rotational vibration was also in line with the VIV in this velocity range. Additionally, as can be seen from Figure 11, $A^*_{1st,4.5}$ = 0.56, which was significantly higher than that when L/D = 4.0. When 5.093 < U_r < 6.091, the A^* also had a tendency to grow again, just like when L/D = 4, but the secondary growth of L/D = 4.5 was less than L/D = 4.0, and it only increased to 0.3 ($A^*_{2nd,4.5}$ = 0.3). For U_r > 6.091, the A^* started to show a downward trend. Further comparing the rotational oscillation responses of L/D = 4.0 and 4.5, it could be found that in the upper and lower branches of VIV, the A^* of L/D = 4.5 was significantly higher than that of L/D= 4.0, indicating that the spacing of the two tandem-coupled cylinders had a significant effect on the arc length responses. Even if the spacing changed slightly, the rotational vibration response could vary considerably.

(3) Further analysis of the arc length response of L/D = 4.0 and 4.5 showed that with continuous increase of the test flow rate, the curve of A^*–U_r had a sudden rise or fall, which meant that each branch of VIV had a clear dividing line. The reason might have been

that the damping ratio of the FIR system at $L/D = 4.0$ and 4.5 belonged to the category of low damping ratio. The study by Khalak and Williamson [52] pointed out that when the damping ratio is small, the VIV response of the cylinder presents three obvious branches.

In summary, when the spacing ratio of the FIR system was 4.0 and 4.5, the flow-induced rotation response was similar to that of VIV response. Compared with the typical VIV rule, the difference was that, after the rotational vibration entered the lower branch of VIV, there was a re-growth region in the arc length response, and this region was no longer significant as the spacing ratio increased. Note that the vibration was also suppressed in this region.

4.1.2. Oscillation Time History and Spectral Characteristics

In order to further study the oscillation characteristics of the FIR system at $L/D = 4.0$ and 4.5, Figures 12 and 13 show the time history and frequency spectra of the rotation angle (θ) at six typical velocities ($U = 0.302$ m/s, 0.483 m/s, 0.519 m/s, 0.592 m/s, 0.665 m/s and 0.882 m/s), respectively. The ordinate in the spectrum was normalized and the dominant oscillation frequency highlighted.

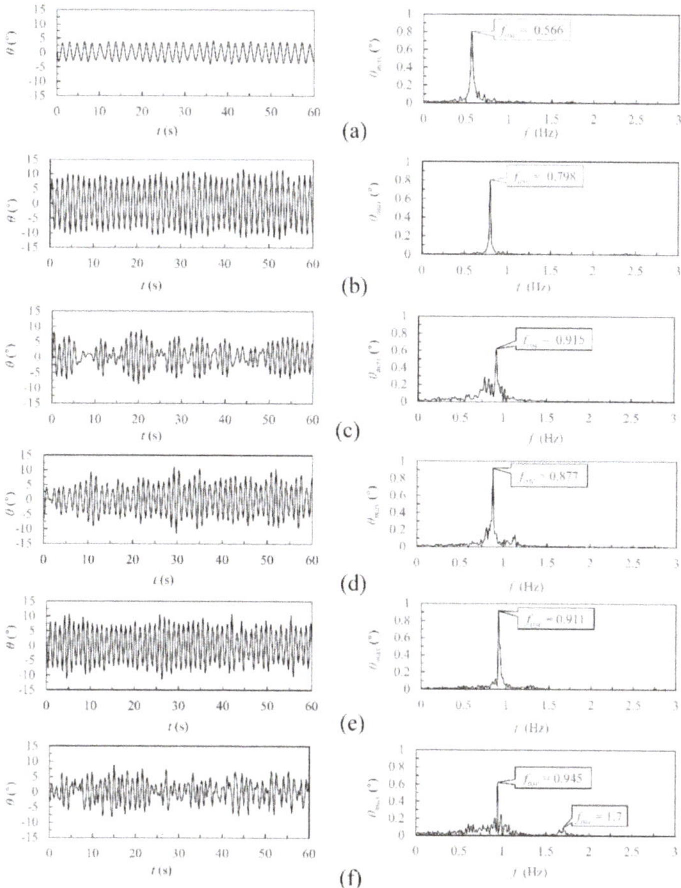

Figure 12. Rotational oscillation time history and spectral characteristics at six typical velocities when $L/D = 4.0$: (**a**) $U = 0.302$m/s ($U_r = 2.631$); (**b**) $U = 0.483$ m/s ($U_r = 4.215$); (**c**) $U = 0.519$ m/s ($U_r = 4.531$); (**d**) $U = 0.592$ m/s ($U_r = 5.165$); (**e**) $U = 0.665$ m/s ($U_r = 5.798$); (**f**) $U = 0.882$m/s ($U_r = 7.699$).

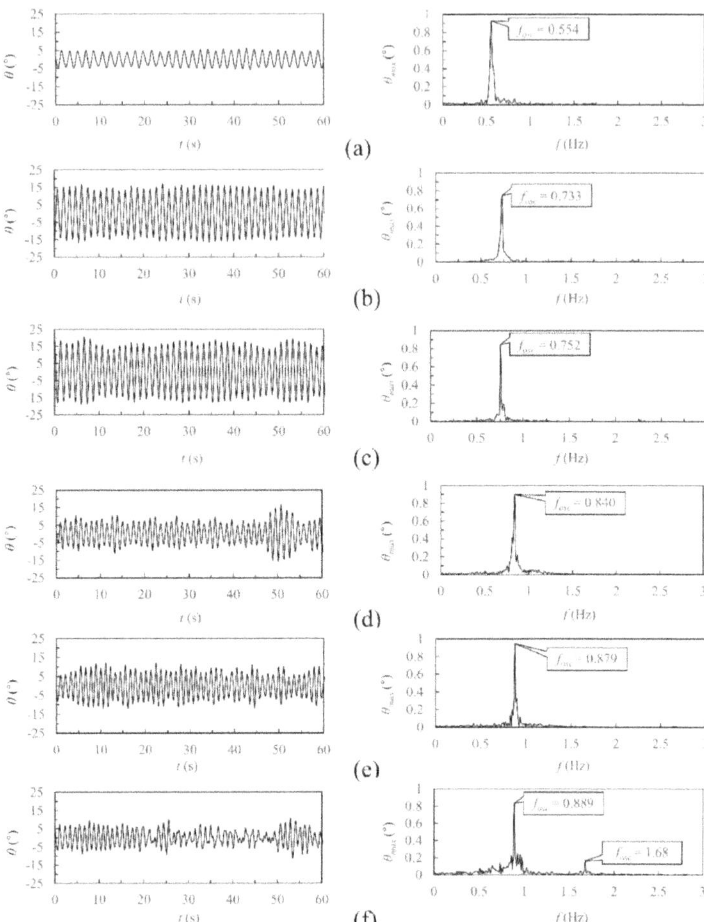

Figure 13. Rotational oscillation time history and spectral characteristics at six typical velocities when $L/D = 4.5$: (**a**) $U = 0.302$ m/s ($U_r = 2.764$); (**b**) $U = 0.483$m/s ($U_r = 4.428$); (**c**) $U = 0.519$ m/s ($U_r = 4.76$); (**d**) $U = 0.592$ m/s ($U_r = 5.426$); (**e**) $U = 0.665$ m/s ($U_r = 6.091$); (**f**) $U = 0.882$ m/s ($U_r = 8.088$).

(1) $U = 0.302$m/s: It can be seen from the time history diagram that the θ of $L/D = 4.0$ and 4.5 was small at this flow rate, reaching only about 3.3° and 4.6°, respectively, and the waveform was relatively uniform. According to the spectrum analysis, the frequency bands of the two cases were narrow, the f_{osc} values were 0.566 Hz and 0.554 Hz, respectively, and there were no frequency doubling characteristics. The results indicated that the rotational vibration was relatively stable and the oscillation energy was low at $U = 0.302$ m/s.

(2) $U = 0.483$ m/s: The waveforms of time history became more uniform and the θ increased remarkably in both cases, increasing to 10.5° and 12.6°, respectively. Additionally, it can be seen from the spectrum diagram that the values of the f_{osc} of the two cases were higher than those of $U = 0.302$ m/s, and the frequency bands became narrower, indicating that the rotational oscillation of the two cases was stable and the vibration energy was concentrated at this velocity.

(3) $U = 0.519$ m/s: For $L/D = 4.0$, the θ began to decrease and the waveform of time history was disordered, the f_{osc} increased to 0.915 Hz but the frequency band became wider, indicating that the rotary oscillation was suppressed and in an unstable state at

U = 0.519 m/s. For L/D = 4.5, the θ was still growing, reaching about 14.2°. The f_{osc} increased to 0.752 Hz and the frequency band was narrow, which suggested that the rotational vibration was not inhibited at this rate and the vibration was still fine. Further analysis of the time history and spectrum diagram with L/D = 4.0 found that the arc length response first increased and then decreased, and the frequency response continued to increase, indicating that the flow-induced rotation characteristics with L/D = 4.0 conformed to the VIV in the range of U = 0.302–0.519 m/s.

(4) U = 0.592 m/s: For L/D = 4.0, the waveform was unstable, but the θ rose slightly compared with that of U = 0.519 m/s. The frequency band was still wide. However, for L/D = 4.5, the waveform of time history became disorganized and the θ was significantly reduced. Additionally, the frequency band became wider and the f_{osc} increased to 0.840 Hz. The results suggested that the rotational vibration with L/D = 4.5 was suppressed at U = 0.592 m/s. Meanwhile, it could also be concluded that for L/D = 4.5, the arc length response also first increased and then decreased when U = 0.302–0.592 m/s, and the dominant frequency always increased, which was also in line with the response law of VIV.

(5) U = 0.665 m/s: For L/D = 4.0, the θ continued to increase, reaching about 7.3°, but did not exceed the value reached at U = 0.483 m/s. Furthermore, compared with U = 0.592 m/s, the frequency band became narrower and the f_{osc} increased to 0.911 Hz. For L/D = 4.5, the θ also increased, but the growth range was smaller than that of L/D = 4.0. The frequency band was narrow and the f_{osc} increased to 0.879 Hz. The results showed that, for L/D = 4.0 and 4.5, after the vibration entered the lower branch of VIV, the arc length response tended to increase again, and the increase of L/D = 4.0 was more significant than that of L/D = 4.5.

(6) U = 0.882 m/s: It was found that the θ reduced in both cases and the waveforms were disordered. Meanwhile, the frequency bands of the two cases became wider, and both had the 2-fold frequency characteristic. This indicated that the rotary oscillations were significantly suppressed, the vibration was unstable and the vibration energy was dispersed at U = 0.882 m/s.

In brief, when L/D = 4.0 and 4.5, the rotation responses generally exhibited the similar VIV law. Differing from the typical VIV response, after the vibration entered the lower branch of VIV, the arc length response would appear as a re-growth region, and with the increase of spacing ratio, this region was no longer significant. The above results were consistent with the discussion of experimental results in Section 4.1.1.

4.2. L/D = 5.0, 5.5 and 6.0

4.2.1. Rotation Response

In this section, the rotation responses at L/D = 5.0, 5.5 and 6.0 are investigated. The complete response results are shown in Figure 14. The specific analyses are as follows:

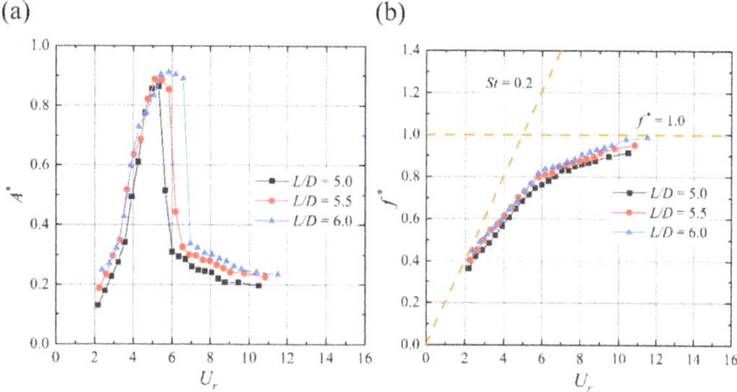

Figure 14. Complete rotation response of FIR system at L/D = 5.0, 5.5 and 6.0: (**a**) Arc length response; (**b**) Frequency response.

(1) On the whole, in the entire test flow velocity range, the rotation responses of these three cases showed the typical VIV response law and there was no re-growth region. The A^* first increased and then decreased sharply, as well as the f^* increasing continuously with increase of U_r, which meant there was entering of the initial branch, the upper branch and the lower branch of VIV, in turn. Furthermore, the rotational oscillation responses of $L/D = 5.0$, 5.5 and 6.0 were characterized by longer arc length. The maximum A^* of all three cases was more than 0.85, among which $L/D = 6.0$ was as high as 0.92, exceeding the A^* of the VIV amplitude response of a single cylinder in the previous study [52]. It could be seen that when the distance between the two cylinders of FIR system increased from 4.5D to 5.0D, the rotation angle θ would increase significantly, which would greatly improve the oscillation level (arc length A).

(2) By carefully comparing the oscillation characteristics of these three cases, it could be seen that, in each branch of VIV, the A^* showed a slight upward trend with the increase of L/D, implying that the case with $L/D = 6.0$ had a larger synchronization interval than the other two cases. Additionally, further analysis of the rotation response of $L/D = 5.0$, 5.5 and 6.0 showed that, when $4.0 < U_r < 5.5$, the rotational oscillation was in the upper branch of VIV, and f^* increased in this flow velocity range. Continuing to increase U_r, the growth rate of f^* would slow down, and in the entire test velocity range, f^* did not exceed 1.0. The reason might have been that the ratio of the moment of inertia J^* of $L/D = 5.0$, 5.5 and 6.0 was still small ($J^* = 1.745$–2.113). Even if the oscillation entered the nonlinear resonance zone when $4.0 < U_r < 5.5$, the vibration frequency was not locked, but still showed an increasing trend. When $U_r > 5.5$, the rate of growth slowed, which was consistent with the research results of Kraghvav [53].

In general, when the spacing ratios of the two tandem-coupled cylinders were 5.0, 5.5 and 6.0, the oscillation responses all showed the typical VIV law and were characterized by high oscillation levels. Additionally, the vibration structure with $L/D = 6.0$ was more suitable for flow-induced motion power generation, because of its the longer arc length and the wider range of synchronization intervals.

4.2.2. Oscillation Time History and Spectral Characteristics

Similarly, in order to further study the vibrational characteristics of the FIR system with $L/D = 5.0$, 5.5 and 6.0, three typical flow velocities ($U = 0.302$ m/s, 0.556 m/s and 0.810 m/s) were selected for each working condition, to analyze the time history and spectral characteristics of the rotation angle θ, as shown in Figures 15–17, respectively.

(1) $U = 0.302$ m/s: At this flow velocity, there was no significant difference in the time history of the three cases. The waveforms were stable and the θ value was low, all maintained at about 5.8°. The frequency bands were narrow and no frequency doubling features appeared. This indicated that the rotational oscillations of these three cases were relatively stable and the vibration energies were small. On the other hand, it could be seen that, when $L/D = 5.0$, 5.5 and 6.0, the FIR system started to vibrate with a smaller flow rate, and when the velocity reached 0.302 m/s, rotational vibration of a lower level would occur.

(2) $U = 0.556$ m/s: The analysis of the time history showed that the waveforms of the three cases were stable. Compared with that of $U = 0.302$ m/s, the θ all increased significantly and could reach about 18.5°. Additionally, as can be seen from the spectral diagram, the f_{osc} of each case was obvious and increased to 0.782 Hz, 0.781 Hz and 0.782 Hz. The rotational oscillations were stable and energies concentrated at this flow velocity.

(3) $U = 0.810$ m/s: The change of time history of the three cases was similar, the θ decreased significantly and the waveforms were disordered. Meanwhile, the f_{osc} of each case continued to increase, the frequency band became wider and there were 2-fold frequency doubling characteristics. The results indicated that, when $U = 0.810$ m/s, the rotational oscillation was suppressed significantly and it was unstable.

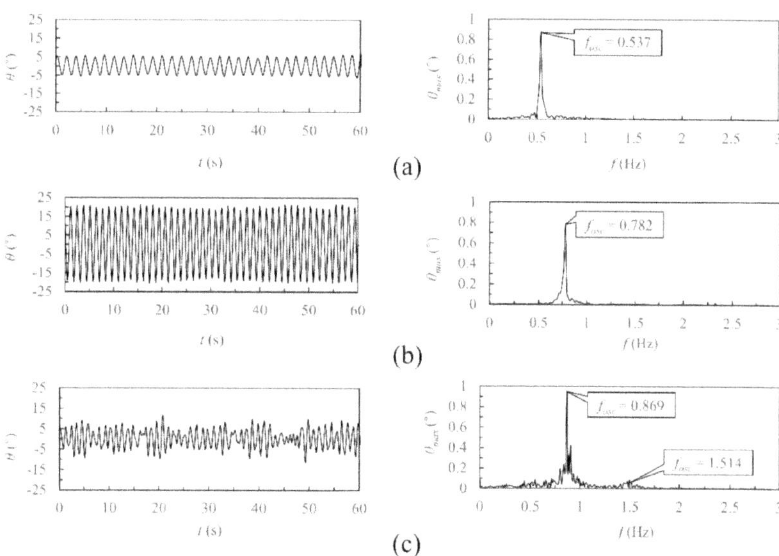

Figure 15. Rotational oscillation time history and spectral characteristics at three typical velocities when L/D = 5.0: (**a**) U = 0.302 m/s (U_r = 2.877); (**b**) U = 0.556 m/s (U_r = 5.302); (**c**) U = 0.810 m/s (U_r = 7.726).

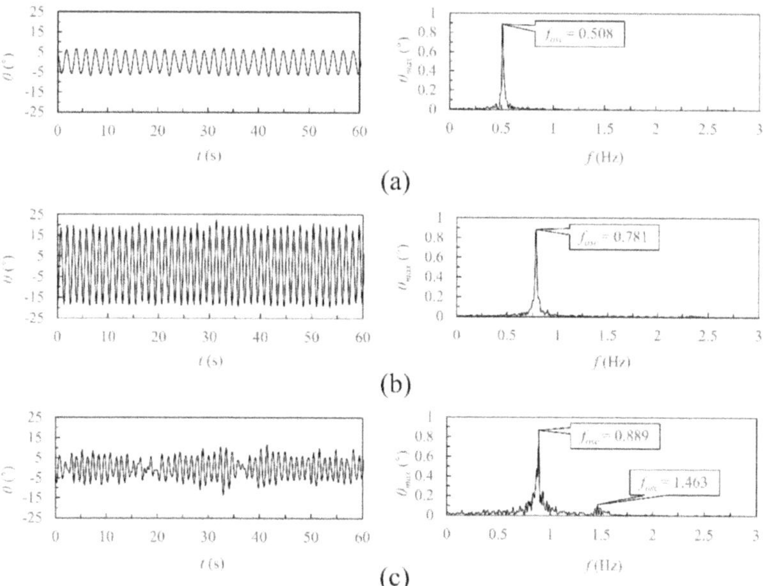

Figure 16. Rotational oscillation time history and spectral characteristics at three typical velocities when L/D = 5.5: (**a**) U = 0.302 m/s (U_r = 2.975); (**b**) U = 0.556 m/s (U_r = 5.483); (**c**) U = 0.810 m/s (U_r = 7.991).

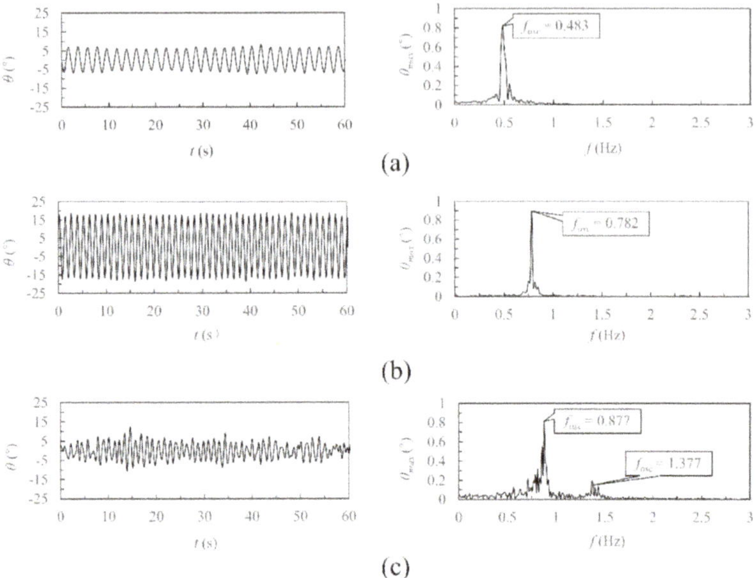

Figure 17. Rotational oscillation time history and spectral characteristics at three typical velocities when L/D = 6.0: (**a**) U = 0.302 m/s (U_r = 3.164); (**b**) U = 0.556 m/s (U_r = 5.83); (**c**) U = 0.810 m/s (U_r = 8.496).

Comprehensive analysis of the oscillation time history and spectrum characteristics of the FIR system with L/D = 5.0, 5.5 and 6.0 showed that, at these three typical velocities, the θ first increased and then decreased, and the f_{osc} continued to increase, which was consistent with the typical response law of VIV. The rotational vibration of these three working conditions could reach a high level because of the larger rotation angle.

4.3. L/D = 7.0, 8.0 and 9.0

4.3.1. Rotation Response

This section analyzes and summarizes the rotation characteristics when L/D = 7.0, 8.0 and 9.0. Figure 18 presents the rotational vibration responses of the FIR system at these three spacing ratios. The specific analysis is as follows:

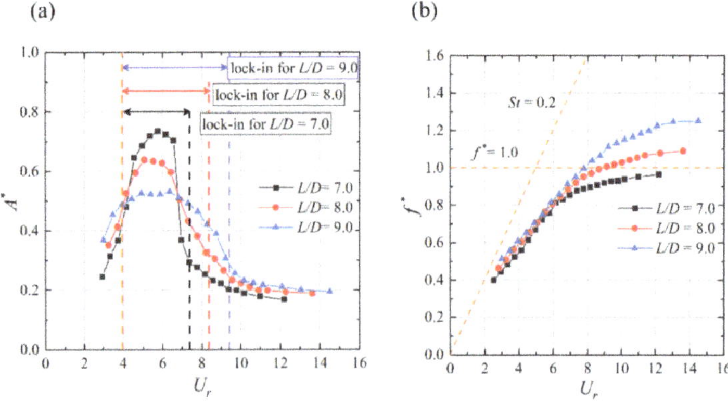

Figure 18. Complete rotation response of FIR system at L/D = 7.0, 8.0 and 9.0: (**a**) Arc length response; (**b**) Frequency response.

(1) On the whole, within the test flow rate range, the rotational responses of these three cases still showed the typical VIV law. With the increase of U_r, the A^* first increased and then decreased, while the f^* presented continuous growth. Additionally, the maximum A^* was 0.53–0.72 at these three spacing ratios, which was lower than the value when $L/D = 5.0$, 5.5 and 6.0, indicating that, when the distance between two cylinders exceeded 0.6 m (corresponding to $L/D = 6.0$), the oscillation level of the FIR system would decrease instead. By analyzing the definition of arc length ratio, A^*, it was found that when $L/D > 6.0$, the θ would be significantly reduced, resulting in the A^* being unable to reach the previous level, which indicated that the distance between the two cylinders should be controlled within a reasonable range to ensure the rotational oscillation reached an ideal level. Additionally, the range of synchronization interval at $L/D = 7.0$, 8.0 and 9.0 was significantly larger than those in the other cases. This meant that the FIR system with $L/D = 7.0$, 8.0 and 9.0 had stronger adaptability to the current velocity and was more favorable to FIM power generation.

(2) Comparing the A^*–U_r curves of the three cases, it could be seen that in the initial and lower branches of VIV, the A^* increased with the increase of L/D, while in the upper branch of VIV, the A^* decreased with the increase of L/D, indicating that the relationship between the size of the synchronization interval in the three cases was as follows: $L/D = 9.0$ was the largest, followed by $L/D = 8.0$, and $L/D = 7.0$, which was the smallest. Further analyzing the frequency responsed of these three working conditions, it was found that when $U_r > 8.5$, the f^* value of $L/D = 8.0$ and 9.0 exceeded 1.0 ($f^* > 1.0$), and f^* continued to increase with increase of U_r. The reason might have been that, when $L/D = 8.0$ and 9.0, the rotational oscillation would also enter the nonlinear resonance zone, but the characteristics of nonlinear resonance zone of these two cases were different from those when $L/D \leq 7.0$. The specific performance was such that the oscillation frequency was no longer locked, but continued to increase with the increase of U_r until $f^* > 1.0$. It should be noted that at this time, the locked interval of arc length response still existed and becames significantly wider. This phenomenon was highly similar to the research results of Kraghvav [53].

In conclusion, when the spacing ratios of two tandem-coupled cylinders were 7.0, 8.0 and 9.0, the rotation responses still presenteds the typical VIV law. Compared with other cases, the rotational oscillation in these three cases were at a medium level, and the maximum A^* was between 0.53 and 0.72. Another notable characteristic of the rotation response was that the range of synchronization interval was larger, and this range tended to increase with the increase of L/D. Furthermore, it should be noted that the longer arc length response could only be generated when the distance between the two cylinders of the FIR system was in a rational range.

4.3.2. Oscillation Time History and Spectral Characteristics

In order to further understand the rotational oscillation characteristics of the FIR system when $L/D = 7.0$, 8.0 and 9.0, herein, four typical velocities ($U = 0.265$ m/s, 0.483 m/s, 0.628 m/s and 0.810 m/s) were selected for each L/D to analyze the time history and spectral characteristics of θ, as shown in Figures 19–21, respectively. The specific analysis is as follows:

(1) $U = 0.265$ m/s: It can be seen from the time-history diagram that the θ of the three cases was small at this velocity. The cases of $L/D = 7.0$ and 8.0 maintained about 4.6°, while the case of $L/D = 9.0$ could reach 5.8°. Furthermore, it can be seen from the spectrogram that no frequency doubling occurred, but the frequency band was wide, which indicated that the vibration of the three cases was in an unstable state and the energy was not concentrated.

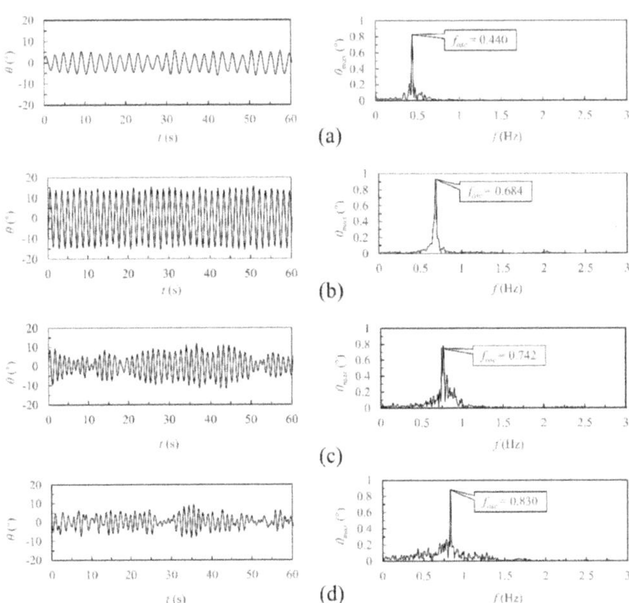

Figure 19. Rotational oscillation time history and spectral characteristics at L/D = 7.0: (**a**) U = 0.265 m/s (U_r = 2.940); (**b**) U = 0.483 m/s (U_r = 5.355); (**c**) U = 0.628 m/s (U_r = 6.965); (**d**) U = 0.810 m/s (U_r = 8.977).

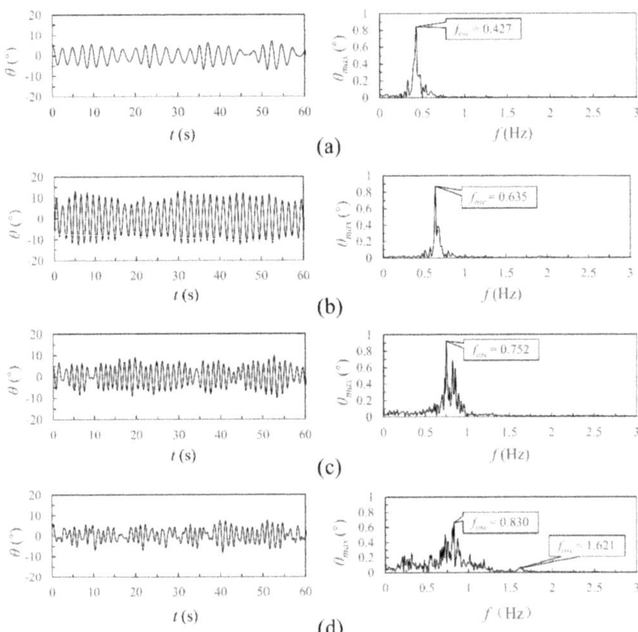

Figure 20. Rotational oscillation time history and spectral characteristics at L/D = 8.0: (**a**) U = 0.265 m/s (U_r = 3.29); (**b**) U = 0.483 m/s (U_r = 5.993); (**c**) U = 0.628 m/s (U_r =7.794); (**d**) U = 0.810 m/s (U_r = 10.046).

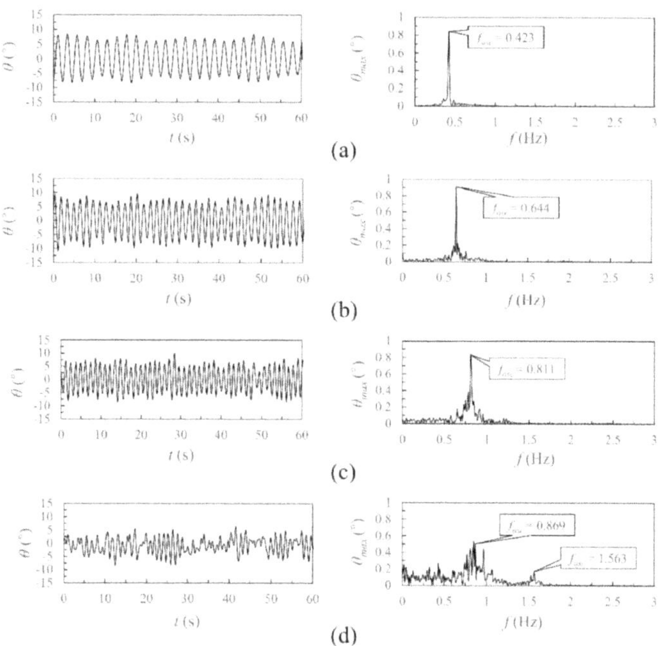

Figure 21. Rotational oscillation time history and spectral characteristics at L/D = 9.0: (**a**) U = 0.265 m/s (U_r = 3.499); (**b**) U = 0.483 m/s (U_r = 6.372); (**c**) U = 0.628 m/s (U_r = 2.288); (**d**) U = 0.810 m/s (U_r = 10.682).

(2) U = 0.483 m/s: Firstly, compared with the time history diagram when U = 0.265 m/s, it could be found that the θ of L/D = 7.0 and 8.0 were greatly improved, increasing to about 11.8° and 8.9°, respectively, while the θ of L/D = 9.0 only increased slightly to 6.7°. Additionally, the waveforms of the three cases at this flow rate were uniform. Then, the analysis of the spectral diagram shows that the f_{osc} of the three cases at this flow velocity were also increased, compared with that at U = 0.265 m/s, the frequency band narrowed, and there was no frequency doubling phenomenon. This showed that the rotational vibration becames stable and the vibrational energy was large.

(3) U = 0.628 m/s: When U increased to 0.628 m/s, it can be seen from the time history diagram that the waveforms of L/D = 7.0 and L/D = 8.0 became disordered, while the waveform of L/D = 9.0 was relatively stable. Meanwhile, the θ of L/D = 7.0 decreased sharply to about 6.1°, while the θ of L/D = 8.0 and 9.0 decreased slightly to about 5.5° and 5.9°, respectively. By analyzing the spectral diagram, it could be found that the f_{osc} of the three cases continued to increase and the frequency band became wider than that of U = 0.483 m/s.

Comprehensive analysis of the time history and frequency spectral characteristics of the FIR system with L/D = 7.0, 8.0 and 9.0 at the above three typical flow rates, indicated that, with the increase of the test flow rate, the θ of the three cases first increased and then decreased, and the f_{osc} kept increasing, which meant that the three cases all showed the response law of VIV. After entering the lower branch of VIV, the suppression effect of L/D = 7.0 was most significant, followed by L/D = 8.0, and was the smallest when L/D = 9.0, which led to the relationship between the range of synchronous intervals of the three cases such that L/D = 9.0 was the largest, L/D = 8.0 was second and L/D = 7.0 was the smallest.

(4) U = 0.810 m/s: When U increased to 0.810 m/s, the θ of L/D = 7.0, 8.0 and 9.0 all decreased to about 3.5° and the waveforms of the time history were all out of

order. Meanwhile, the frequency bands became wider and there were the characteristics of frequency doubling, indicating that, after the vibration response of these three cases entered the lower branch of VIV, there was no "re-growth region", and the vibration was still very unstable.

In conclusion, when the spacing ratios of the FIR system were 7.0, 8.0 and 9.0, the rotational oscillation responses presented the typical VIV law, and the range of the synchronization interval increased with increase of L/D. The result corresponds to the rotation response shown in Figure 18a, which proved the authenticity of the experimental results obtained.

5. Conclusions

In this paper, an experimental investigation on the flow-induced rotation (FIR) of two mechanically tandem-coupled cylinders at different spacing ratios (L/D) was carried out with the help of a self-circulating water tank. The Reynolds number ranged from 20,105 to 96,575. The complete rotation responses of the FIR system with L/D = 4.0–9.0 were analyzed emphatically. The major conclusions are as follows:

(1) For L/D = 4.0 and 4.5: The rotation response of the FIR system is similar to the VIV response, on the whole. The rotational oscillation is at a low level and the maximum arc length ratio (A^*) is less than 0.56. Compared with the typical VIV response, the difference is that the arc length response appears as a re-growth region with the increase of velocity after entering the lower branch of VIV, but the oscillation is still suppressed in this region. As L/D increases, the re-growth region is no longer significant.

(2) For L/D = 5.0, 5.5 and 6.0: The rotation responses of the FIR system present the typical VIV response, and the oscillation reaches a high level with the maximum A^* exceeding 0.85. Additionally, the vibration structure with L/D = 6.0 is more suitable for flow-induced motion power generation because of the longer arc length and the wider range of synchronization interval.

(3) For L/D = 7.0, 8.0 and 9.0: The rotation responses still show the VIV law. The oscillation is at the medium level, and the maximum A^* is between 0.53 and 0.72. Compared with the other cases, the oscillation response of these three cases have a wider range of synchronization interval, and the range shows an increasing trend with the growth of the spacing ratio.

In this paper, the flow-induced rotation characteristics of two tandem-coupled cylinders at different spacing ratios were investigated. Nevertheless, the two key parameter $c_{\text{tangent},up}$ and $c_{\text{tangent},down}$ in Equation (22) were not actually measured in the existing experiments. A torque sensor will be introduced to measure these two parameters in future research. Additionally, the more definite influence of the term containing $\left(\dot{\theta}\right)^2$ in Equation (18) on the FIR system will also be a focus of subsequent research. Finally, a generator will be installed, on the basis of the present experimental device, to evaluate the power generation level of the FIR system. Meanwhile, the new system's dynamic equation can be further analyzed and solved after considering the generator damping.

Author Contributions: Conceptualization, F.L. and X.Y.; methodology, F.L., W.F., X.Y., D.R. and N.S.; software, W.F., X.Y., X.W. and D.Y.; validation, W.F., X.Y. and D.R.; formal analysis, F.L., W.F. and W.F.; investigation, W.F. and X.Y.; resources, F.L. and X.Y.; data curation, F.L., W.F. and X.Y.; writing—original draft preparation, W.F.; writing—review and editing, F.L., X.Y., D.R., N.S., X.W. and D.Y.; visualization, W.F. and X.Y.; supervision, F.L. and X.Y.; project administration, F.L. and X.Y.; funding acquisition, F.L. and X.Y. All authors have read and agreed to the published version of the manuscript.

Funding: This research was funded by the National Natural Science Foundation of China, grant number 51909190.

Institutional Review Board Statement: Not applicable.

Informed Consent Statement: Not applicable.

Data Availability Statement: The data presented in this study are available on request from the corresponding author. The data are not publicly available, due to privacy.

Acknowledgments: The authors would like to thank the peer reviewers and editors for their hard work and constructive feedback, which made a significant contribution to improving the paper.

Conflicts of Interest: The authors declare no conflict of interest.

References

1. Mohamed, A.; Zaki, S.A.; Shirakashi, M.; Ali, M.S.; Samsudin, M.Z. Experimental investigation on vortex-induced vibration and galloping of rectangular cylinders of varying side ratios with a downstream square plate. *J. Wind Eng. Ind. Aerodyn.* **2021**, *211*, 104563. [CrossRef]
2. Hong, K.S.; Shah, U.H. Vortex-induced vibrations and control of marine risers: A review. *Ocean Eng.* **2018**, *152*, 300–315. [CrossRef]
3. Xu, W.; Ji, C.; Sun, H.; Ding, W.; Bernitsas, M.M. Flow-induced vibration of two elastically mounted tandem cylinders in cross-flow at subcritical Reynolds numbers. *Ocean Eng.* **2019**, *173*, 375–387. [CrossRef]
4. Liu, G.; Li, H.; Qiu, Z.; Leng, D.; Li, Z.; Li, W. A mini review of recent progress on vortex-induced vibrations of marine risers. *Ocean Eng.* **2020**, *195*, 106704. [CrossRef]
5. Leng, D.; Liu, D.; Li, H.; Jin, B.; Liu, G. Internal flow effect on the cross-flow vortex-induced vibration of marine risers with different support methods. *Ocean Eng.* **2022**, *257*, 111487. [CrossRef]
6. Hwang, Y.C.; Kim, S.; Kim, H.K. Cause investigation of high-mode vortex-induced vibration in a long-span suspension bridge. *Struct. Infrastruct. Eng.* **2020**, *16*, 84–93. [CrossRef]
7. Khan, H.H.; Islam, M.D.; Fatt, Y.Y.; Janajreh, I.; Alam, M.M. Flow-induced vibration on two tandem cylinders of different diameters and spacing ratios. *Ocean Eng.* **2022**, *258*, 111747. [CrossRef]
8. Yan, B.; Ren, H.; Li, D.; Yuan, Y.; Li, K.; Yang, Q.; Deng, X. Numerical Simulation for Vortex-Induced Vibration (VIV) of a High-Rise Building Based on Two-Way Coupled Fluid-Structure Interaction Method. *Int. J. Struct. Stab. Dyn.* **2022**, *22*, 2240010. [CrossRef]
9. Bearman, P.W. Vortex shedding from oscillating bluff bodies. *Annu. Rev. Fluid Mech.* **1984**, *16*, 195–222. [CrossRef]
10. Sarpkaya, T. A critical review of the intrinsic nature of vortex-induced vibrations. *J. Fluids Struct.* **2004**, *19*, 389–447.
11. Allen, J.J.; Smits, A.J. Energy harvesting eel. *J. Fluids Struct.* **2001**, *15*, 629–640. [CrossRef]
12. Taylor, G.W.; Burns, J.R.; Kammann, S.A.; Powers, W.B.; Welsh, T.R. The energy harvesting eel: A small subsurface ocean/river power generator. *IEEE J. Ocean. Eng.* **2001**, *26*, 539–547. [CrossRef]
13. Pobering, S.; Schwesinger, N. A novel hydropower harvesting device. In Proceedings of the International Conference on MEMS, NANO and Smart System, Banff, AB, Canada, 25–27 August 2004.
14. Bernitsas, M.M.; Raghavan, K.; Ben-Simon, Y.; Garcia, E.M.H. VIVACE (Vortex Induced Vibration Aquatic Clean Energy): A new concept in generation of clean and renewable energy from fluid flow. *J. Offshore Mech. Arct. Eng.* **2008**, *130*, 041101. [CrossRef]
15. Bernitsas, M.M.; Ben-Simon, Y.; Raghavan, K.; Garcia, E.M.H. The VIVACE converter: Model tests at high damping and Reynolds number around 105. *J. Offshore Mech. Arct. Eng.* **2009**, *131*, 011102. [CrossRef]
16. Raghavan, K.; Bernitsas, M.M. Experimental investigation of Reynolds number effect on vortex induced vibration of rigid circular cylinder on elastic supports. *Ocean Eng.* **2011**, *38*, 719–731. [CrossRef]
17. Lee, J.H.; Bernitsas, M.M. High-damping, high-Reynolds VIV tests for energy harnessing using the VIVACE converte. *Ocean Eng.* **2011**, *38*, 1697–1712. [CrossRef]
18. Shao, N.; Xu, G.; Liu, F.; Yan, X.; Wang, X.; Deng, H.; Zheng, Z. Experimental Study on the Flow-Induced Motion and Hydrokinetic Energy of Two T-section Prisms in Tandem Arrangement. *Appl. Sci.* **2020**, *10*, 1136. [CrossRef]
19. Liu, M.; Wang, H.; Shao, F.; Jin, X.; Tang, G.; Yang, F. Numerical investigation on vortex-induced vibration of an elastically mounted circular cylinder with multiple control rods at low Reynolds number. *Appl. Ocean Res.* **2022**, *118*, 102987. [CrossRef]
20. Chen, W.; Ji, C.; Xu, D.; Alam, M.M. Three-dimensional direct numerical simulations of two interfering side-by-side circular cylinders at intermediate spacing ratios. *Appl. Ocean Res.* **2022**, *123*, 103162. [CrossRef]
21. King, R.; Johns, D.J. Wake interaction experiments with two flexible circular cylinders in flowing water. *J. Sound Vib.* **1976**, *45*, 259–283. [CrossRef]
22. Zdravkovich, M.M. Review of flow interference between two circular cylinders in various arrangements. *J. Fluids Eng.* **1977**, *99*, 618–633. [CrossRef]
23. Zdravkovich, M.M. Flow induced oscillations of two interfering circular cylinders. *J. Sound Vib.* **1985**, *101*, 511–521. [CrossRef]
24. Zdravkovich, M.M. The effects of interference between circular cylinders in cross flow. *J. Fluids Struct.* **1987**, *1*, 239–261. [CrossRef]
25. Igarashi, T. Characteristics of the flow around two circular cylinders arranged in tandem: 1st report. *Bull. JSME* **1981**, *24*, 323–331. [CrossRef]
26. Bokaian, A.; Geoola, F. Wake-induced galloping of two interfering circular cylinders. *J. Fluid Mech.* **1984**, *146*, 383–415. [CrossRef]
27. Bokaian, A.; Geoola, F. Wake displacement as cause of lift force on cylinder pair. *J. Eng. Mech.* **1985**, *111*, 92–99. [CrossRef]
28. Yao, X.; Chen, Q.; Xu, W. Experimental research on two circular cylinders in tandem arrangements at critical Reynolds numbers. *J. Vib. Eng.* **1994**, *7*, 17–22.

29. Mahir, N.; Rockwell, D. Vortex formation from a forced system of two cylinders. Part I: Tandem arrangement. *J. Fluids Struct.* **1996**, *10*, 473–489. [CrossRef]
30. Mahir, N.; Rockwell, D. Vortex formation from a forced system of two cylinders. Part II: Side-by-side arrangement. *J. Fluids Struct.* **1996**, *10*, 491–500. [CrossRef]
31. Brika, D.; Laneville, A. Wake interference between two circular cylinders. *J. Wind Eng. Ind. Aerodyn.* **1997**, *72*, 61–70. [CrossRef]
32. Meneghini, J.R.; Saltara, F.; Siqueira, C.L.R.; Ferrarijr, J.A. Numerical simulation of flow interference between two circular cylinders in tandem and side-by-side arrangements. *J. Fluids Struct.* **2001**, *15*, 327–350. [CrossRef]
33. Lin, J.C.; Yang, Y.; Rockwell, D. Flow past two cylinders in tandem: Instantaneous and averaged flow structure. *J. Fluids Struct.* **2002**, *16*, 1059–1071. [CrossRef]
34. Zhou, Y.; Yiu, M.W. Flow structure, momentum and heat transport in a two-tandem-cylinder wake. *J. Fluid Mech.* **2006**, *548*, 17–48. [CrossRef]
35. Prasanth, T.K.; Mittal, S. Flow-induced oscillation of two circular cylinders in tandem arrangement at low Re. *J. Fluids Struct.* **2009**, *25*, 1029–1048. [CrossRef]
36. Prasanth, T.K.; Mittal, S. Vortex-induced vibration of two circular cylinders at low Reynolds number. *J. Fluids Struct.* **2009**, *25*, 731–741. [CrossRef]
37. Assi, G.R.S.; Meneghini, J.R.; Aranha, J.A.P.; Bearman, P.W.; Casaprima, E. Experimental investigation of flow-induced vi-bration interference between two circular cylinders. *J. Fluids Struct.* **2006**, *22*, 819–827. [CrossRef]
38. Assi, G.R.S.; Bearman, P.W.; Meneghini, J.R. On the wake-induced vibration of tandem circular cylinders: The vortex interaction excitation mechanism. *J. Fluid Mech.* **2010**, *661*, 365–401. [CrossRef]
39. Bao, Y.; Huang, C.; Zhou, D.; Tu, J.; Han, Z. Two-degree-of-freedom flow-induced vibrations on isolated and tandem cylinders with varying natural frequency ratios. *J. Fluids Struct.* **2012**, *35*, 50–75. [CrossRef]
40. Ji, C.; Chen, W.; Huang, J.; Xu, W. Numerical investigation on flow-induced vibration of two cylinders in tandem arrangements and its coupling mechanisms. *Acta Mech. Sin.* **2014**, *46*, 862–870.
41. Chen, W.; Ji, C.; Williams, J.; Xu, D.; Yang, L.; Cui, Y. Vortex-induced vibrations of three tandem cylinders in laminar crossflow: Vibration response and galloping mechanism. *J. Fluids Struct.* **2018**, *78*, 215–238. [CrossRef]
42. Qin, B.; Alam, M.M.; Ji, C.; Liu, Y.; Xu, S. Flow-induced vibrations of two cylinders of different natural frequencies. *Ocean Eng.* **2018**, *155*, 189–200. [CrossRef]
43. Arionfard, H.; Nishi, Y. Flow-induced vibrations of two mechanically coupled pivoted circular cylinders: Characteristics of vibration. *J. Fluids Struct.* **2018**, *80*, 165–178. [CrossRef]
44. Arionfard, H.; Nishi, Y. Flow-induced vibration of two mechanically coupled pivoted circular cylinders: Vorticity dynamics. *J. Fluids Struct.* **2018**, *82*, 505–519. [CrossRef]
45. Zhu, H.; Tang, T.; Alam, M.M.; Song, J.; Zhou, T. Flow-induced rotation of a circular cylinder with a detached splitter plate and its bifurcation behavior. *Appl. Ocean Res.* **2022**, *122*, 103150. [CrossRef]
46. Chen, L.; Wu, G. Vortex-induced three-degree-of-freedom vibration of a piggyback circular cylinder system. *Appl. Ocean Res.* **2022**, *123*, 103145. [CrossRef]
47. Ding, L.; Bernitsas, M.M.; Kim, E.S. 2-D URANS vs. experiments of flow induced motions of two circular cylinders in tandem with passive turbulence control for 30,000 <Re <105,000. *Ocean Eng.* **2013**, *72*, 429–440.
48. Kim, E.S.; Bernitsas, M.M. Performance prediction of horizontal hydrokinetic energy converter using multiple-cylinder synergy in flow induced motion. *Appl. Energy* **2016**, *170*, 92–100. [CrossRef]
49. Sun, H.; Ma, C.; Kim, E.S.; Nowakowski, G.; Mauer, E.; Bernitsas, M.M. Hydrokinetic energy conversion by two rough tandem-cylinders in flow induced motions: Effect of spacing and stiffness. *Renew. Energy* **2017**, *107*, 61–80. [CrossRef]
50. Williamson, C.H.K.; Govardhan, R. Vortex-induced vibrations. *Annu. Rev. Fluid Mech.* **2004**, *36*, 413–455. [CrossRef]
51. Shao, N.; Lian, J.; Xu, G.; Liu, F.; Deng, H.; Ren, Q.; Yan, X. Experimental investigation of flow-induced motion and energy conversion of a T-section prism. *Energies* **2018**, *11*, 2035. [CrossRef]
52. Khalak, A.; Williaamson, C.H.K. Fluid forces and dynamics of a hydroelastic structure with very low mass and damping. *J. Fluids Struct.* **1997**, *11*, 973–982. [CrossRef]
53. Raghavan, K. Energy Extraction from a Steady Flow Using Vortex Induced Vibration. Ph.D. Thesis, University of Michigan, Ann Arbor, MI, USA, 2007.

MDPI

Article

Identification of Shark Species Based on Their Dry Dorsal Fins through Image Processing

Luis Alfredo Carrillo-Aguilar [1], Esperanza Guerra-Rosas [2], Josué Álvarez-Borrego [1,*], Héctor Alonso Echavarría-Heras [1] and Sebastián Hernández-Muñóz [3,4]

[1] Centro de Investigación Científica y de Educación Superior de Ensenada (CICESE), Baja California, Carretera Ensenada-Tijuana No. 3918, Zona Playitas, Ensenada 22860, Baja California, Mexico

[2] Facultad de Ingeniería, Arquitectura y Diseño, Universidad Autónoma de Baja California, Km. 103 Carretera Tijuana-Ensenada, Ensenada 22860, Baja California, Mexico

[3] Biomolecular Laboratory, Center for International Programs and Sustainability Studies, Universidad Veritas, San José 10105, Costa Rica

[4] Sala de Colecciones, Facultad de Ciencias del Mar, Universidad Católica del Norte, Coquimbo 1781421, Chile

* Correspondence: josue@cicese.mx

Abstract: Shark populations worldwide have suffered a decline that has been primarily driven by overexploitation to meet the demand for meat, fins, and other products for human consumption. International agreements, such as CITES, are fundamental to regulating the international trade of shark specimens and/or products to ensure their survival. The present study suggests algorithms to identify the dry fins of 37 shark species participating in the shark fin trade from 14 countries, demonstrating high sensitivity and specificity of image processing. The first methodology used a non-linear composite filter using Fourier transform for each species, and we obtained 100% sensitivity and specificity. The second methodology was a neural network that achieved an efficiency of 90%. The neural network proved to be the most robust methodology because it supported lower-quality images (e.g., noise in the background); it can recognize shark fin images independent of rotation and scale, taking processing times in the order of a few seconds to identify an image from the dry shark fins. Thus, the implementation of this approach can support governments in complying with CITES regulations and in preventing illegal international trade.

Keywords: CITES; shark fins; image processing; Fourier transform; neural network; non-linear composite filter

Citation: Carrillo-Aguilar, L.A.; Guerra-Rosas, E.; Álvarez-Borrego, J.; Echavarría-Heras, H.A.; Hernández-Muñóz, S. Identification of Shark Species Based on Their Dry Dorsal Fins through Image Processing. *Appl. Sci.* **2022**, *12*, 11646. https://doi.org/10.3390/app122211646

Academic Editors: Enjin Zhao, Hao Qin and Lin Mu

Received: 6 September 2022
Accepted: 15 November 2022
Published: 16 November 2022

1. Introduction

The increased human exploitation and habitat deterioration over the last half-century has decreased shark populations worldwide [1,2]. Consequently, more than one-third of chondrichthyan species (sharks, rays, and chimeras, hereafter referred to as 'sharks') are threatened with extinction due to a myriad of human-caused threats; however, observed population declines are driven primarily by overexploitation in largely unregulated and unmonitored target and bycatch fisheries worldwide [3]. A global catch assessment estimated that approximately 100 million sharks are caught annually worldwide, including illegal, unreported, and unregulated catch [4]. The reassessment of 1199 species by the International Union for Conservation of Nature (IUCN) Red List reveals that almost 400 chondrichthyan species are jeopardized with extinction [5].

International efforts to improve the management and conservation of sharks have focused on the use of multilateral environmental agreements, such as the Convention on International Trade in Endangered Species of Wild Fauna and Flora (CITES), to ensure that products derived from shark and ray species are traded legally and sustainably [6]. At present, CITES has listed 46 shark and ray species in the Appendices, and the participating

184 countries worldwide should monitor and control trading of shark products to ensure sustainability, legality, and traceability from international trade operations [7].

Economic globalization and exploitation of sharks have strengthened the demand and supply of domestic and international markets for sharks and ray products (mainly meat and fins) [8]. Shark meat markets have remained stable over the last decade, with Brazil, Spain, Uruguay, and Italy accounting for 57% of the average global shark meat imports [9–11]. In contrast, Hong Kong and mainland China are major worldwide trade and consumption centers for seafood, where shark fins are considered a prized cultural treasure and luxury food items, such as sharkfin soup—which is served on formal and special occasions [12].

Unfortunately, international trade data for sharks and their derivative products are rarely collected at the species level, hampering the monitoring of shark species or their derivative parts, such as fins and meat [8]. This represents a major challenge for the implementation of effective monitoring, enforcement, and requirements of countries—referred to as parties—in meeting their obligations under CITES [13].

To achieve compliance with domestic and international regulations for the shark fin trade, there are several accessible identification tools to aid in the implementation of CITES trade controls for listed species, both domestically and at various points along the supply chain (i.e., software iSharkFin version number 1.4, fin guides, and genetic approaches). First, the bioinformatics tool, iSharkFin, developed by FAO and the University of Vigo, was designed to identify 39 species from wet shark fins [14,15]; however, some limitations need to be considered when using this software, including the misidentification of CITES-listed species, particularly when dry fins are analyzed because of the discordance between iSharkFin results, visual diagnostic characteristics, and genetic identification. Currently, this is the only software that is working.

Second, several visual shark fin identification (ID) guides can provide users with a fast and cheap tool for the identification of unprocessed fins from CITES species based on the morphological characteristics of certain fin types, such as the shape and coloration patterns [16–20]; however, the effectiveness of fin ID guides is highly dependent on the training and expertise of users in identifying fins from morphologically distinct species, such as *Sphyrna lewini* and *S. zygaena* [20].

Lastly, advances in molecular approaches that are typically used for the identification of shark and ray species, or their derivative products, in markets have made them more accessible than ever before because these assays can be performed quickly in basic laboratories and are relatively inexpensive. Two widely used genetic tools used to identify body parts at the species level, such as meat and fins, are (a) DNA barcoding (using the COI or ND2 mitochondrial genes [21–26]) and (b) multiplex PCR assays based on the nuclear ribosomal DNA internal transcribed spacer (ITS2) [27–32]. Nevertheless, in Latin American countries, due to financial and logistical restrictions for molecular analysis—such as the salary for a technician—dedicated molecular labs, and validation of a genetic tool for law enforcement systems and courts, DNA techniques are implemented as workflows for domestic inspections from importation, exportation, and re-exportation. As a result, there is an urgent need for a robust tool that can aid in the identification of shark fins in CITES enforcement contexts.

Here, we provide computer techniques and digital correlation systems that offer an accuracy-based solution for image processing, because we can determine the object position to identify the problem image. This first model (non-linear composite filter) has been self-developed and the second model (neural network using the Local Binary Pattern) is a Matlab tool.

Most filters do not function efficiently when the problem image has small distortions, different sizes, rotations, or illumination. Therefore, in recent years, numerous efforts have been made to develop distortion-invariant systems using linear and non-linear filters [33]. Correlation filters were used to identify different species. For example, ceratium was identified with 90% efficiency, independent of images with different rotation sizes [34].

Subsequently, different shrimp tissues were identified to detect hypodermal necrosis and hematopoietic infection virus (IHHN) [35].

In addition, three different approaches involving molecular, morphometric, and image processing were implemented to identify wet and dry dorsal fins in two CITES-listed species (Isurus oxyrinchus and Lamna nasus) and a blue shark (Prionace glauca) from the Chilean shark fin market. The results showed that morphometric analysis lacked the accuracy to discriminate among species, whereas DNA-based identification and image processing were 100% successful [9].

In this study, we used two different image-processing approaches: a non-linear composite filter using the Fourier transform and a neural network to identify the species of origin of 37 dry dorsal fins sourced from 14 countries using photos of the global shark fin trade.

2. Methodology

2.1. General Information about the Image Database

The database used in this project was shark fin photos from the international fin trade established in the project "Enhancing the morphological tools to identify illegal shark fins traded in central America" financed by the Shark Conservation Fund in 2008. Part of this project includes 1029 photos of dry dorsal fins from 37 commercially important shark species taken from 14 different countries: United States, Mexico, Belize, Guatemala, Costa Rica, El Salvador, Panamá, Colombia, Ecuador, Perú, Chile, South Africa, Hong Kong, and Fiji. The database includes two groups (CITES-listed and non-CITES-listed). The CITES-listed species are very important because most of the shark populations are in critical danger, however, there are shark species that are not CITES-listed but are as important as the ones who are CITES-listed; that is why we decided to merge the two groups. The dry shark fin database was identified. Figure 1 shows four different dry shark fin species. The first one corresponds to *Sphyrna lewini* (A), the second to *Sphyrna zygaena* (B), the third to *Carcharodon carcharias* (C), and the last to *Trianodon obesus* (D). The photos were classified by species because we are interested in the population aspects of sharks and rays, including the genetic diversity, connectivity, and morphological supporting tools that can prevent the illegal trafficking of shark products in international trade in Latin America. To validate the use of the algorithms, all the shark fin photos were previously visually identified by shark fin identification experts [18–20], based on their knowledge and published fin field guides, and in particular, the experience training international workshop for government agencies who enforce international trade regulations of CITES-listed species. We compared the photos using two approaches: (i) a non-linear composite filter using Fourier transform and (ii) a neural network applied to test species identification from the dry shark fins. The images can be rotated or scaled. The algorithms were realized in Matlab language.

Figure 1. Dry shark fins from the first dry fin shark species up to the last dry fin shark species. (**A**) is *Sphyrna lewini*, (**B**) is *Sphyrna zygaena*, (**C**) is *Carcharodon carcharias* and (**D**) is *Trianodon obesus*. Database without noise in the background.

We created two databases for the neural network. The first dataset includes 1029 dry dorsal fins with a white background (without noise) and the second dataset contains 4438 dry dorsal fins with noise in the background (random variation of brightness or color information in the background of an image) (Figure 2). We gathered the second dataset of 4438 by removing the background of the first dataset of 1029 photos.

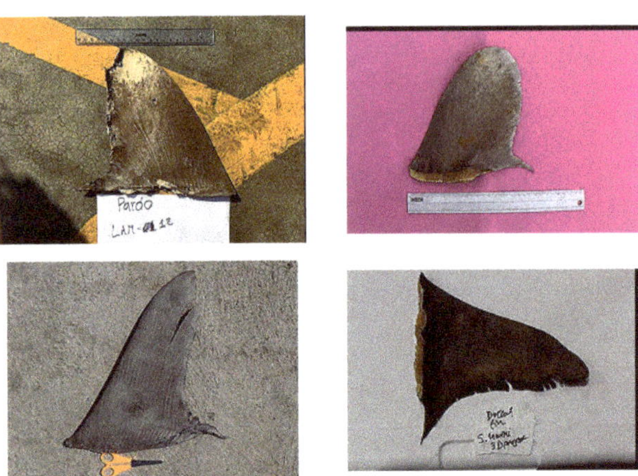

Figure 2. Dry dorsal fins shark database with noise in the background. In the four images we can see different objects, lines, and colors, which may hinder correct identification.

2.2. Non-Linear Compositive Filter

In this section, we present a detailed description of the non-linear composite filters. Figure 3 shows the steps of the non-linear composite filter. In step (A), on the left, there is an input training set (information of the species we want to recognize), I_n, defined by:

$$I_n = \{f_1(x,y), f_2(x,y), \ldots, f_n(x,y)\} \quad (1)$$

where $f_i(x,y)$ for $i = 1,2,\ldots,n$ is a two-dimensional function that represents a digitalized dry shark fin image. In this step, we have n dry shark fin photos, where each one is represented with $f_i(x,y)$.

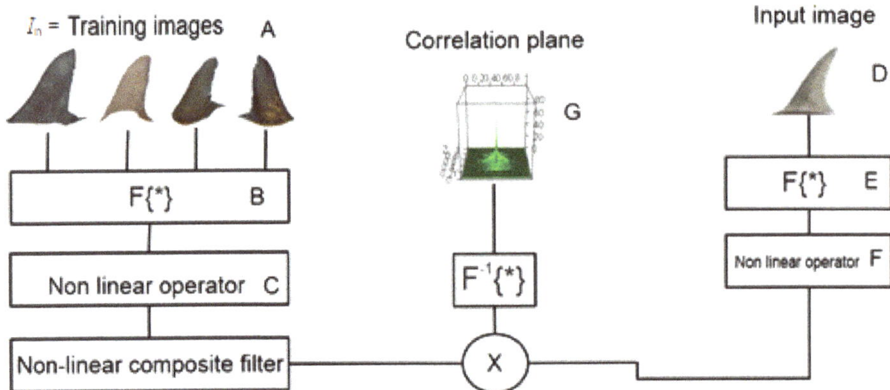

Figure 3. Steps to obtain the non-linear composite filter.

Then, the fast Fourier transform (FFT) was applied to each of the images of the dry shark fins, and because the FFT is a linear integral, we took the total FFT:

$$F(u,v) = \sum_{i=1}^{n} F_i(u,v), \tag{2}$$

where $F_i(u,v)$ represents the Fourier transform for each image in the training set; however, n is the total of dry shark fins in this set, and u and v are the frequency components (Step B).

Furthermore, $F(u,v)$ can be written like:

$$F(u,v) = |F|^k \exp(i\varphi) \tag{3}$$

where k is a non-linear operator ($0 \leq k < 1$) and φ is the phase (Step C).

With the k value selected, we get a better signal in both images. In this case, we choose $k = 0$; for this reason, the non-linear composite filter was realized with filters of phase only. The same procedure is applied to the input images (Steps D, E, and F). The results obtained from both the training and input images were multiplied to obtain a correlation plane [36] (Step G). If we have a single peak in the correlation plane, it means that we get a correct identification.

2.3. Neural Network

The second methodology consists of a neural network [37]. We used the local binary pattern function to obtain a vector of 59 elements for each image [38]. This algorithm is a simple and efficient descriptor that describes the textures (edges, corners, spots, and flat regions), and it is invariant to rotation and scale [39]. Furthermore, the Levenberg–Marquardt method is used [40]. The neural network consists of the following steps.

The neurons are simple information processors. The output layer comprises neurons that receive signals from the environment $(x1, x2, x3, \ldots, x59)$. In this case, the input layer was the texture vector of the image. The hidden layer has three elements (error, weight, and sigmoid function). The errors and weights were random values. The sigmoid function transforms negative values into 0 and positive values are represented by 1; it is one of the most widely used non-linear activation functions. The mathematical expression is as follows:

$$y = \frac{1}{1+e^{-x}} \tag{4}$$

where y, $(0 \leq y \leq 1)$ is the output and x is the real input value in the sigmoid function (logistic function).

The output layer consisted of 38 neurons, and each neuron was a dry shark dorsal fin.

Figure 4 shows the steps of a neural network with one hidden layer, which are described below. A three-layer neural network was used in this study. The first layer is an input layer containing 59 elements. The hidden layer was a single layer with 20 neurons, and the third layer was the output layer with 38 output values. Each of these outputs corresponds to a species of dry shark dorsal fin. Each of these features was assigned a random weight and an error value. In the hidden layer, the weight values are summed and the error is subtracted. The obtained value was affected by the sigmoid function. This procedure was performed to obtain the value in the output layer.

Of the 38 outputs, 37 belonged to each shark species studied in this study and one control group. This control group was created such that when a dry dorsal fin was identified and did not belong to any of the 37 species, the network would place it in the control group and thus avoid a possible error when identifying it with another species. These neural network steps are repeated as a cycle. In each neural network, 80% of the images were randomly selected for training, 10% were randomly selected for testing, and 10% were randomly selected for validating data. The photos of fins with different backgrounds that were used in the training of the neural network are not the same as those used to perform the validation and testing of the network. This procedure is performed until the global minimum value of the error function is obtained. The purpose of testing is to compare the outputs from the neural network against targets in an independent set, and the purpose of the validation set is to fine-tune the hyperparameters of the model and is considered a part of the training of the model [41]. Generally, 80% are used for training, 10% are used for testing, and 10% are used for validation in neural networks.

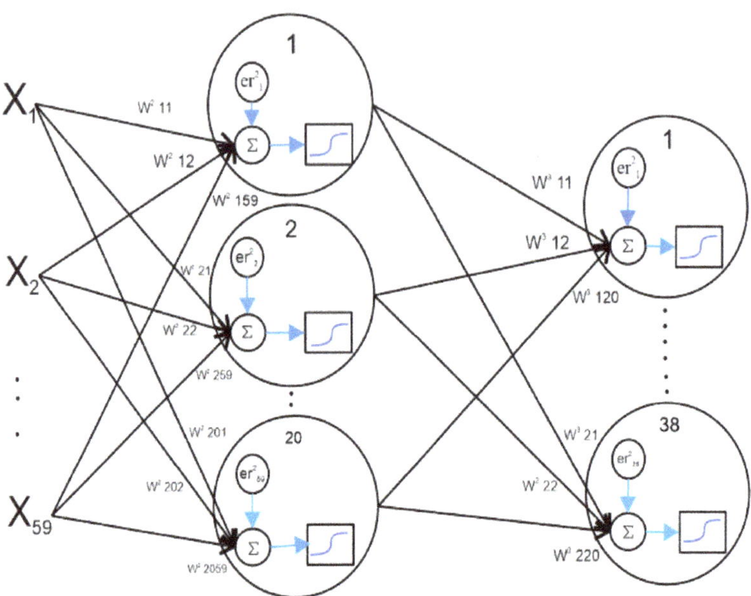

Figure 4. Scheme of a neural network with a hidden layer.

Finally, the percentages for sensitivity and specificity were applied to each result to determine the effectiveness of each methodology.

Sensitivity was defined as the proportion of individuals correctly identified as belonging to Species 1. The mathematical expression is as follows:

$$\frac{TP}{TP+TN} \tag{5}$$

where TP corresponds to true positives and TN corresponds to true negatives.

Specificity was defined as the proportion of correctly identified individuals that did not belong to Species 1.

$$\frac{TN}{TN+FP} \tag{6}$$

where TN corresponds to true negatives and FP corresponds to false positives.

3. Results

3.1. Non-Linear Composite Filter

The numerical simulations performed for the non-linear composite filters provided the most representative results for identifying dry shark fins from the CITES-listed and non-listed species (n = 37). Table 1 shows that the species-specific composite filters developed for the 37 shark species showed excellent identification of CITES-listed and non-listed species (n = 37), with 100% sensitivity and specificity. In addition, the optimal value of the non-linear operator (k) was found to be 0.

Table 1. Sensitivity and specificity percentage of each of the dry dorsal fin species of sharks using the non-linear composite filter.

Scientific Name	Dry Dorsal Fins	% Sensitivity	% Specificity
Sphyrna lewini	262	100	100
Sphyrna zygaena	92	100	100
Sphyrna mokarran	16	100	100
Lamna nasus	9	100	100
Carcharodon carcharias	16	100	100
Carcharhinus longimanus	22	100	100
Carcharhinus falciformis	101	100	100
Alopias vulpinus	24	100	100
Alopias pelagicus	75	100	100
Alopias superciliosus	98	100	100
Isurus oxyrinchus	49	100	100
Isurus paucus	3	100	100
Rhincodon typus	4	100	100
Carcharhinus acronotus	4	100	100
Carcharhinus brachyurus	2	100	100
Carcharhinus brevipinna	22	100	100
Carcharhinus isodon	5	100	100
Carcharhinus leucas	19	100	100
Carcharhinus limbatus	51	100	100
Carcharhinus obscurus	27	100	100
Carcharhinus perezi	3	100	100
Carcharhinus plumbeus	1	100	100
Carcharhinus sealei	6	100	100
Carcharhinus signatus	2	100	100
Carcharhinus taurus	8	100	100
Galeocerdo cuvier	33	100	100
Ginglymostoma cirratum	2	100	100
Ginglymostoma unami	4	100	100
Mustelus lunulatus	3	100	100
Mustelus mustelus	5	100	100
Negaprion acutidens	2	100	100
Negaprion brevirostris	2	100	100
Prionace glauca	39	100	100
Rhizoprionodon acutus	1	100	100
Rhizoprionodon longurio	10	100	100
Sphyrna tiburo	5	100	100
Trianodon obesus	2	100	100

3.2. Neural Network

Four experiments were conducted using a neural network that varied the number of neurons in the hidden layer, species, and noise in the images. The second experiment was the best neural network because we obtained a 90% efficiency with 20 neurons in the hidden layer. Efficiencies between 84% and 88% were obtained in the other runs. The experiments were conducted as follows.

Table 2 shows the results from the first experiment with 37 shark species with a white background and one control group using ten neurons in the hidden layer. We repeated the neural network 15 times to determine the optimal neural network efficiency. The epochs are the number of cycles that the neural network performed to reach the global minimum value of the error function. Efficiency is a relative value that shows the ratio between the achieved result and the used resource. In this experiment, the best neural network achieved an efficiency of 88.9%.

Table 2. First experiment. Fifteen runs of the neural network with 37 species, 1 control group (38 "species"), and 10 neurons in the hidden layer.

Species	Neural Network	Layer	Epochs	Time	% Efficiency	Neurons
38	1	1	128	44 min	88.6	10
38	2	1	127	43 min	88.4	10
38	3	1	127	20 min	86.1	10
38	4	1	122	49 min	85.8	10
38	5	1	138	44 min	87.4	10
38	6	1	125	43 min	83.8	10
38	7	1	128	77 min	86.3	10
38	8	1	128	46 min	88.9	10
38	9	1	128	47 min	88.6	10
38	10	1	127	22 min	86.1	10
38	11	1	127	43 min	88.4	10
38	12	1	122	44 min	85.8	10
38	13	1	138	37 min	87.4	10
38	14	1	125	40 min	83.8	10
38	15	1	128	60 min	86.3	10

Table 3 shows the second experiment with 37 shark species with a white background and one control group. We obtained an efficiency of 90% (shown in yellow) for the four neural networks.

Table 3. Second experiment. Fifteen runs of the neural network with 37 species, 1 control group (38 "species"), and 20 neurons in the hidden layer.

Species	Neural Network	Layer	Epochs	Time	% Efficiency	Neurons
38	1	1	179	90 min	87.8	20
38	2	1	259	120 min	87.6	20
38	3	1	166	82 min	87.8	20
38	4	1	128	76 min	86.3	20
38	5	1	151	89 min	88.8	20
38	6	1	150	78 min	90.6	20
38	7	1	201	99 min	90.3	20
38	8	1	161	83 min	84.9	20
38	9	1	179	94 min	87.8	20
38	10	1	259	134 min	87.6	20
38	11	1	166	92 min	87.8	20
38	12	1	151	78 min	88.8	20
38	13	1	150	66 min	90.6	20
38	14	1	201	89 min	90.3	20
38	15	1	161	72 min	84.9	20

Table 4 shows the sensitivity and specificity percentage of each dry dorsal fin shark species using the neural network with 90% efficiency. In this table, we show the 100% sensitivity for *Carcharhinus plumbeus, Ginglymostoma cirratum* and *Negaprion acutidens*. *Carcharhinus limbatus* had a sensitivity of 56.43%; this was the lowest percentage of all species.

The rest had between 65.34% and 99.04% sensitivity. The specificity was between 98.05% and 100%.

Table 4. Sensitivity and specificity percentage of each dry dorsal fin shark species using the neural network with 90% efficiency.

Scientific Name	Dry Dorsal Fins without Noise	% Sensitivity	% Specificity
Sphyrna lewini	262	92.4%	98.05%
Sphyrna zygaena	92	66.96%	99.72%
Sphyrna mokarran	16	78.21%	99.46%
Lamna nasus	9	91.34%	99.72%
Carcharodon carcharias	16	88.10%	99.4%
Carcharhinus longimanus	22	87.25%	99.86%
Carcharhinus falciformis	101	65.34%	99.24%
Alopias vulpinus	24	88.46%	99.72%
Alopias pelagicus	75	83%	99.37%
Alopias superciliosus	98	90.26%	99.70%
Isurus oxyrinchus	49	71.15%	99.16%
Isurus paucus	3	98.05%	99.89%
Rhincodon typus	4	98.07%	100%
Carcharhinus acronotus	4	96.15%	99.94%
Carcharhinus brachyurus	2	97.05%	99.91%
Carcharhinus brevipinna	22	92.85%	99.86%
Carcharhinus isodon	5	98%	100%
Carcharhinus leucas	19	85.57%	99.67%
Carcharhinus limbatus	51	56.43%	99.91%
Carcharhinus obscurus	27	89.74%	99.45%
Carcharhinus perezi	3	95.14%	99.59%
Carcharhinus plumbeus	1	100%	99.83%
Carcharhinus sealei	6	95.28%	99.91%
Carcharhinus signatus	2	97.05%	99.80%
Carcharhinus taurus	8	95.37%	99.61%
Galeocerdo cuvier	33	82.23%	99.89%
Ginglymostoma cirratum	2	100%	99.78%
Ginglymostoma unami	4	97.11%	100%
Mustelus lunulatus	3	98.05%	99.89%
Mustelus mustelus	5	99.04%	99.89%
Negaprion acutidens	2	100%	99.80%
Negaprion brevirostris	2	99.01%	99.91%
Prionace glauca	39	91.59%	99.64%
Rhizoprionodon acutus	1	97.02%	99.94%
Rhizoprionodon longurio	10	94.54%	99.91%
Sphyrna tiburo	5	97.14%	99.86%
Trianodon obesus	2	99.01%	99.89%
Random group	80	93.91%	99.91%

We performed a third experiment based on the first two experiments. Nine shark species had only five images, which is why they were not considered in this experiment. There were 27 dry shark fin species with a white background and one control group.

Table 5 shows the results of the third experiment, with 26 species and one control group. Here, we have three neural networks with an 89% efficiency.

Table 5. Third experiment. Fifteen runs of the neural network with 27 species and 20 neurons in the hidden layer.

Species	Neural Network	Layer	Epochs	Time	% Efficiency	Neurons
27	1	2	168	21 min	87.7	20
27	2	2	207	27 min	89.7	20
27	3	2	169	21 min	89.6	20
27	4	2	209	26 min	87.5	20
27	5	2	181	23 min	90.1	20
27	6	2	176	22 min	89.1	20
27	7	2	216	29 min	87	20
27	8	2	191	25 min	88.4	20
27	9	2	181	38 min	87.2	20
27	10	2	246	52 min	89.4	20
27	11	2	164	35 min	88.4	20
27	12	2	198	41 min	87.6	20
27	13	2	287	51 min	88.2	20
27	14	2	168	26 min	88.5	20
27	15	2	227	30 min	87.3	20

Table 6 shows the fourth experiment with 37 species and one control group. Here, we have two neural networks with an 89% efficiency. In this experiment, there was a good percentage because the number of dry shark fins increased for each species.

Table 6. Fourth experiment. Fifteen runs of the neural network with 37 species, 1 control group (38 "species"), and 20 neurons in the hidden layer.

Species	Neural Network	Layer	Epochs	Time	% Efficiency	Neurons
38	1	1	186	108	80.1	20
38	2	1	186	110	83.1	20
38	3	1	147	48	84.5	20
38	4	1	154	52	89	20
38	5	1	148	49	81.3	20
38	6	1	137	45	86.1	20
38	7	1	130	46	82.9	20
38	8	1	147	48	84.5	20
38	9	1	147	49	84.5	20
38	10	1	147	48	84.5	20
38	11	1	147	49	84.5	20
38	12	1	194	66	82.9	20
38	13	1	150	53	83.4	20
38	14	1	154	55	89	20
38	15	1	147	49	84.5	20

The final experiment (Table 6) was performed using a database of dry shark fin images with and without background noise to increase the number of dry fins in each species. From this database, 4438 images of the dry dorsal fins of sharks were obtained.

Table 7 shows the sensitivity and specificity of each dry dorsal fin shark species using a neural network with 89% efficiency. In this table, we show the 100% sensitivity for

Carcharhinus plumbeus. *Sphyrna zygaena* had a sensitivity of 66.37%; this was the lowest percentage of all species. The rest had between 75% and 99% sensitivity. The specificity was between 98% and 99%.

Table 7. Sensitivity and specificity percentage of each dry dorsal fin shark species using the neural network with 89% efficiency. This table represents the database of dry shark fins with noise in the background of the images.

Scientific Name	Dry Dorsal Fins with Noise	% Sensitivity	% Specificity
Sphyrna lewini	459	91.28%	98.65%
Sphyrna zygaena	113	66.37%	99.35%
Sphyrna mokarran	104	82.69%	99.66%
Lamna nasus	101	90.09%	99.63%
Carcharodon carcharias	101	96.03%	99.68%
Carcharhinus longimanus	105	92.38%	99.66%
Carcharhinus falciformis	130	77.69%	99.25%
Alopias vulpinus	113	86.72%	99.63%
Alopias pelagicus	138	78.98%	99.40%
Alopias superciliosus	173	86.70%	99.45%
Isurus oxyrinchus	116	81.03%	99.22%
Isurus paucus	103	98.05%	99.81%
Rhincodon typus	105	98.05%	99.92%
Carcharhinus acronotus	100	95%	99.89%
Carcharhinus brachyurus	101	95.04%	99.94%
Carcharhinus brevipinna	105	74.28%	99.61%
Carcharhinus isodon	102	92.15%	99.89%
Carcharhinus leucas	109	79.81%	99.58%
Carcharhinus limbatus	102	75.49%	99.23%
Carcharhinus obscurus	104	88.46%	99.33%
Carcharhinus perezi	104	96.15%	99.58%
Carcharhinus plumbeus	102	100%	99.94%
Carcharhinus sealei	104	92.30%	99.79%
Carcharhinus signatus	103	97.08%	99.79%
Carcharhinus taurus	108	88.88%	99.79%
Galeocerdo cuvier	117	71.79%	99.92%
Ginglymostoma cirratum	101	96.03%	99.71%
Ginglymostoma unami	100	99%	99.79%
Mustelus lunulatus	100	94%	99.92%
Mustelus mustelus	100	94%	99.94%
Negaprion acutidens	100	97%	99.92%
Negaprion brevirostris	100	98%	99.89%
Prionace glauca	100	81%	99.30%
Rhizoprionodon acutus	100	99%	99.89%
Rhizoprionodon longurio	100	85%	99.87%
Sphyrna tiburo	100	92%	99.92%
Trianodon obesus	100	96%	99.89%
Random group	115	85.21%	99.63%

4. Discussion

The results obtained in this study show that the non-linear composite phase filter can successfully correlate (100%) with 37 different species of dry shark dorsal fins. In this context, the results obtained were similar to those obtained using species-specific composite filters to identify the dry fins (dorsal fins, right-sided pectoral fins, and caudal fins) of three shark species: *Prionace glauca*, *Isurus oxyrinchus*, and *Lamna nasus*. A 100% identification was recorded among the fins of each species analyzed [36]; however, in the study of [36], an inverse Gaussian filter was used to enhance the high frequencies, and the technique in [34] was used to have rotation invariance and the confidence level was calculated (95.4%). Only a phase filter was used in this study, and the percentages for the sensitivity and specificity were calculated.

A non-linear compound filter uses this value, k, as the non-linear operator. By changing the value to 1, we obtain a classically matched filter that has the advantage of optimizing the output when the input signal (image problem) is degraded by additive white noise [33]. When $k = 0$, we have a phase-only filter that maximizes the light efficiency in an optical system; moreover, when $k = -1$, we have an inverse filter that minimizes the correlation energy criteria. This last filter produces a narrower peak in the output correlation plane if the reference image and problem image are the same [34]. When the non-linear operator modifies the Fourier transform of the problem and reference images, we consider that we have a non-linear processor. The intermediate values of (0.1, 0.2, 0.3, ... , 0.9) allow us to vary the characteristics of the processor, such as the discrimination capacity and its variance to illumination [42]. It is essential to consider that in these results with the non-linear composite filter, the non-linear filter law (when k is different from zero) was discarded because when varying the value between 0 and 1, it was found that the best correlation peak was at $k = 0$. This represents a correlation using a phase-only filter [33].

The disadvantages of using the non-linear composite phase filter are as follows. Suppose we use n filters corresponding to n species. In this case, the algorithm takes a long time to process hundreds of problem images that contain different species; however, this disadvantage does not occur when using neural networks, since identifying a fin photo takes between 0.18 s to 0.48 s and information from all species has already been integrated.

The percentage of efficiency in the tables corresponds to the confusion matrix. The diagonal of this matrix indicates the number of fins correctly identified and the percentages outside of this diagonal shows all the fins that were not correctly identified.

The first neural network with a white background obtained 90% efficiency. *Sphyrna lewini* had 92% efficiency from 262 images, *Sphyrna zygaena* had 66% efficiency from 92 images, and *Sphyrna mokarran* had 78% efficiency from 16 images. In the second neural network with the noise in the background, we obtained 89% efficiency. *Sphyrna lewini* had 91% efficiency from 459 images, *Sphyrna zygaena* had 66% efficiency from 113 images, and *Sphyrna mokarran* had 82% efficiency from 104 images. This indicates that 66% of the images were correctly identified as belonging to *Sphyrna zygaena*. The low percentage of sensitivity is because there is less information from the images in both species; therefore, a more significant number of images is needed to obtain a more robust model. However, there are some species with 94–100% sensitivity, such as *G. unami*, *N. brevirostris*, and *N. acutidens*, which have high percentages because they do not look like the rest of the other species.

Having more variability in the database for each species will benefit the algorithm because it holds more information for each species and has a higher sensitivity percentage.

The local binary pattern function is essential because it is a texture classifier (that focuses on edges, corners, spots, and flat regions). It is designed to tolerate noise and handle grayscale, rotation, and scale-invariant images [38]. Therefore, our database is composed of photos in which some of the images are rotated in different directions to create a robust algorithm.

The advantage of using more than one layer and, in each layer, using more than 20 neurons is that we might obtain better efficiency, but, as a consequence, the neural network will take longer for training. Therefore, it will be better if we increase the number of images in each species to have better efficiency.

The neural network can be replicated to identify wet dorsal fins, as well as wet and dry pectoral fins. This is the first step in creating a tool for CITES agents to use to prevent international trade in the Asian market. Even so, building capacities for the implementation of CITES species is highly recommended in Latin American and global countries. Nevertheless, algorithmic tools must be provided to government agencies and inspectors in order to prevent international trade. Updating the identification of CITES species and non-CITES species with algorithms from machine learning systems could be salvageable in the future in order to conserve the remaining shark populations, which have been in decline since 1950 due to overfishing.

5. Conclusions

All species were identified using a non-linear composite filter with 100% sensitivity and specificity. Although a perfect percentage was obtained, this was not the best methodology for the following reasons: (1) It is not rotation- or scale-invariant; and (2) the filter takes a long time to identify a problem image (fin photo) because the problem image must be correlated with each image in the database. It can be made invariant if a non-linear composite phase filter is fed with hundreds of rotated and scaled images of the species to be identified. This filter can also be fed images with different illumination levels and fragmented images.

The best methodology for identifying dry dorsal fins for this study is the neural network, primarily because of the short time required to identify a species. The sensitivity and specificity of the studied species can be increased when the network is fed hundreds or thousands of images. Two high percentages were obtained in this study: 90% with images without a background and 89% with images with noise in the background. This methodology supports noisy images and is invariant to scale and rotation. In addition, it takes between 0.18 s and 0.48 s to identify a problem image (fin photo).

If we do not understand the problem impacting the shark populations in the following years, we would be responsible for driving all shark species to extinction because of a lack of conscience.

Author Contributions: Methodology, L.A.C.-A., H.A.E.-H. and S.H.-M.; Software, E.G.-R.; Validation, L.A.C.-A.; Formal analysis, J.Á.-B.; Investigation, L.A.C.-A., E.G.-R. and H.A.E.-H.; Data curation, S.H.-M.; Visualization, E.G.-R.; Supervision, J.Á.-B.; Project administration, S.H.-M.; Funding acquisition, J.Á.-B. and H.A.E.-H. All authors have read and agreed to the published version of the manuscript.

Funding: This research was funded by Centro de Investigación Científica y de Educación Superior de Ensenada (CICESE), Baja California, grant number F0F181.

Institutional Review Board Statement: Not applicable.

Informed Consent Statement: Not applicable.

Acknowledgments: Luis Alfredo Carrillo Aguilar hold a Master degree in Centro de Investigación Científica y de Educación Superior de Ensenada (CICESE) supported by CONACYT scholarship. SHARK CONSERVATION FUNDING We thank Debra Abercrombie for reviewing this article.

Conflicts of Interest: The authors declare no conflict of interest related to this study.

References

1. Julia, K.B.; Ransom, A.M.; Daniel, G.K.; Boris, W.; Shelton, J.H.; Penny, A.D. Collapse and conservation of shark populations in the northwest Atlantic. *Science* **2003**, *299*, 389–392. [CrossRef]
2. Peter, W.; Myers, R.A. Shifts in open-ocean fish communities coinciding with the commencement of commercial fishing. *Ecology* **2005**, *86*, 835–847. [CrossRef]
3. Rafael, M.; Imanol, M.; William, D.; Catherine, S.; Nicholas, K.D.; Kent, E.C.; Beth, P.; Nadia, D.R.; Caroline, P.; Craig, H.T.; et al. Monitoring extinction risk and threats of the world's fishes based on the sampled red list index. *Rev. Fish Biol. Fish.* **2022**, *32*, 975–991. [CrossRef]
4. Boris, W.; Brendal, D.; Lisa, K.; Christine, A.W.P.; Demian, C.; Michael, R.H.; Steven, T.K.; Samuel, H.G. Global catches, explotation rates, and rebuilding options for sharks. *Mar. Policy* **2013**, *40*, 194–204.
5. Nicholas, D.K.; Pacoureau, N.; Rigby, C.L.; Pollom, R.A.; Jabado, R.W.; Ebert, D.A.; Finucci, B.; Pollock, C.M.; Cheok, J.; Derrick, D.H.; et al. Overfishing drives over one-third of all sharks and rays toward a global extinction crisis. *Curr. Biol.* **2021**, *31*, 4773–4787.
6. CITES. 2019. Appendices I, II and III (Valid from 26 November 2019). Available online: www.cites.org/eng/app/appendices.php (accessed on 7 May 2022).
7. Amanda, C.J.V.; Yvonne, J.C.; Sarah, F.L.; Susan, L.; Felix, C. The role of CITES in the conservation of marine fishes subject to international trade. *Fish Fish. Curr.* **2014**, *15*, 563–592.
8. Alyson, P.; Kelly, M.; Emily, K.; Audrey, C.; Daniel, K.; Stefania, V.; Kim, F. *CITES and the Sea: Trade in Commercially Exploited CITES-Listed Marine Species*; FAO Fisheries and Aquaculture Technical Paper No. 666; FAO: Rome, Italy, 2021.

9. Felix, D.; Shelley, C. *State of the Global Market for Shark Products*; FAO Fisheries and Aquaculture Technical Paper No. 666; FAO: Rome, Italy, 2015; Volume 31, pp. 4773–4787.
10. Nicola, O.; Glenn, S. *An Overview of Major Shark Traders Catchers and Species, State of the Global Market for Shark Products*; TRAFFIC: Cambridge, UK, 2019.
11. Bianca, R.S.; Rodrigo, B.; Nathalie, G.; Carolina, C. Brazil can protect sharks worldwide. *Science* **2021**, *373*, 633. [CrossRef]
12. Kwok, H.S.; Allen Nicola, T. From boat to bowl: Patterns and dynamics of shark fin trade in Hong Kong—Implications for monitoring and management. *Mar. Policy* **2017**, *81*, 330–339.
13. Cardeñosa, D.; Fields, A.T.; Babcock, E.A.; Zhang, H.; Feldheim, K.; Shea, S.K.H.; Fischer, G.A.; Chapman, D.D. CITES-listed sharks remain among the top species in the contemporary fin trade. *Conserv. Lett.* **2018**, *11*, e12457. [CrossRef]
14. Lindsay, J.M.; Marone, B. *SharkFin Guide: Identifying Sharks from Their Fins*; FAO: Rome, Italy, 2017.
15. Monica, B.; Frederik, H.M.; Jenny, L.G.; Lindsay, M.J.; Melany, V.M.; Carlotta, M.; Elisa, P.C.; Jürgen, H.; Castor, G. Performance of iSharkFin in the identificationof wet dorsal fins from priority shark species. *Ecol. Inform.* **2022**, *68*, 101514.
16. Hideki, N.; Toru, K. *Identification of Eleven Sharks Caught by Tuna Long-Line Using Morphological Characters of Their Fins*; Information Paper of the FAO Technical Working Group on the Conservation and Management of Sharks; Food and Agriculture Organization: Rome, Italy, 2000; p. 11.
17. Anonymous. *Characterization of Morphology of Shark Fin Products: A Guide of the Identification of Shark Fin Caught by Tuna Long-Line Fishery*; Fisheries Agency of Japan: Tokyo, Japan, 2016; p. 24.
18. Debra, A.L.; Demian, C.D.; Simon, J.B.G.; John, C.K. *Visual Identification of Fins from Common Elasmobranchs in the Northwest Atlantic Ocean*; NMFS-SEFSC-643; Fundación Mundo Azul: Guatemala, Guatemala, 2013; p. 51.
19. Christopher, A.C.; Sebastian, H.M.; Elisa, A.M. *Guía de Identificación de Aletas de Tiburones en Guatemala Incluidas en el Apéndice II de CITES*; Fundación Mundo Azul: Guatemala, Guatemala, 2018.
20. Sebastian, H.M.; Maike, H.; Debra, A.L. *Guía de Identificación de Aletas de Tiburones en el Perú*, 1st ed.; Oceana: Lima, Peru, 2022. [CrossRef]
21. Ward, R.D.; Zemlak, T.S.; Innes, B.H.; Last, P.R.; Hebert, P.D. DNA barcoding Australia's fish species. *Philos. Trans. R. Soc. B Biol. Sci.* **2005**, *360*, 1847–1857. [CrossRef]
22. Ward, R.D.; Zemlak, T.S.; Innes, B.H.; Last, P.R.; Hebert, P.D. DNA sequence-based approach to the identification of shark and ray species and its implications for global elasmobranch diversity and parasitology. *Bull. Am. Mus. Nat. Hist.* **2012**, *367*, 1–262.
23. Yang, Z.; Zhongze, W.; Chunguang, Z.; Zhibin, M.; Zhigang, J.; Jie, Z. DNA barcoding of Mobulid Ray Gill Rakers for Implementing CITES on Elasmobranch in China. *Sci. Rep.* **2016**, *6*, 37567. [CrossRef]
24. Diego, C.; Andrew, F.; Debra, A.; Kevin, F.; Stanley, S.K.H.; Demian, C.D. A multiplex PCR mini-barcode assay to identify processed shark products in the global trade. *PloS ONE* **2016**, *12*, e0185368.
25. Dirk, S.; Andrea, B.A.; Rebekah, H.L.; Paul, H.; Robert, H.; Mahmood, S.S. DNA analysis of traded shark fins and mobulid gill plates reveals a high proportion of species of conservation concern. *Sci. Rep.* **2017**, *7*, 9505. [CrossRef]
26. Grace, W.C.B.; Hoi, Y.W.; Kwang, T.S.; Pang, C.S. Rapid detection of CITES-listed shark fin species by loop-mediated isothermal amplification assay with potential for field use. *Sci. Rep.* **2020**, *10*, 4455. [CrossRef]
27. Mahmood, S.; Shelley, C.; Melissa, P.; Lisa, N.; Nancy, K.; Michael, S. Genetic identification of pelagic shark body parts for conservation and trade monitoring. *Conserv. Biol.* **2002**, *16*, 1036–1047. [CrossRef]
28. Demian, C.; Debra, L.A.; Cristhophe, J.D.; Ellen, K.P. A streamlined, bi-organelle, multiplex PCR approach to species identification: Application to global conservation and trade monitoring of the great white shark, Carcharodon carcharias. *Conserv. Genet.* **2003**, *4*, 415–425. [CrossRef]
29. Jennifer, M.; Elle, P.K.; Clarke, S.; Colin, N.; Russ, H.; Mahmood, S. Genetic tracking of basking shark products in international trade. *Anim. Conserv.* **2007**, *10*, 199–207. [CrossRef]
30. Debra, L.A. Efficient PCR-Based Identification of Shark Products in Global Trade: Applications for the Management and Conservation of Commercially Important Mackerel Sharks (Family Lamnidae), Thresher Sharks (Family Alopiidae) and Hammerhead Sharks (Family Sphyrnidae). Master's Thesis, Nova Southeastern University, Fort Lauderdale, FL, USA, 2004. Available online: http://nsuworks.nova.edu/occ_stuetd/131 (accessed on 31 January 2004).
31. Debra, L.A.; Clarke, S.; Mahmood, S. Global-scale genetic identification of hammerhead sharks: Application to assessment of the international fin trade and law enforcement. *Conserv. Genet.* **2005**, *6*, 775–788. [CrossRef]
32. Diego, C.; Jessica, Q.; Kwok, H.S.; Demian, C.D. Multiplex real-time PCR assay to detect illegal trade of CITES-listed shark species. *Sci. Rep.* **2018**, *8*, 16313. [CrossRef]
33. Ángel, C.B. Non-Linear Pattern Recognition Invariant to Position, Rotation, Scale and Image Noise. Ph.D. Thesis, Department of Engineering, UABC University, Ensenada, Baja California, México, 2010.
34. José, P.P.; Josué, A.B. Optical-digital system applied to the identification of five phytoplankton species. *Mar. Biol.* **2020**, *132*, 357–365. [CrossRef]
35. Josué, A.B.; María Cristína, C.S. Detection of IHHN virus in shrimp tissue by digital color correlation. *Aquaculture* **2020**, *194*, 1–9. [CrossRef]
36. Sebastián, H.; Cristian, G.E.; Josué, A.B.; Teresa, G.M.; Pilar, H. A multidisciplinary approach to identify pelagic shark fins by molecular, morphometric and digital correlation data. *Hidrobiologica* **2020**, *20*, 71–80.

37. Aaron, L.L.J.; Esperanza, G.R.; Josué, A.B. Multi-class diagnosis of skin lesions using the fourier spectral information of images on additive color model by artificial neural network. *IEEE Access* **2021**, *9*, 35207–35216. [CrossRef]
38. Abdolhossein, F.; Ahmad Reza, N.N. Noise tolerant local binary pattern operator for efficient texture analysis. *Pattern Recognit. Lett.* **2020**, *33*, 1093–1100. [CrossRef]
39. Ojala, T.; Pietikainen, M.; Maenpaa, T. Multiresolution Gray-Scale and Rotation Invariant Texture Classification with Local Binary Patterns. *IEEE Trans. Pattern Anal. Mach. Learn.* **2002**, *24*, 971–987. [CrossRef]
40. Cortés, O.C. Application of the Levenberg-Marquardt Method and the Conjugate Gradient in the Estimation of the Heat Generation of a Hot Plate Apparatus with Guard. Master's Thesis, Department of Mechanical Engineering, Cenidet University, Cuernavaca, Mexico, 2004.
41. Available online: https://www.google.com/search?q=testing+and+validating+data+in+a+neural+network&sxsrf=ALiCzsYuSZj_AwopgyDhspEsc4lzkOzmgQ%3A1667342819650&ei=46FhY8myJ8KlkPIPzNCWgA0&oq=testing+and+validating+data+in+a+ne&gs_lcp=Cgxnd3Mtd2l6LXNlcnAQARgAMgUIIRCgATIFCCEQoAEyBQghEKABMgQIIRAVMggIIRAWEB4QHTIICCEQFhAeEB0yCAghEBYQHhAdMggIIRAWEB4QHTIICCEQFhAeEB0yCAghEBYQHhAdOgQIIxAnOgoIABCxAxCDARBDOgsIABCABBCxAxCDAToRCC4QgAQQsQMQgwEQxwEQ0QM6BAgAEEM6CwguEIAEELEDEIMBOgsILhCABBDHARDRAzoLCC4QsQMQgwEQ1AI6BwgAELEDEEM6EQguEIAEELEDEMcBENEDENQCOggIABCABBCxAzoFCAAQgAQ6CwgAELEDEIMBEMkDOgUILhCABDoICAAQgAQQywE6CQgAEIAEEA0QEzoGCAAQHhANOgYIABAWEB5KBAhBGABKBAhGGABQAFiXd2D6mAFoAXAAeAGAAXqIAawakgEFMjguMTCYAQCgAQHAAQE&sclient=gws-wiz-serp (accessed on 1 November 2022).
42. Bahram, J. Non-linear joint power spectrum based optical correlation. *Appl. Opt.* **2021**, *28*, 2358–2367. [CrossRef]

 *applied sciences*

MDPI

Article

Underwater Sparse Acoustic Sensor Array Design under Spacing Constraints Based on a Global Enhancement Whale Optimization Algorithm

Lening Wang [1], Hangfang Zhao [1,2,3,*] and Qide Wang [4]

1 College of Information Science and Electronic Engineering, Zhejiang University, Hangzhou 310027, China
2 Key Laboratory of Ocean Observation Imaging Testbed of Zhejiang Province, Zhejiang University, Hangzhou 310027, China
3 The Engineering Research Center of Oceanic Sensing Technology and Equipment, Ministry of Education, Beijing 100816, China
4 The State Key Laboratory of Computer-Aided Design and Computer Graphics, Zhejiang University, Hangzhou 310027, China
* Correspondence: hfzhao@zju.edu.cn

Abstract: Sparse arrays with low cost and engineering complexity are widely applied in many fields. However, the high peak sidelobe level (PSLL) of a sparse array causes the degradation of weak target detection performance. Particularly for the large size of underwater low-frequency sensors, the design problem requires a minimum spacing constraint, which further increases the difficulty of PSLL suppression. In this paper, a novel swarm-intelligence-based approach for sparse sensor array design is proposed to reduce PSLL under spacing constrains. First, a global enhancement whale optimization algorithm (GEWOA) is introduced to improve the global search capability for optimal arrays. A three-step enhanced strategy is used to enhance the ergodicity of element positions over the aperture. In order to solve the adaptation problem for discrete array design, a position decomposition method and a V-shaped transfer function are introduced into off-grid and on-grid arrays, respectively. The effectiveness and superiority of the proposed approach is validated using experiments for designing large-scale low-frequency arrays in the marine environment. The PSLL of the off-grid array obtained by GEWOA was nearly 3.8 dB lower than that of WOA. In addition, compared with other intelligent algorithms, the on-grid array designed using GEWOA had the lowest PSLL.

Keywords: sparse array design; underwater acoustic sensor array; peak sidelobe level; intelligence optimization algorithm

Citation: Wang, L.; Zhao, H.; Wang, Q. Underwater Sparse Acoustic Sensor Array Design under Spacing Constraints Based on a Global Enhancement Whale Optimization Algorithm. *Appl. Sci.* **2022**, *12*, 11825. https://doi.org/10.3390/app122211825

Academic Editors: Lin Mu, Enjin Zhao and Hao Qin

Received: 6 October 2022
Accepted: 17 November 2022
Published: 21 November 2022

1. Introduction

Sparse sensor arrays are widely used in observation and communication systems, such as radar, sonar, satellite, etc. [1–3]. With fewer array elements, such arrays obtain better array performance, including the narrow main lobe width and the PSLL compared with equally spaced arrays. In addition, such arrays have clear advantages in terms of the installation cost and system complexity.

Generally, the design problems of sparse sensor arrays are mainly divided into two types [4]: (1) The element positions are optimized to minimize the PSLL with a fixed number of array elements and aperture [5]. (2) The array beampattern is designed to match the desired beampattern by optimizing the array element positions and weights while reducing the number of array elements [6,7]. Some arrays, such as large towed arrays and the seabed array, are prone to shift in array position and change in array configuration during operation, which causes a mismatch between the real positions and the design ones.

For example, the towed array is flexible with oil-filled polyurethane tubing. When the array is towed, it is difficult to keep the array straight; therefore, the real positions of the elements deviate from the designed positions. For the seabed array mounted at the bottom of sea, the uncertainties of the element positions and the entire array configuration are induced by ship drift and subsurface currents when the array deploys to the seafloor from the layout vessel. We focus on the first type problem under the marine environment to design a random sparse array with low PSLL to enhance the detection of weak targets [8].

The existing methods to solve the first problem are mainly divided into two categories: deterministic optimization methods and stochastic optimization methods. Deterministic optimization methods, including minimum redundancy array [9], coprime array [10], and difference coarray [11], are easy to implement; however, the designed arrays are generally suboptimal. Stochastic optimization methods use intelligent algorithms to find the element positions that satisfy the design requirements. Genetic algorithm (GA) [12], particle swarm optimization (PSO) algorithm [13], and simulated annealing algorithm [14] have been successfully applied to sparse sensor array design.

However, these intelligent algorithms converge slowly and easily fall into local optima. Some improved algorithms [15] have been proposed to solve the problem for array design—for example, a series of swarm intelligence optimization algorithms mimicking swarm behavior, including the wolf pack algorithm (WPA) [16], bat algorithm (BA) [17], and cuckoo search (CS) algorithm [18], which show good search accuracy and quick convergence speed. Although the stochastic optimization method suffers from global performance degradation during the design process of large-scale arrays, this method outperforms the deterministic optimization method in terms of the array performance as the calculation capacity increases.

The whale optimization algorithm (WOA) [19]—a swarm intelligence algorithm—has been applied successfully to sparse sensor array design for both fixed aperture and non-fixed aperture and has shown its advantages in sidelobe suppression and null steering for small- and medium-scale arrays [20,21]. Combining the salp swarm algorithm (SSA) and WOA, a novel swarm algorithm was proposed to improve its convergence accuracy for conformal antenna array design [22] and for dual-band aperture-coupled antenna design [23].

Nevertheless, the global search capability of WOA decreases with the expansion of the search range in large-scale array design. Researchers have attempted to improve the algorithm from several aspects in successive steps. The initialization strategy played an important role [24]. Chaos initialization improves population diversity due to its ergodic and random nature, and the chaotic particle algorithm was demonstrated to successfully reduce the PSLL of sparse sensor arrays [25].

With the development of intelligent algorithms, more chaotic initialization algorithms, including the chaotic invasive weed optimization algorithm [26], chaotic cuckoo search algorithm [27], and chaotic sparrow search algorithm [28], have been applied to sparse array design for improving global search capability. These success examples lead us to believe that WOA embedded with chaotic initialization will improve the performance of large-scale array design.

In the next step, search strategy optimization can be achieved by introducing adaptive weights [29] and search path planning [30]. WOA with adaptive weights had a good global search capability [31,32]. Search path planning is another effective method. Levy flight strategy, as a random search path strategy with global search capability, was introduced to WOA [33]. Although the method has not been widely applied to array design within WOA, it has been successful within other swarm algorithms for linear and circular array design, including the differential evolutionary algorithm [34], CS algorithm [27], InvasiveWeed Optimization [35], and biogeography-based optimization approach [36]. Unfortunately, the Levy flight strategy destroys the original whale spiral search, which causes the convergence speed to slow down; hence, a new search strategy should be introduced to maintain the speed.

In the third step to produce a new population, it was found that opposition-based learning (OBL) can also improve the search accuracy of WOA [37]. The combined strategy of adaptive weights with OBL was applied to high-dimensional global optimization problems [38]. However, it is easy to fall into the inverse local optimum, since the search range of the inverse solution is limited. A new learning algorithm should be developed to jump out of local minima. In short, the existing methods still have some shortcomings in the large-scale array design, particularly for the insufficient global search capability. To tackle this issue, GEWOA with a three-step enhanced global search strategy is introduced in this paper.

Additionally, the sparse sensor array design is a discrete problem, while the WOA is suitable for the solution of continuous variables. Therefore, WOA needs to be customized for discrete problems. Continuous variables discretization is generally achieved by transfer functions. If the discrete array element is on an integral multiple of the half wavelength, then this is an on-grid problem, otherwise it is an off-grid problem. The former is a discrete optimization problem with popular S-shaped and V-shaped transfer functions [39,40]. The latter uses binary discrete continuous variables with the ability to achieve lower PSLL [41,42]. Considering the physical limit of the array element size, the array design needs to constrain the minimum spacing between adjacent array elements while minimizing the PSLL.

A novel approach based on GEWOA for sparse sensor array design is proposed to reduce PSLL. A three-step enhanced global search strategy is introduced in GEWOA to improve the global search capability. In the initial stage, chaotic initialization is embedded in GEWOA, which enhances the ergodicity of the algorithm. In the search stage, the conventional spiral strategy is replaced by the Archimedean spiral strategy (ASS) to avoid premature algorithm results. In the offspring selection stage, refraction learning (RL) is used to obtain the inverse solution of the offspring, which avoids falling into local optima.

Moreover, to solve the adaptation problem for the discrete array design based on GEWOA, a position decomposition method is applied to make the element positions of off-grid arrays continuous, and a V-shaped transfer function is introduced into the on-grid array design to map the search position into the discrete grid points to obtain the desired array. The effectiveness of the proposed method is validated on the design tasks for sparse sensor arrays. To recap, the main contributions of this paper can be summarized as follows:

(1) A novel GEWOA with a three-step enhanced global search strategy is proposed for sparse sensor array design. Due to the strong global search capability, the arrays obtained by GEWOA show good performance in PSLL.
(2) A position decomposition method and a V-shaped transfer function are introduced into off-grid arrays and on-grid arrays, respectively, to make GEWOA suitable for discrete array design.
(3) The effectiveness of the proposed method is validated on design tasks for large-scale low-frequency sensor arrays in the marine environment. Additionally, comparisons with other representative methods, such as GA, PSO, GWO, and WOA, also demonstrate the superiority of our method.

The rest of this paper is organized as follows: Section 2 presents the problem description and the background. Section 3 introduces the proposed GEWOA. In Section 4, the effectiveness and superiority of the proposed method are demonstrated. Finally, Section 5 concludes this paper.

2. Problem Description

Let us consider that a sparse linear array has N isotropic radiative elements, which are distributed along the x-axis at $\mathbf{D} = [d_0, d_1, \ldots, d_i, \ldots, d_{N-1}], i \in [0, N-1]$. The array aperture is L. The beampattern of the array is given by:

$$P = \sum_{i=1}^{N} w_i e^{\mathrm{j}\frac{2\pi}{\lambda} d_i \sin(\theta - \theta_0)} \tag{1}$$

where $\mathrm{j} = \sqrt{-1}, \lambda$ denotes the wavelength of the operating frequency, w_i denotes the complex excitation of the ith element in the array, θ is the steering angle, and θ_0 represents the desired beam direction. The schematic of the array elements distribution is shown in Figure 1. Sparse arrays can be divided into on-grid arrays and off-grid arrays according to the positions of the array elements. The grid points are set at positions that are integer multiples of half the wavelength.

The black solid circles in Figure 1 are the actual positions of the array elements. The red hollow circles are grid points. If all array elements fall on the grid points, it is an on-grid array. Otherwise, it is an off-grid array. The PSLL of the off-grid array could be lower than that of on-grid array. In other words, an off-grid array can use fewer elements than an on-grid array to meet the same PSLL. However, the on-grid array is easier to process with less computation charge. Therefore, different design methods can be selected with a trade-off between the PSLL and computation charge.

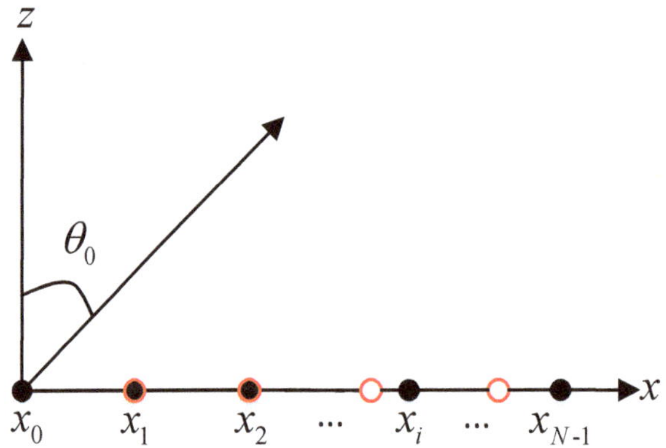

Figure 1. Schematic of the array element distribution.

In this paper, we focus on the layout problem of large-scale sparse sensor arrays, and all weights are set to 1 ($w_n = 1$ for all elements). The beampattern of a uniform array is a Dirichlet function with a PSLL of about -13 dB. However, the positions of the sparse array can be distributed nonuniformly. Assuming that the weight vector of a uniform array is $\tilde{\mathbf{w}}$, then the weight vector $\mathbf{w}$ of the sparse array can be expressed as $\mathbf{w} = \tilde{\mathbf{w}} \odot \mathbf{t}$, where $\mathbf{t} = [1, 0, 1, \ldots, 0, 1]$ can be considered a tapering vector, which consists of 1 and 0, with 1 for element and 0 for no element, and $\odot$ denotes the Hadamard product.

The tapering operation enables the side lobes to be lower than those of the uniform array. The optimal positions by minimizing PSLL are filled with 1 in vector $\mathbf{t}$. The beampattern of a sparse array obtained through Equation (1) is the convolution of the Dirichlet function with the spatial spectrum of weight vector $\mathbf{w}$, which is based on the array positions. Hence, a sparse sensor array with a low PSLL can be obtained by optimizing the array element positions. Considering the physical limit of the sensors' size, the minimum spacing of the elements must be restricted. Therefore, the array design problem can be formulated as follows:

$$\min_{\mathbf{X}} \mathbf{PSLL}(P) \quad \text{s.t.} \begin{cases} x_i - x_j \geq d_c > 0 \\ i, j \in \mathbf{Z}, 0 \leq j \leq i \leq N-1 \\ x_0 = 0, x_{N-1} = L \end{cases} \tag{2}$$

where x_i and x_j denote the position of the ith and jth element, respectively. d_c is the minimum distance between the adjacent elements. $\mathbf{Z}$ denotes the set of positive integers.

3. Methodology

3.1. Overview

The flowchart of the proposed method is shown in Figure 2. For off-grid array design, the position decomposition is used to make the element positions continuous. This strategy scales the array aperture L to the new aperture L^*, and then the proposed GEWOA is performed to obtain the optimized design result. For on-grid array design, GEWOA is directly performed to update the positions of the population, and then a V-shaped transfer function is used to map the search positions into the discrete grid points to obtain the desired array.

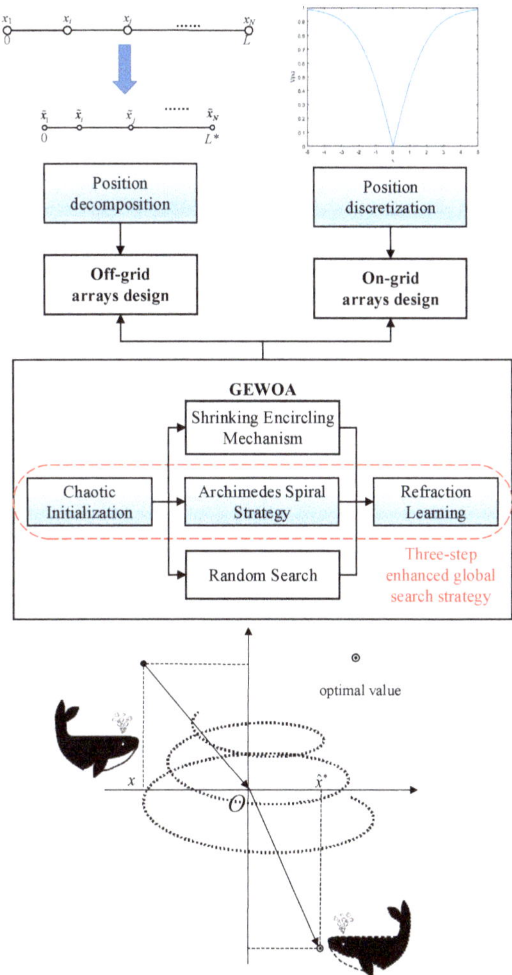

Figure 2. Flowchart of the GEWOA method.

In GEWOA, a three-step enhanced global search strategy is introduced to improve the global search capability of the array element positions as shown in the red dashed box. Chaotic initialization is used to initialize the population. Next, the position updating process is executed, where the three strategies, including the shrinking encircling mechanism, ASS, and random search, are selected according to preset rules.

After the position updating, RL is performed to seek the inverse solution to optimize the offspring. Next, the best offspring is selected by calculating the PSLL according to the objective function. Finally, the iterative optimization process is continued until the convergence condition is satisfied or the maximum number of iterations is reached.

3.2. Global Enhancement Whale Optimization Algorithm

GEWOA is a swarm intelligence optimization algorithm inspired by the foraging process of humpback whales. It is inspired by WOA and continues to follow its strategy execution process. The execution process simulates the behavior of whales attacking prey through a shrinking encircling mechanism and spiral updating position. In addition, a random search strategy is performed to avoid falling into a local solution [19]. However, the global search capability of the original strategy decreases as the search range increases. A three-step enhanced global search strategy, including chaotic initialization, ASS, and RL, is proposed to improve the search performance in GEWOA.

3.2.1. Chaotic Initialization

Chaos is a phenomenon that can traverse all states without repetition in a certain range [24]. We embed the chaos model to calculate the initial value aiming at increases the diversity of the population. Conventional chaotic mapping functions include Logistic mapping, Tent mapping, Kent mapping, Henon mapping, and Sin mapping.

After setting the initial value $\widetilde{s}_0$ and the parameters, the chaotic sequence $\widetilde{\mathbf{S}}(\widetilde{s}_1, \widetilde{s}_2, \ldots, \widetilde{s}_{N-2})$ can be obtained by using the equations in Table 1, where $\alpha \in (0,4]$ is the control parameter of Logistic mapping. $\beta = 1.4, \zeta = 0.3$ are the control parameters of Henon mapping, which allows the equation to enter a chaotic state and generates a chaotic sequence with strong randomness.

$\mu \in (0,1)$ is the the control parameter of Kent mapping, $\eta \in (0,2]$ is the control parameter of Tent mapping, and $\gamma \in [0,1]$ is the control parameter of Sin mapping. Then, the positions of the initial population $\widetilde{\mathbf{X}}(\widetilde{x}_1, \widetilde{x}_2, \ldots, \widetilde{x}_{N-2})$ can be obtained by:

$$\widetilde{x}_n = \widetilde{s}_n(ub - lb) + lb, n = 0, 1, 2, \ldots, N-1 \tag{3}$$

Table 1. Chaotic mapping equations.

Name	Chaotic Map Equation	Parameters
Logistic	$\widetilde{s}_{n+1} = \alpha\widetilde{s}_n(1-\widetilde{s}_n)$	α
Henon	$\widetilde{s}_{n+1} = 1 - \beta\widetilde{s}_n^2 + \zeta\widetilde{s}_n$	β, ζ
Kent	$\widetilde{s}_{n+1} = \begin{cases} \widetilde{s}_n/\mu & 0 < \widetilde{s}_n \leq 0.5 \\ (1-\widetilde{s}_n)/(1-\mu) & 0.5 < \widetilde{s}_n \leq 1 \end{cases}$	μ
Tent	$\widetilde{s}_{n+1} = \begin{cases} \eta\widetilde{s}_n & \widetilde{s}_n < 0.5 \\ \eta(1-\widetilde{s}_n) & \widetilde{s}_n \geq 0.5 \end{cases}$	η
Sin	$\widetilde{s}_{n+1} = \gamma\sin(\pi\widetilde{s}_n)$	γ

In this paper, different chaotic initialization strategies are simulated to determine the best one as the optimal initialization strategy that has the lowest PSLL. Compared with the randomly distributed population in WOA, chaotic initialization overcomes the shortcoming of falling into a local solution to a certain extent.

For a swarm of whales with M individuals, the search space of each whale position is N-dimensional. The position of the jth whale in the population after the qth iteration is $\mathbf{X}(q) = (x_{0,j}, x_{1,j}, \ldots, x_{i,j}, \ldots, x_{N-1,j}), i \in [0, N-1], j \in [0, M-1]$. Here, $\mathbf{X} \in [lb, ub]$, ub is the upper boundary of the search range, and lb is the lower boundary of the search range.

3.2.2. Shrinking Encircling Mechanism

The optimal position is regarded as the position of a target prey, and other individuals surround the optimal position. Such behavior is called the shrink encircling mechanism, and its expressions are as follows:

$$\widetilde{\mathbf{D}} = \mid \mathbf{C}\mathbf{X}^*(q) - \mathbf{X}(q) \mid \tag{4}$$

$$\mathbf{X}(q+1) = \mathbf{X}^*(q) - \mathbf{A} \cdot \widetilde{\mathbf{D}} \tag{5}$$

where q is the number of iterations, and $\mathbf{X}^*(q)$ is the optimal position. $\mathbf{X}(q)$ denotes the individual position in the population after qth iterations. $\mathbf{X}(q+1)$ is the updated position. $\cdot$ denotes the multiplication of the corresponding elements. $\mathbf{A}$ and $\mathbf{C}$ are coefficient vectors:

$$\mathbf{A} = 2\mathbf{a} \cdot \mathbf{r} - \mathbf{a} \tag{6}$$

$$\mathbf{C} = 2\mathbf{r} \tag{7}$$

where $\mathbf{a}$ decreases linearly from 2 to 0, and $\mathbf{r}$ is a random vector in $[0,1]$. It is worth noting that the individuals move to the optimal position by adjusting $\mathbf{A}$ and $\mathbf{C}$.

3.2.3. Random Search

To avoid falling a local optimum, an individual is randomly selected as a target prey, and the positions of the population are updated by:

$$\widetilde{\mathbf{D}} = \mid \mathbf{C}\mathbf{X}_{rand}(q) - \mathbf{X}(q) \mid \tag{8}$$

$$\mathbf{X}(q+1) = \mathbf{X}_{rand}(q) - \mathbf{A} \cdot \widetilde{\mathbf{D}} \tag{9}$$

where $\mathbf{X}_{rand}$ is a randomly selected whale position.

3.2.4. Archimedean Spiral Strategy

Whales swim in a spiral shape while shrinking towards their prey with the spiral position update model:

$$\mathbf{X}(q+1) = \mathbf{X}^*(q) + \mathbf{D}_{p}e^{bl}\cos(2\pi l) \tag{10}$$

where $\mathbf{D}_p$ denotes the distance between candidate whales and their prey, b is a constant used to control the logarithmic spiral shape, and l is a random number between $[-1,1]$.

The spiral position update method of WOA follows the logarithmic spiral model, which is characterized by equal spiral angles but unequal pitches. Such a method can find the optimal value quickly; however, it likely misses the intermediate optimal solution. Archimedean spiral curve provides a better idea to improve the original spiral search strategy, which avoids the reduced accuracy caused by long search steps. Each position dimension of a whale is generated by (11) instead of (10).

$$\mathbf{X}(q+1) = \mathbf{X}^*(q) + (\mathbf{D}_p + b \cdot l)e^{bl}\cos(2\pi l) \tag{11}$$

Figure 3 shows the Archimedean spiral curve and the logarithmic spiral curve, where the blue solid line is the Archimedean spiral curve, and the red dashed line is the logarithmic spiral curve. It can be seen that the Archimedean spiral curve is characterized by equal pitch, and the search step can be set by setting the pitch.

In the process of WOA, the strategy selection probability is set to 50% to choose encircling prey or the spiral updating position:

$$\mathbf{X}(q+1) = \begin{cases} \mathbf{X}^*(q) - \mathbf{A} \cdot \widetilde{\mathbf{D}}, p < 0.5 \\ \mathbf{X}^*(q) + (\mathbf{D}_p + b \cdot l)e^{bl}\cos(2\pi l), p \geq 0.5 \end{cases} \tag{12}$$

where p is a random value between (0, 1).

If p is less than 50%, encircling prey is chosen. Otherwise, the Archimedean spiral strategy is chosen, and the whale position is updated by (11). In encircling prey, the prey is selected to pass by $\mathbf{A}$. If $|\mathbf{A}| < 1$, the optimal individual in the population is selected as the prey to execute encircling prey by (4) and (5). If $|\mathbf{A}| \geq 1$, an individual is randomly selected as the prey, and the positions of population are updated by (8) and (9). By following the above three mechanisms, the whale swarm can gradually approach the optimal position.

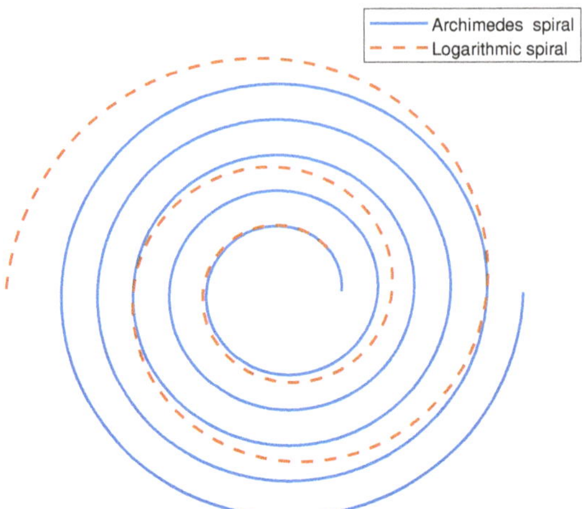

Figure 3. The curves of an Archimedes spiral and a logarithmic spiral.

3.2.5. Refraction Learning

RL is an inverse solution solving method that introduces the refraction principle of light into opposition-based learning (OBL). The superior inverse solution is screened out by OBL, and it is used to replace the existing solution for the next iteration. This strategy has a higher probability of reaching the global optimal solution [37]. However, the conventional OBL is not adjustable in the search range and cannot prevent the inverse solution from falling into the local optimal solution. To fix the defect, a scaling factor is introduced to control the search range according to the refraction principle.

Figure 4 shows the process of RL, where the center point of the search interval $[x_{\min}, x_{\max}]$ is denoted as O. The light propagates from $P(x, y)$ to the origin O in one medium and enters another medium. Refraction occurs after it enters a new medium, and the refraction point is $\hat{P}^*$. The distance $|\overrightarrow{PO}|$ from the incident point to the origin O is h, and the distance $|\overrightarrow{O\hat{P}^*}|$ between the origin O and the refraction point is h^*. According to Shell's Law, the relationship between the angle of incidence α and that of refraction β satisfies the shell formula: $n = \sin\alpha / \sin\beta$, where n is the refractive index.

Definition 1. *The projection $\hat{x}^*$ of the refraction point $\hat{P}^*$ on the x-axis is defined as the inverse solution of x based on the RL.*

According to the shell formula, we find

$$\begin{aligned} \sin\alpha &= ((x_{min} + x_{max})/2 - x)/h \\ \sin\beta &= (\hat{x}^* - (x_{min} + x_{max})/2)/h^* \end{aligned} \quad (13)$$

Assuming $k = h/h^*$, (13) can be rewritten as

$$\hat{x}^* = (x_{min} + x_{max})/2 + (x_{min} + x_{max})/(2kn) - x/kn \quad (14)$$

The inverse solution $\hat{x}^*$ can be solved using (14), where k is the scaling factor, which is used to adjust the positions of the inverse solutions on the x-axis. n is a fine-tuning factor. It can be seen that, assuming $n = 1, k = 1$, (14) can be simplified as $\hat{x}^* = -x$, which is the expression for OBL. It can be seen that OBL is a special case of RL. The refractive index of light from air to water is about 0.75. Assuming that $n = 0.75$ is fixed, the change in incident angle causes the change in the projection point x^*.

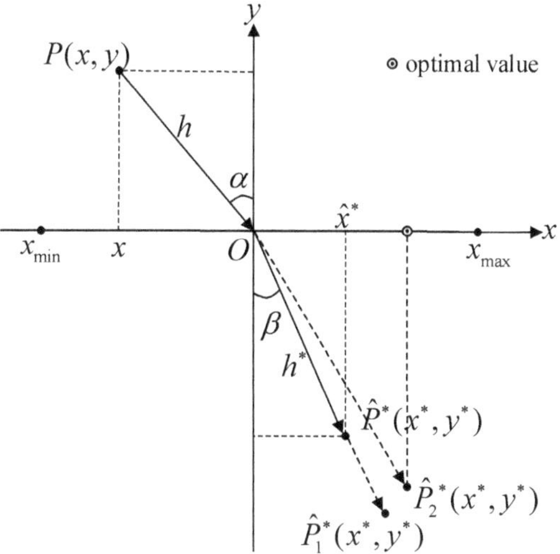

Figure 4. The process of refraction learning.

The refraction point can be moved from point $\hat{P}^*$ to point $\hat{P}_1^*$ by adjusting k, which is closer to the optimal point. If the $\hat{P}_1^*$ does not reach the optimal value, the optimal solution can be reached by adjusting the incidence angle and moving the refraction point to the $\hat{P}_2^*$. In order to improve the projection range of the inverse solution, we set k in a linearly decreasing trend:

$$k = k_{min} + (k_{max} - k_{min})t/t_{max} \tag{15}$$

The N-dimensional positions of a whale after position updating are denoted as:

$$\widetilde{\mathbf{X}} = (\widetilde{x}_0, \widetilde{x}_1, \ldots, \widetilde{x}_i, \ldots, \widetilde{x}_{N-1}) \tag{16}$$

where $x_{i,j} \in [x_{min}, x_{max}], i = 0, 1, \ldots, N-1$. x_{min} and x_{max} are the minimum and maximum values in the search range, respectively. The inverse solution $\hat{\mathbf{X}}^*$ obtained by RL is

$$\hat{\mathbf{X}}^* = (\hat{x}_0^*, \hat{x}_1^*, \ldots, \hat{x}_i^*, \ldots, \hat{x}_{N-1}^*) \tag{17}$$

If $\mathbf{PSLL}(\hat{\mathbf{X}}^*) < \mathbf{PSLL}(\widetilde{\mathbf{X}})$, $\hat{\mathbf{X}}^*$ is chosen as the best candidate solution. Otherwise, $\widetilde{\mathbf{X}}$ is chosen as the best candidate solution.

3.3. Off-Grid Arrays Design Based on GEWOA

For off-grid arrays, considering the limit of the array element size, the minimum spacing d_c is constrained. This means that the search range is not the whole aperture L. To solve this problem, the positions $\mathbf{D}(d_0, d_1, \ldots, d_{N-1})$ are decomposed. Since the positions of the two array elements x_0 and x_{N-1} are fixed, only the positions $\mathbf{D}^*(d_1, d_2, \ldots, d_{N-2})$ of the $N-2$ array elements need to be designed. Thus, a position decomposition method is applied to make $\mathbf{D}^*$ continuous:

$$\mathbf{D}^* = \widetilde{\mathbf{X}} + \begin{bmatrix} d_c \\ 2d_c \\ \vdots \\ (N-2)d_c \end{bmatrix} = \begin{bmatrix} \widetilde{x}_1 + d_c \\ \widetilde{x}_2 + 2d_c \\ \vdots \\ \widetilde{x}_{N-2} + (N-2)d_c \end{bmatrix} \tag{18}$$

In (19), $\widetilde{x}_1 \leq \widetilde{x}_2 \leq, \ldots, \leq \widetilde{x}_{N-2}, d_1 > d_c, d_{N-1} < L - d_c$. The solution for the array position $\mathbf{D}^*$ is converted to the solution for $\widetilde{\mathbf{X}}$ by the the position decomposition method. Here, the search range of $N-2$ array elements becomes $L - 2d_c$. In addition, $(N-3)d_c$

needs to be subtracted because the spacing between adjacent array elements is not less than d_c. Therefore, the search aperture becomes L^*:

$$L^* = L - 2d_c - (N-3)d_c = L - (N-1)d_c \quad (19)$$

Through the above position decomposition, the discontinuous problem can be converted into a continuous problem. Then, GEWOA can be performed to achieve off-grid array design (Algorithm 1).

Algorithm 1: Off-grid array design based on GEWOA.

Input: Initial chaotic sequences **S**, maximum iterations t_{Max}, and the minimum spacing d_c
Output: Optimal individual whale position
Determine the new search aperture L^* using position decomposition;
Initialize $\widetilde{\mathbf{X}}$ according to search aperture L^* using chaotic sequence **S**;
Find the individual position corresponding to the optimal fitness value $\mathbf{X}^*$;
while $t < t_{Max}$ *or* **PSLL** *does not meet demand* **do**
  Set the parameters $\mathbf{a}, \mathbf{r}, l, p$, and calculate $\mathbf{A}, \mathbf{C}$.;
  Update $\widetilde{\mathbf{X}}(t)$. **if** $p < 0.5$ **then**
    if $|\mathbf{A}| < 1$ **then** Enforcing shrinking encircling mechanism and updating position according to (4) and (5) ;
    else Random search using (8) and (9);
  else
    Archimedes spiral strategy using (11);
  end
  Set k and n and use refraction learning to find the inverse solution $\hat{\mathbf{X}}^*(t)$;
  If $\mathbf{PSLL}(\widetilde{\mathbf{X}}) < \mathbf{PSLL}(\hat{\mathbf{X}}^*)$, choose $\widetilde{\mathbf{X}}$ as a new population; Otherwise, choose $\hat{\mathbf{X}}^*$ as a new population; $t = t + 1$
end

The algorithmic details of the proposed method for off-grid array design based on GEWOA are summarized in Algorithm 2. First, position decomposition is performed to obtain the search range L^*. Secondly, GEWOA is performed to search the optimal position. In the initial stage of GEWOA, the population initialization is performed by using chaos sequences to obtain $\widetilde{\mathbf{X}}$, and the maximum number of iterations t_{max} is set. The optimal whale in the initialized population is selected by (2), and then the iterative optimization is performed.

The parameters $\mathbf{a}, \mathbf{r}, l, p$ of each whale are set, and $\mathbf{A}, \mathbf{C}$ are calculated for choosing different strategies of position updating. If $p > 0.5$, ASS for position updating is performed. Else, if $|\mathbf{A}| < 1$, a shrinking encircling mechanism is executed. Otherwise, a random search is used. After the position updating, the inverse solution $\hat{\mathbf{X}}^*$ is generated by using RL. The best solution is selected from the new offspring and its inverse solution. The iterative optimization process is continued until the convergence condition is satisfied or the maximum number of iterations is reached. Finally, the optimal array positions are output by (19).

3.4. On-Grid Arrays Design Based on GEWOA

For on-grid arrays, the positions are discrete on half wavelength grid points. Therefore, the V-shaped transfer function is used to discretize the output of GEWOA. The V-shaped transfer function maps the whale position to the [0,1] interval. When the mapped values fall in different intervals, discrete values are selected. The expression of the V-shaped transfer function is:

$$V(x_{ij}(t)) = \left| \frac{e^{x_{ij}(t)} - 1}{e^{x_{ij}(t)} + 1} \right| \\ x'_{ij} = \begin{cases} G & 0 \leq V(x_{ij}(t)) < r_1 \\ G-1 & r_1 \leq V(x_{ij}(t)) < r_2 \\ \cdots & \\ 0 & r_M \leq V(x_{ij}(t)) < 1 \end{cases} \tag{20}$$

where x' is the converted discrete value. $r_1, r_2, \ldots, r_{k-1}$ are random numbers between $(0,1)$, which satisfy $0 < r_1 < r_2 < \ldots < r_{k-1} < 1$. When $0 \leq V(x_{ij}(t)) < r_1$, the whale position is mapped to G. $G = \mathbf{floor}(2 * L/\lambda)$ is the largest grid point (integer multiple of half wavelength), where $\mathbf{floor}()$ is a function to find the nearest integer less than or equal to the parameter. The algorithmic details of the proposed method for on-grid array design based on GEWOA are summarized in Algorithm 2.

Algorithm 2: On-grid array design based on GEWOA.

Input: Initial chaotic sequences $\widetilde{\mathbf{S}}$ and maximum iterations t_{Max}
Output: Optimal individual whale position
Initialize $\widetilde{\mathbf{X}}$ using chaotic sequence $\mathbf{S}$;
Mapping $\widetilde{\mathbf{X}}$ to discrete grid points $\mathbf{G}$ using V-shaped transfer function;
Find the individual position corresponding to the optimal fitness value $\mathbf{X}^*$;
while $t < t_{Max}$ *or* **PSLL** *does not meet demand* **do**
  Set the parameters $\mathbf{a}, \mathbf{r}, l, p$, and calculate $\mathbf{A}, \mathbf{C}$.;
  Update $\widetilde{\mathbf{X}}(t)$. **if** $p < 0.5$ **then**
    if $|\mathbf{A}| < 1$ **then** Enforcing shrinking encircling mechanism and updating position according to (4) and (5) ;
    else Random search using (8) and (9);
  else
    Archimedes spiral strategy using (11);
  end
  Set k and n and use refraction learning to find the inverse solution $\hat{\mathbf{X}}^*(t)$;
  Mapping $\widetilde{\mathbf{X}}$ and $\hat{\mathbf{X}}^*(t)$ to discrete positions $\mathbf{X}'$ and $\hat{\mathbf{X}}'$ using V-shaped transfer function;
  If $\mathbf{PSLL}(\mathbf{X}') < \mathbf{PSLL}(\hat{\mathbf{X}}')$, choose $\mathbf{X}'$ as a new population; Otherwise, choose $\hat{\mathbf{X}}'$ as a new population; $t = t + 1$
end

4. Experiment

In this section, first, two kinds of numerical experiments on sparse-sensor-array-design tasks are performed to validate the effectiveness and superiority of the proposed method. (1) For off-grid array design, the proposed three-step enhanced global search strategy in GEWOA is simulated step by step to test the effectiveness of each strategy. Furthermore, the designed results are compared with those of WOA.

(2) For on-grid array design, the proposed GEWOA is compared with some other intelligence algorithm-based design methods, including GA, PSO, and WOA, in terms of PSLL. Then, the array designed using the proposed method is verified in the marine environment, and the beampattern obtained by the array based on GEWOA is analyzed.

The basic design requirements in numerical experiments are as follows. The array aperture is 200 m, the number of array elements is 40, and the array operating frequency is 300 Hz with a wavelength of 5 m. The objective function is shown in (2), where PSLL is chosen as the evaluation criterion. The number of whales in the population $M = 30$, the dimension of search $N = 40$, the number of iterations $t_{max} = 1000$.

The minimum spacing of the array d_c is set to 2 m for off-grid arrays, and the array elements are set at half-wavelength grid points for on-grid arrays. In addition, we also conduct numerical experiments for large-scale low-frequency sensor array design, and the design conditions are as follows: the array operating frequency is 300 Hz, the number of array elements is 256, and the array aperture is 1280 m.

4.1. Results of Off-Grid Arrays

4.1.1. Performance of Chaotic Initialization

In this part, different chaotic initialization methods are used to design the off-grid sensor array, and the convergence speed and beampattern performance are analyzed. Five chaotic mapping functions, including Logistic, Henon, Kent, Sin, and Tent, are tested under two termination conditions.

Under the first termination condition, which depends on the maximum number of iterations t_{max}, the search accuracy of five chaotic initialization methods is compared. The parameter settings and the obtained PSLLs are shown in Table 2, where the PSLL is the minimum value among 10 experiments. The beampatterns of the obtained array are shown in Figure 5. The iterative process of five chaotic initialization methods are shown in Figure 6. It can be seen that the search accuracy is improved after chaotic initialization compared to the original WOA (denoted as Origin in Figures 5 and 6). In addition, the Kent initialization method yields the array with the lowest PSLL.

Table 2. PSLLs and the numbers of iterations for five chaotic initialization methods.

Chaotic Initialization	Parameter Settings	PSLL (dB)	Iterations
Logistic	$\alpha = 3.9, \tilde{s}_0 = 0.5$	−19.96	659
Henon	$\beta = 1.4, \zeta = 0.3, \tilde{s}_0 = 0$	−19.63	451
Kent	$\mu = 0.625, \tilde{s}_0 = 0.45$	−19.98	800
Sin	$\eta = 0.867, \tilde{s}_0 = 0.5$	−19.30	733
Tent	$\gamma = 1.41, \tilde{s}_0 = 0.1$	−19.77	488
Origin	-	−18.65	897

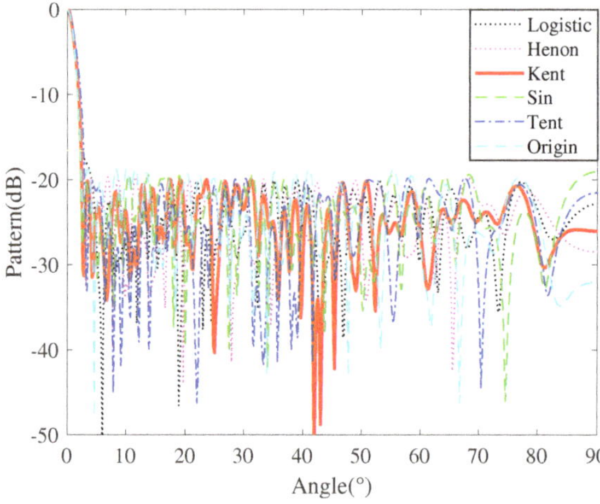

Figure 5. Beampatterns of the obtained array based on five chaotic initialization methods.

Under the second termination condition that the algorithm stops if the PSLL remains unchanged beyond 100 iterations, the convergence speed of five chaotic initialization methods is compared. The numbers of iterations are shown in Table 2. It can be seen that Origin<Kent<Sin<Tent<Logistic<Henon in terms of the convergence speed.

Here, the Kent initialization method is adopted in GEWOA because it has the best search accuracy and faster convergence speed compared with the original method. Considering that the search accuracy is the most important criterion for array design, the selection is reasonable.

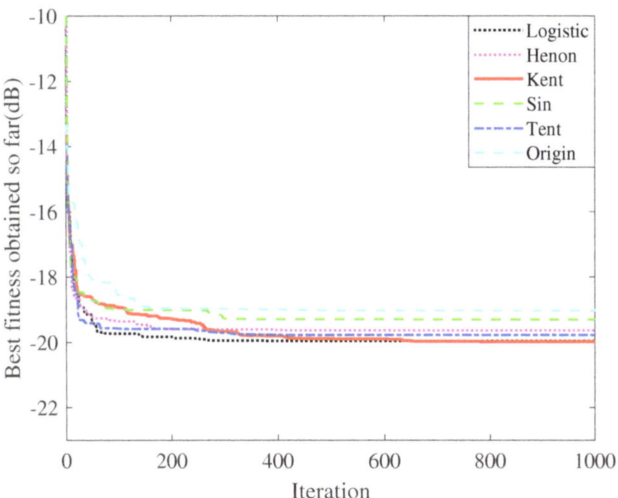

Figure 6. The iterative process of five chaotic initialization methods.

4.1.2. Performance of Archimedean Spiral Strategy

In this part, we compare three search path planning strategies, including the original spiral position update, ASS, and Levy flight strategy. The Archimedean spiral spacing is set to 10 m. The beampatterns and the iterative process are shown in Figures 7 and 8, respectively. It can be seen that the PSLL of ASS is −21.02 dB, while the PSLL of Levy flight strategy is −20.76 dB. After adding ASS, the PSLL is 2.37dB lower than the original strategy. ASS improves the original strategy, and has a lower PSLL than the Levy flight strategy.

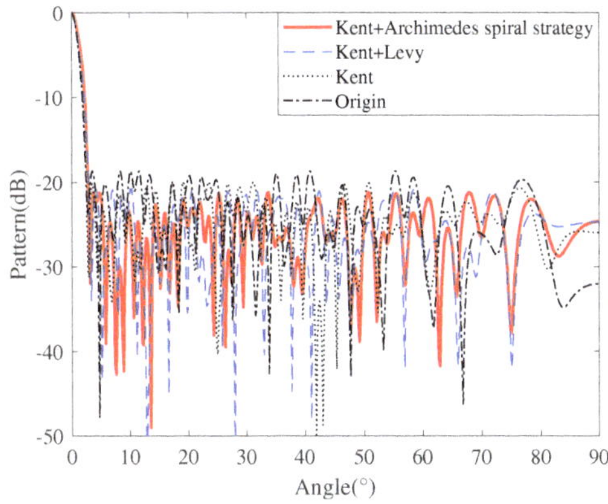

Figure 7. Beampatterns under three search path planning strategies.

4.1.3. Performance of Refraction Learning

On the basis of Kent chaotic initialization and ASS, RL is used to further optimize the offspring to obtain an array with the lower PSLL. The setting parameters of RL are $k_{max} = 1.2, k_{min} = 0.8, t_{max} = 10$. The beampatterns of 40-element off-grid arrays and the iterative process before and after RL optimization are shown in Figures 9 and 10. It can be seen that the PSLL of array without RL optimization is −21.02 dB, and the PSLL of array with RL optimization is −22.47 dB. Compared with WOA, the PSLL obtained by GEWOA is reduced by 3.81 dB.

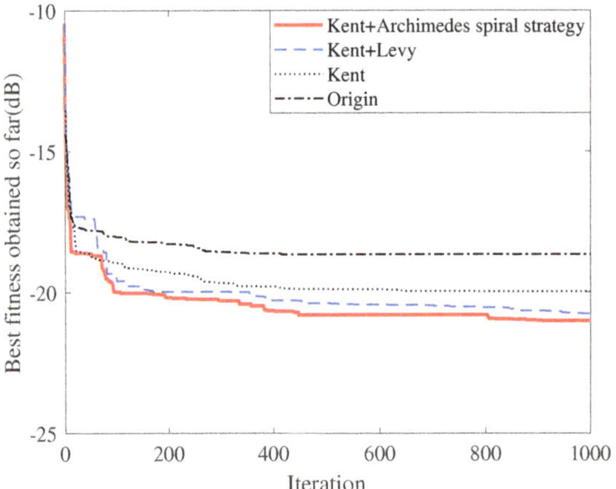

Figure 8. The iterative process under three search path planning strategies.

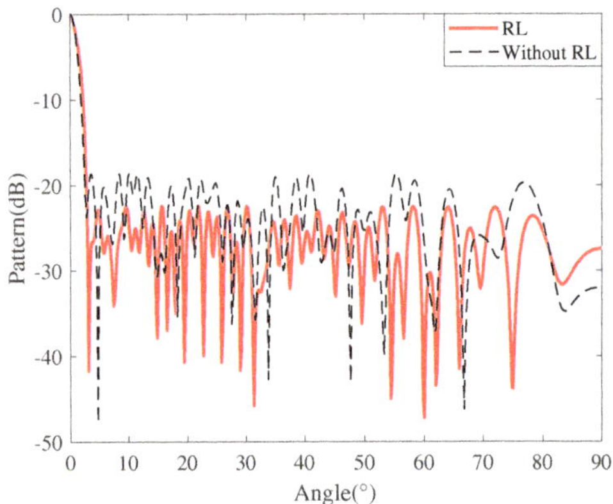

Figure 9. Beampatterns with and without RL.

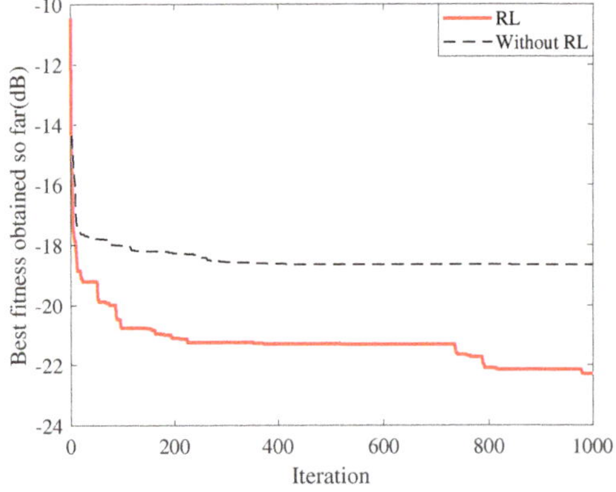

Figure 10. The iterative process with and without RL.

In order to test the influence of RL parameter on the search accuracy, experiments with different RL parameters are performed. The three sets of RL parameters are (1) $k = 1$; (2) $k_{max} = 1.2, k_{min} = 0.8, t_{max} = 5$; and (3) $k_{max} = 1.2, k_{min} = 0.8, t_{max} = 10$. The array beampatterns and iterative process are shown in Figures 11 and 12, where the PSLLs of three sets of RL parameters are −21.47, −22.30, and −22.47 dB. The PSLLs of the designed arrays improve as the search range of the inverse solution is expanded.

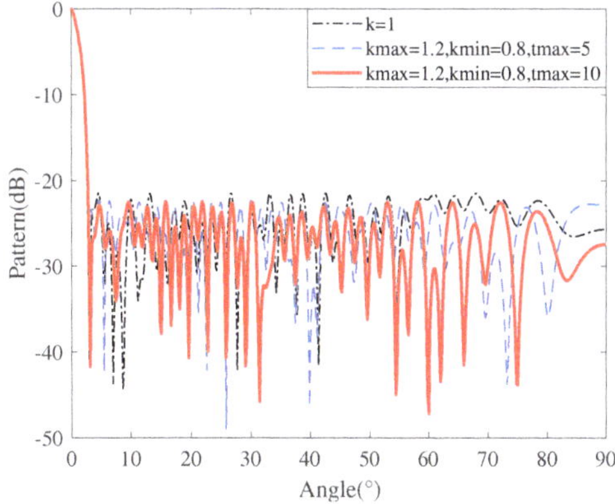

Figure 11. Beampatterns of off-grid arrays (40-element) under three sets of RL parameters.

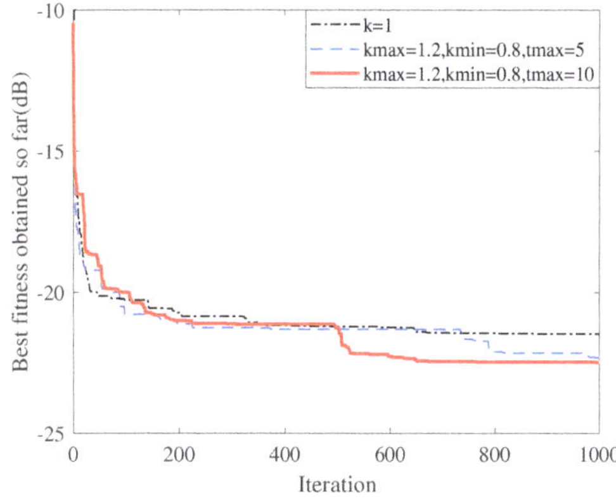

Figure 12. The iterative process under three sets of RL parameters.

4.2. Results of On-Grid Arrays

For on-grid arrays, the grids are set according to the half wavelength $\lambda/2 = 2.5$ m, and the number of grid points is 81. The proposed GEWOA discretized by a V-shaped transfer function is used to design on-grid arrays. In order to validate the superiority of the proposed GEWOA, three intelligence algorithms, including GA, PSO, and WOA, are compared with GEWOA.

The beampatterns of 40-element on-grid arrays and iterative process are shown in Figures 13 and 14, respectively. The PSLLs of GA, PSO, WOA, and GEWOA are −14.90, −15.10, −15.66, and −17.14 dB, respectively. It can be seen that the on-grid array designed using GEWOA obtains the lowest PSLL, which proves the advantage of the GEWOA in terms of the global search capability.

The proposed array design method, including the off-grid array design method and on-grid array design method, are successfully applied to the large-scale low-frequency arrays (256-element) design, and the obtained arrays have low PSLL. The beampatterns of the designed arrays and iterative process are shown in Figures 15 and 16, respectively. The results show that the PSLLs of the off-grid array and the on-grid array obtained by the proposed methods are −27.71 and −21.47 dB, which verifies the feasibility of the algorithm in large-scale low-frequency array design.

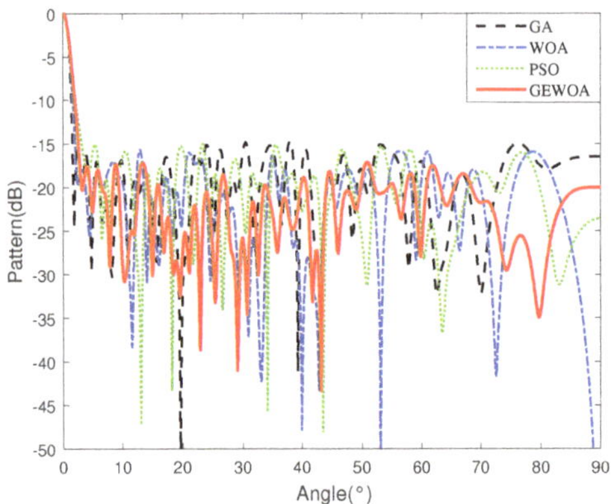

Figure 13. Beampatterns of on-grid arrays (40-element) designed using four intelligence algorithms.

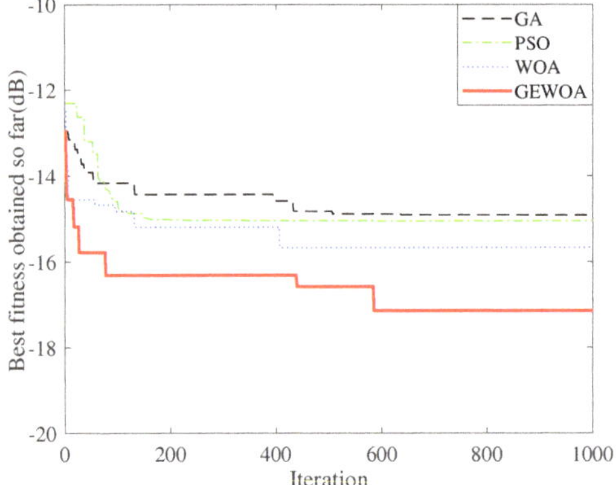

Figure 14. The iterative process of four intelligence algorithms.

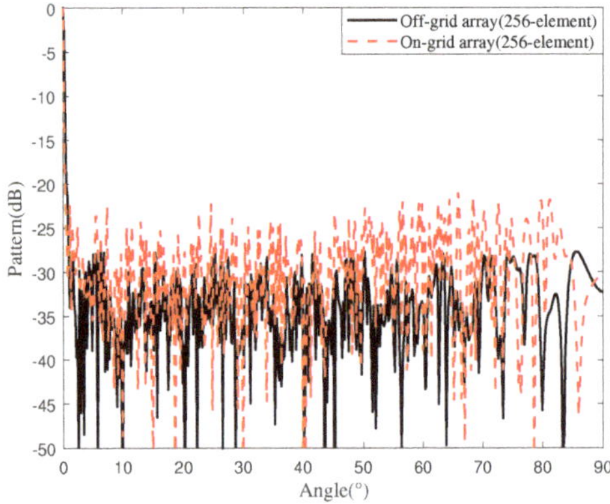

Figure 15. Beampattern of the on-grid large-scale low-frequency senor array (256-element) obtained using the proposed method.

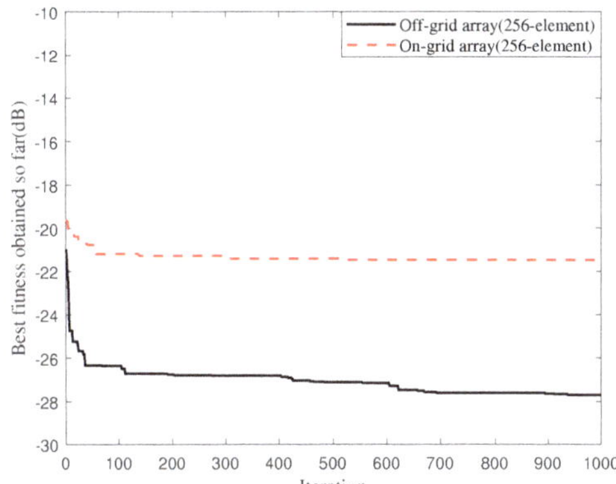

Figure 16. The iterative process of the on-grid large-scale low-frequency senor array (256-element) based on the proposed method.

In order to compare the performance of the proposed method with typical configuration arrays, such as coprime arrays and nested arrays, three different arrays were designed at the same aperture and the same number of elements. The 40-element on-grid array was designed using the proposed method. The nested array has three levels of nesting. The first level is a 20-element uniform array with half-wavelength element spacing. The second level of nesting is a 10-element uniform array with a wavelength element spacing. The third level of nesting is a 10-element uniform array with two times wavelength element spacing.

To ensure the same aperture, the 40-element coprime array was designed, and its elements were arranged on three and four times half-wavelength grids. Figure 17 shows the beampatterns of the three arrays. It can be seen that the PSLL of the array designed using GEWOA are −14.79 dB, and the PSLLs of the coprime array and the nested array are −5.82 and −6.02 dB, respectively. The main lobe widths of three arrays are approximately the same. However, the sparse array designed using GEWOA has a lower PSLL, which proves the advantage of the proposed method compared with the determined methods.

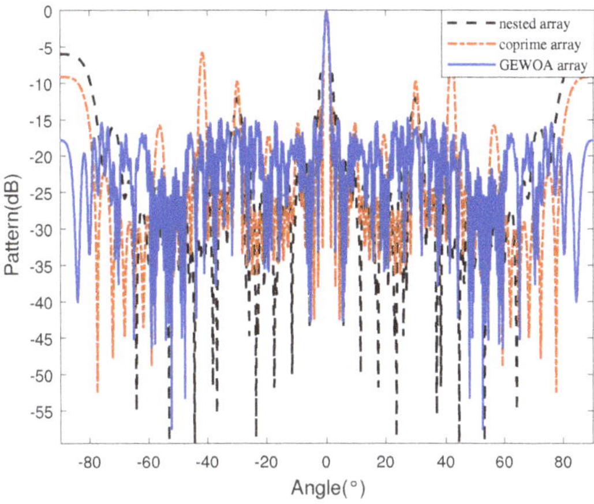

Figure 17. Beampatterns of the coprime array, nested array, and GEWOA array.

To summarize, the above experimental results show that the proposed GEWOA with a three-step enhanced global search strategy had a good global search capability, and the array designed using GEWOA had a lower PSLL when compared with the other algorithms and deterministic methods. In addition, the proposed array design method was successfully applied to the large-scale low-frequency array design, and the obtained array had a low PSLL.

4.3. Discussion and Analysis

To demonstrate the superiority of the proposed method, we compared the performance of the sparse array designed using GEWOA with those of arrays obtained by traditional deterministic and hybrid methods, including the GWO array [23] and the different coarray (DC) designed using the method in [11]. Array design via these methods was performed at the same aperture.

The number of array elements is 40, and the array aperture is 200 m. The beampatterns of the designed arrays are shown in Figure 18, and their PSLLs are shown in Table 3, which indicates the lowest PSLL of our method. To further illustrate the repeatability of our method, we conducted 50 Monte Carlo simulations of the algorithms under the same conditions.

The results show that our method had a 96% probability of making PSLL below −17 dB. We compared the time efficiency of these methods on a CPU Intel E3-1220 v6 @ 3.00 GHz, and −15 dB was chosen as the convergence threshold. The computation times of GEWOA, GWO, and DC were 14.39, 20.54, and 122.41 s, and it can be seen that our method achieved the best computational efficiency.

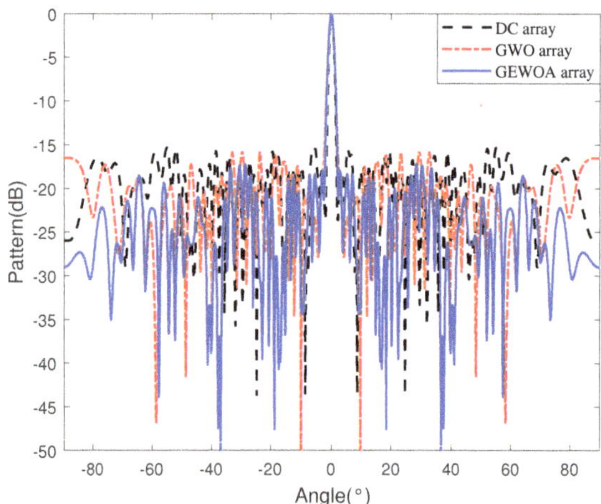

Figure 18. Beampatterns of sparse arrays designed using different methods.

Table 3. The PSLLs of arrays designed using different algorithms.

Arrays	DC Array	GWO Array	GEWOA
PSLL(dB)	−15.34 dB	−15.82 dB	−17.58 dB
Threshold Time(s)	122.41 s	20.54 s	14.39 s

4.3.1. Mutual Coupling Analysis

Due to the limitations of the natural frequencies, the mode shapes of frequencies, and inertial effects, the low-frequency acoustic sensors, such as transducers are usually large [43]. For engineering implementation, the minimum spacing of the array elements is constrained. The distance between adjacent element of the off-grid array designed using GEWOA can be less than half the wavelength, which results in the array element falling in the radiated sound field of that adjacent element.

The mutual coupling phenomenon occurs between adjacent array elements. This increases with the decrease of array element spacing, and directly affects the setting of the minimum spacing of the array element. According to the mutual radiation impedance model [44], the mutual radiation impedance X_{AB} of an array element in the radiation field of another array element is

$$X_{AB} = \rho c S k a^2 / d_{AB}. \quad (21)$$

where ρ is the medium density. c is the speed of sound. S is the surface area of the array element. a is the radius of the sensor element. d_{AB} is the distance of two elements. k is the wave number assuming the resonant frequencies of the array element is 3.5 kHz. The radius of the array element is 0.4 m. The curve of the radiation impedance changing with the array element spacing is shown in Figure 19.

It can be seen that $d/\lambda \geq 0.4$, mutual-impedance is less than its self-impedance. In the off grid array design, the operating frequency is 300 Hz, and the minimum spacing of array elements is set to 2 m due to engineering requirements. Since the minimum spacing is greater than 0.4 times the wavelength of the resonant frequency, the mutual radiation effect is less than the self-radiation, and it can be disregarded.

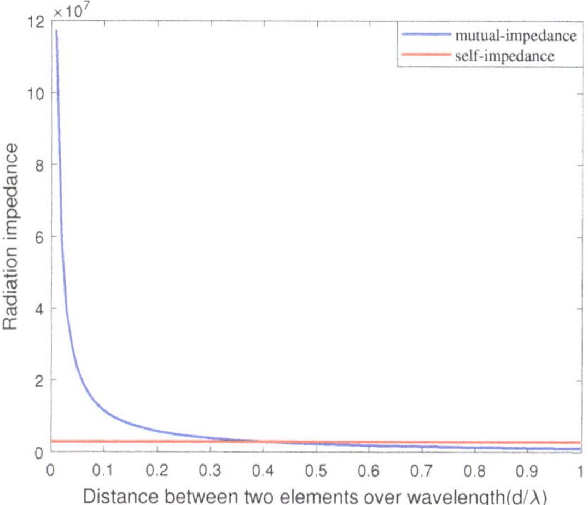

Figure 19. Curve of the radiation impedance changing with the array element spacing.

4.3.2. Array Gain Analysis

Array gain is an important indicator of array performance. Assuming that the sparse arrays receive a 300 Hz sinusoidal signal with a duration of 1 s. The array gains of the 40-element off-grid array and on-grid array designed using GEWOA are calculated to verify the feasibility of the proposed method under the Gaussian white noise background. The theoretical value of the array gain for the 40-element array is 16.02 dB.

The change trend of the array gain with the input signal-to-noise ratio (SNR_{in}) is shown in Figure 20. It can be seen that the array gain decreases at a lower input SNR. When the SNR_{in} is −30 dB, the array gain of both arrays decreases by about 1.5 dB compared with the theoretical value. When the SNR_{in} is higher than −20 dB, the array gain of both arrays decreases by, at most, 0.3 dB compared with the theoretical value, which proves the feasibility of the designed array based on the proposed method from the perspective of array gain.

4.3.3. Position Uncertainty Analysis

To verify the effect of array performance when the array element position is varied, Gaussian perturbation is added to the element position of the on-grid array designed using GEWOA in Section 4.2 to compare the effect of the PSLL performance. The outer layer of most fiber optic towing arrays is the polyurethane tubing, which is filled with oil. This connection limits the offset of the array element position. Assume that the position perturbation with Gaussian distribution $\mathcal{N}\ (0, (\frac{\lambda}{12})^2)$, which can guarantee 99.73% probability within a quarter-wavelength error range.

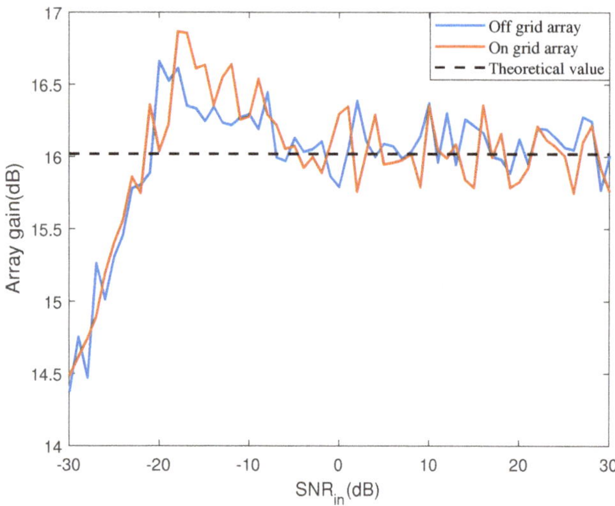

Figure 20. Curve of the radiation impedance changing with the array element spacing.

The beampattern of the sparse array designed using GEWOA before and after adding the position perturbation is shown in Figure 21. It can be seen that the PSLL of the designed array is −17.58 dB, while the PSLL of the array is −15.87 dB in the case of perturbation of all array elements. We performed 100 experiments using Monte Carlo simulations. The average value of PSLL for the position perturbation case is 16.23 dB. A quarter-wavelength Gaussian position perturbation causes the PSLL of the array to degrade by about 1.35 dB.

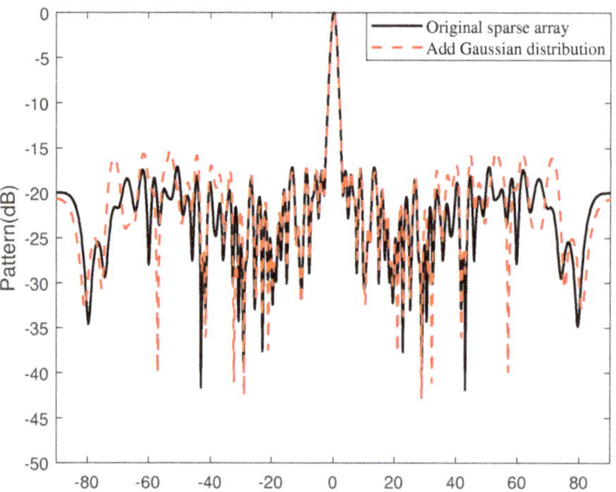

Figure 21. The beampattern after position perturbation.

4.4. Experiment in the Marine Environment

To verify the effectiveness of the proposed method, the performance of the designed arrays in the marine environment was analyzed. Due to the long production cycle and high investment of large-scale low-frequency sensor arrays, 40-element on-grid arrays were designed for the experiments. We extracted 40 elements from an 81-element uniform array to obtain the required arrays. The positions of the arrays designed using GA, PSO, WOA, and GEWOA are shown in Figure 22.

All array elements are set on the grid points. The beampatterns of designed arrays are shown in Figure 23. The PSLLs and main lobe widths of the beampatterns are shown in Table 4. By analyzing the performance of the sparse-sensor array designed based on GEWOA in the marine environment and comparing with the performance of the arrays obtained by GA, PSO, and WOA, the effectiveness of the algorithm was verified. The experimental conditions were as follows: the experimental array was a uniform hydrophone array with a half wavelength of 0.4167 m as shown in Figure 24. The design frequency of the array was 1800 Hz, and the number of array elements was 81.

The experimental data were from the 2020 South China Sea Experiment. The designed arrays were used to receive the transmit signal. The received data of 40-element on-grid arrays obtained by GA, PSO, WOA, and GEWOA were extracted from the 81-element uniform array. The relative azimuth angle between the sound source and the array was 15°. The frequency of the transmitted signal was 1800 Hz. Its duration was 1 s, and the transmission period was 40 s.

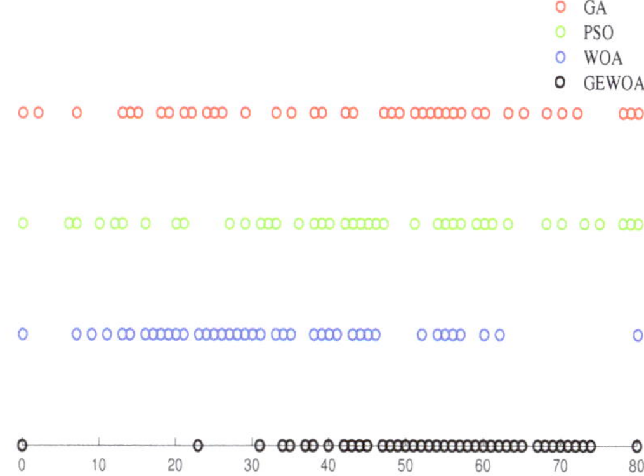

Figure 22. The positions of arrays designed using GA, PSO, WOA, and GEWOA.

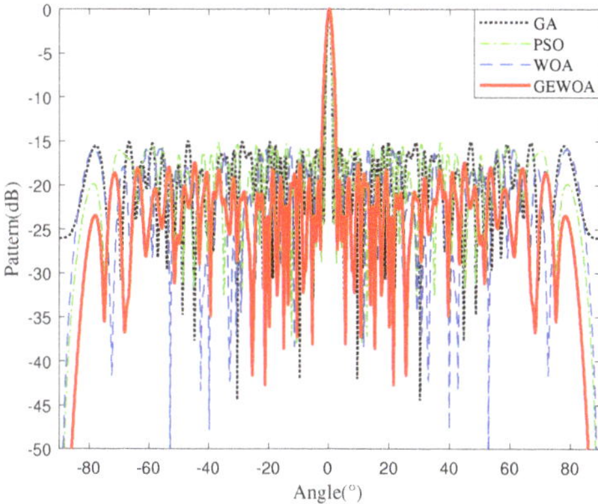

Figure 23. Beampatterns of the on-grid arrays (40-element) designed using GA, PSO, WOA, and GEWOA.

Figure 24. Profile display of equipment in the marine environment.

The beampatterns were obtained by conventional beamforming processing of the direct wave signals. The PSLLs of the beampatterns were analyzed to evaluate the array performance. The beampatterns obtained by the 40-element on-grid array designed using GA, PSO, WOA, and GEWOA are shown in Figure 25. The PSLLs and main lobe widths of the beampatterns in marine environment are shown in Table 4. It can be seen that the PSLLs of the beampatterns in the marine environment are degraded, due to the deviation of the array position from the designed ideal position.

The main lobe widths vary relatively little. The PSLL of the array designed using GEWOA is −17.10 dB in the marine environment. According to Table 4, it can be seen that compared to GA, the main lobe width expanded by nearly 0.06 degrees, which was only increased by 4.5% in the marine experiments. However, the PSLL decreased by 3.9 dB, which is a decrease of nearly 29.5%.

Compared with WOA, the main lobe width expanded by 0.01 degree, which is an increase of only 0.7%, while the PSLL was reduced by 2.5 dB, which is a decrease of 17.1%. The PSLL of GEWOA was 3.9 dB lower than that of the traditional GA algorithm with a small change in the performance of the main lobe width. In comparison with PSLLs obtained using other arrays, the array designed using GEWOA had the lowest PSLL.

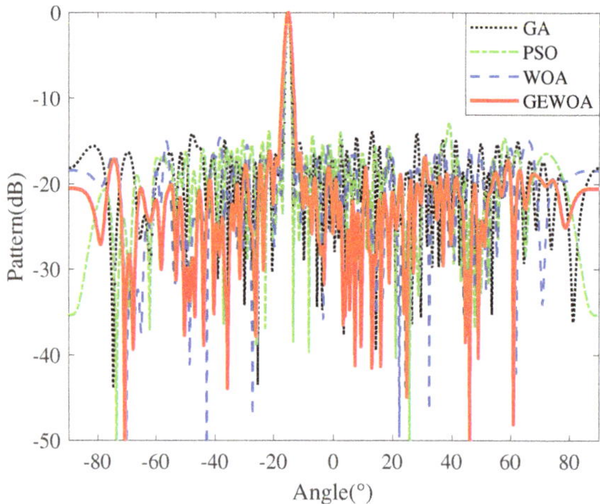

Figure 25. Beampatterns of the on-grid arrays (40-element) designed using four intelligence algorithms in the marine environment.

Table 4. The PSLLs of arrays designed using different algorithms.

Algorithms	GA	PSO	WOA	GEWOA
PSLL (dB)	−15.03	−15.12	−16.03	−17.85
Main lobe width (°)	1.32	1.32	1.36	1.38
PSLL in marine environment(dB)	−13.20	−13.95	−14.60	−17.10
Main lobe width in marine environment (°)	1.33	1.34	1.38	1.39

5. Conclusions

In this paper, we proposed a novel approach based on GEWOA for sparse sensor array design to suppress PSLL under spacing constrains. A three-step enhanced global search strategy was introduced into GEWOA to improve the global search capability. In the initial stage, chaotic initialization was embedded in GEWOA to enhance the ergodicity of the algorithm.

In the search stage, the conventional spiral strategy was replaced by the ASS strategy to avoid premature algorithm results. In the offspring selection stage, RL was used to obtain the inverse solution of the offspring, which prevents falling into local optima. Moreover, in order to solve the adaptation problem for discrete array design based on GEWOA, a position decomposition method and a V-shaped transfer function were introduced into the off-grid arrays and on-grid arrays, respectively.

The effectiveness and superiority of the proposed method were validated by experiments on large-scale low-frequency sparse sensor array design tasks. The experimental results show that the proposed GEWOA with a three-step enhanced global search strategy had a good global search capability, and the array designed using GEWOA had the lowest PSLL compared with other intelligent algorithms. Additionally, the array designed using the proposed method was further verified in the marine environment, where the proposed GEWOA still achieved the lowest PSLL. In the future, we will attempt to expand our application of the proposed method on 2-D or 3-D arrays.

Author Contributions: Conceptualization, L.W. and Q.W.; methodology, L.W.; software, L.W.; validation, L.W., H.Z. and Q.W.; formal analysis, L.W.; investigation, L.W.; resources, Q.W.; data curation, H.Z.; writing—original draft preparation, L.W.; writing—review and editing, H.Z. and Q.W.; visualization, L.W.; supervision, H.Z.; project administration, H.Z.; funding acquisition, H.Z. All authors have read and agreed to the published version of the manuscript.

Funding: This research received no external funding.

Institutional Review Board Statement: Not applicable.

Informed Consent Statement: Not applicable.

Data Availability Statement: The data presented in this study are available on request from the corresponding author.

Conflicts of Interest: The authors declare no conflict of interest.

References

1. Sayin, A.; Hoare, E.G.; Antoniou, M. Design and Verification of Reduced Redundancy Ultrasonic MIMO Arrays Using Simulated Annealing & Genetic Algorithms. *IEEE Sens. J.* **2020**, *20*, 4968-4975. [CrossRef]
2. Liu, J.; Shi, F.; Sun, Y.; Li, P. An ADS-Based Sparse Optimization Method for Sonar Imaging Sensor Arrays. *Appl. Sci.* **2020**, *10*, 3176. [CrossRef]
3. Wang, Y.; Wu, W.; Zhang, X.; Zheng, W. Transformed nested array designed for DOA estimation of non-circular signals: Reduced sum-difference co-array redundancy perspective. *IEEE Commun. Lett.* **2020**, *24*, 1262–1265. [CrossRef]
4. Liu, F.; Liu, Y.; Han, F.; Ban, Y.L.; Jay Guo, Y. Synthesis of Large Unequally Spaced Planar Arrays Utilizing Differential Evolution With New Encoding Mechanism and Cauchy Mutation. *IEEE Trans. Antennas Propagat.* **2020**, *68*, 4406–4416. [CrossRef]
5. He, X.; Alistarh C.; Podilchak, S.K. Optimal MIMO Sparse Array Design Based on Simulated Annealing Particle Swarm Optimization. In Proceedings of the 2022 16th European Conference on Antennas and Propagation (EuCAP), Madrid, Spain, 27 March–1 April 2022; pp. 1–5. [CrossRef]

6. Shen, H.O.; Wang, B.H.; Li, L.J. Effective approach for pattern synthesis of sparse reconfigurable antenna arrays with exact pattern matching. *Iet Microw. Antenna Propagat.* **2016**, *10*, 748–755. [CrossRef]
7. Li, Z.; Cai, J.J.; Hao, C. Synthesis of sparse linear arrays using reweighted gridless compressed sensing. *IET Microw. Antenna Propagat.* **2021**, *15*, 1945–1959. [CrossRef]
8. Sun, S.; Zhang, Y.D. 4D Automotive Radar Sensing for Autonomous Vehicles: A Sparsity-Oriented Approach. *IEEE J Sel. Top. Signal Process.* **2021**, *15*, 879–891. [CrossRef]
9. Ebrahimi, M.; Modarres-Hashemi, M.; Yazdian, E. An Efficient Method for Sparse Linear Array Sensor Placement to Achieve Maximum Degrees of Freedom. *IEEE Sens. J.* **2021**, *21*, 20788–20795. [CrossRef]
10. Hameed, K.; Tu, S.; Ahmed, N.; Khan, W.; Armghan, A.; Alenezi, F.; Alnaim, N.; Qamar, M.S.; Basit, A.; Ali, F. DOA Estimation in Low SNR Environment through Coprime Antenna Arrays: An Innovative Approach by Applying Flower Pollination Algorithm. *Appl. Sci.* **2021**, *11*, 7985. [CrossRef]
11. Wang, L.; Zhao, H. Difference Coarray Design based on Genetic Algorithm and Convex Optimization. In Proceedings of the 2021 OES China Ocean Acoustics (COA) Conference, Harbin, China, 14–17 July 2021; pp. 791–795.
12. Chen, K.; He, Z.; Han, C. A modified real GA for the sparse linear array synthesis with multiple constraints. *IEEE Trans. Antennas Propagat.* **2006**, *54*, 2169–2173. [CrossRef]
13. Yu, X.; Cui, G.; Yang, S. Coherent unambiguous transmit for sparse linear array with geography constraint. *IET Radar Sonar Navig.* **2017**, *11*, 386–393. [CrossRef]
14. Li, J.X.; Ren, S.; Guo, C.J. Synthesis of Sparse Arrays Based On CIGA (Convex Improved Genetic Algorithm). *J. Microw. Optoelectron. Electromagn.* **2020**, *19*, 444–456. [CrossRef]
15. Yang, X.; Li, Y.; Liu, F.; Lan, T.; Teng, L.; Sarkar, T.K. Antenna position optimization method based on adaptive genetic algorithm with self-supervised differential operator for distributed coherent aperture radar. *IET Radar Sonar. Navig.* **2021**, *15*, 677–685. [CrossRef]
16. Wang, T.; Xia, K.W.; Tang, H.L.; Zhang, S.W.; Sandrine, M. A Modified Wolf Pack Algorithm for Multiconstrained Sparse Linear Array Synthesis. *Int. J. Antennas Propag.* **2020**, *5*, 1–12. [CrossRef]
17. Pan, Y.; Zhang, J. Synthesis of linear symmetric antenna arrays using improved bat algorithm. *Microw. Opt. Technol. Lett.* **2020**, *62*, 2383–2389. [CrossRef]
18. Wang, R.Q.; Jiao, Y.C. Synthesis of Sparse Linear Arrays With Reduced Excitation Control Numbers Using a Hybrid Cuckoo Search Algorithm with Convex Programming. *IEEE Antennas Wirel. Propag. Lett.* **2020**, 428–432. [CrossRef]
19. Mirjalili, S.; Andrew, L. The whale optimization algorithm. *Adv. Eng. Softw.* **2016**, *95*, 51–67. [CrossRef]
20. Zhang, C.; Fu, X.; Peng, S.; Wang, Y. Linear unequally spaced array synthesis for sidelobe suppression with different aperture constraints using whale optimization algorithm. In Proceedings of the 2018 13th IEEE Conference on Industrial Electronics and Applications (ICIEA), Wuhan, China, 31 May–2 June 2018; pp. 69–73.
21. Zhang, C.; Fu, X.; Ligthart, L.P.; Peng, S.; Xie, M. Synthesis of Broadside Linear Aperiodic Arrays With Sidelobe Suppression and Null Steering Using Whale Optimization Algorithm. *IEEE Antennas Wirel. Propag. Lett.* **2018**, *17*, 347–350. [CrossRef]
22. Prabhakar, D.; Moturi, S. Side lobe pattern synthesis using hybrid SSWOA algorithm for conformal antenna array. *Eng. Sci. Technol.* **2019**, *22*, 1169–1174. [CrossRef]
23. Boursianis, A.D.; Papadopoulou, M.S.; Salucci, M.; Polo, A.; Kostas, P.E. Emerging Swarm Intelligence Algorithms and Their Applications in Antenna Design: The GWO, WOA, and SSA Optimizers. *Appl. Sci.* **2021**, *11*, 8330. [CrossRef]
24. Fuertes, G.; Vargas, M.; Alfaro, M.; Soto-Garrido, R.; Sabattin, J.; Peralta, M.A. Chaotic genetic algorithm and the effects of entropy in performance optimization. *Chaos Interdiscip. J. Nonlinear Sci.* **2019**, *29*, 013132. [CrossRef] [PubMed]
25. Li, X.; Duan, B.; Zhou, J.; Song, L.; Zhang, Y. Planar Array Synthesis for Optimal Microwave Power Transmission with Multiple Constraints. *IEEE Antennas Wirel. Propag. Lett.* **2017**, *16*, 70–73. [CrossRef]
26. Wu, H.; Liu, C.; Xie, X. Pattern Synthesis of Planar Nonuniform Circular Antenna Arrays Using a Chaotic Adaptive Invasive Weed Optimization Algorithm. *Math. Probl. Eng.* **2014**, 1–13. [CrossRef]
27. Sun, G.; Liu, Y.; Chen, Z.; Liang, S.; Wang, A.; Zhang, Y. Radiation Beam Pattern Synthesis of Concentric Circular Antenna Arrays Using Hybrid Approach Based on Cuckoo Search. *IEEE Trans. Antennas Propagat.* **2018**, *66*, 4563–4576. [CrossRef]
28. Liang, Q.; Chen, B.; Wu, H.; Ma, C.; Li, S. A Novel Modified Sparrow Search Algorithm with Application in Side Lobe Level Reduction of Linear Antenna Array. *Wirel. Commun. Mob. Comput.* **2021**, *2014*, 1–25. [CrossRef]
29. Yang, X.; Liu, J.; Liu, Y.; Xu, P.; Yu, L.; Zhu, L.; Chen, H.; Deng, W. A Novel Adaptive Sparrow Search Algorithm Based on Chaotic Mapping and T-Distribution Mutation. *Appl. Sci.* **2021**, *11*, 11192. [CrossRef]
30. Yao, J.; Sha, Y.; Chen, Y.; Zhang, G.; Hu, X.; Bai, G.; Liu, J. IHSSAO: An Improved Hybrid Salp Swarm Algorithm and Aquila Optimizer for UAV Path Planning in Complex Terrain. *Appl. Sci.* **2022**, *12*, 5634. [CrossRef]
31. Puri, H.; Chaudhary, J.; Bingi, K.; Sivaramakrishnan, U.; Panga, N. Design of Adaptive Weighted Whale Optimization Algorithm. In Proceedings of the 2021 IEEE Madras Section Conference (MASCON), Chennai, India, 27–28 August 2021; pp. 1–6.
32. Luan, F.; Cai, Z.; Wu, S.; Jiang, T.; Yang, J. Improved Whale Algorithm for Solving the Flexible Job Shop Scheduling Problem. *Mathematics 2019*, *7*, 384. [CrossRef]
33. Zhou, Y.; Ling, Y.; Luo, Q. Lévy Flight Trajectory-Based Whale Optimization Algorithm for Global Optimization. *IEEE Access* **2017**, *5*, 6168–6186. [CrossRef]

34. Dai, D.; Yao, M.; Ma, H.; Jin, W.; Zhang, F. An Effective Approach for the Synthesis of Uniformly Excited Large Linear Sparse Array. *IEEE Antennas Wirel. Propag. Lett.* **2018**, *17*, 377–380. [CrossRef]
35. Zheng, T.; Liu, Y.; Geng, S. IWORMLF: Improved Invasive Weed Optimization With Random Mutation and Lévy Flight for Beam Pattern Optimizations of Linear and Circular Antenna Arrays. *IEEE Access* **2020**, *8*, 19460–19478. [CrossRef]
36. Liu, L.; Wang, A.; Sun, G.; Zheng, T.; Yu, C. An improved biogeography-based optimization approach for beam pattern optimizations of linear and circular antenna arrays. *Int. J. Numer. Model. Electron. Netw. Devices Fields* **2021**, *34*, 1–33. [CrossRef]
37. Alamri, H. Opposition-based Whale Optimization Algorithm. *Adv. Sci. Lett.* **2018**, *24*, 7461–7464. [CrossRef]
38. Fan, Q.; Chen, Z.; Li, Z.; Xia, Z.; Wang, D. A new improved whale optimization algorithm with joint search mechanisms for high-dimensional global optimization problems. *Eng. Comput.* **2021**, *37*, 1851–1878. [CrossRef]
39. He, Y.; Li, Y.; Liu, X. A novel discrete whale optimization algorithm for solving knapsack problems, *Appl. Intell.* **2020**, *50*, 3350–3366. [CrossRef]
40. Mohammed, A.K.; Altamir, S.A.; Algamal, Z. Improving whale optimization algorithm for feature selection with a time-varying transfer function. *Numer. Algebr. Control. Optim.* **2020**, *11*, 87–98. [CrossRef]
41. Abdel-Basset, M.; El-Shahat, D.; Sangaiah, A.K. A modified nature inspired meta-heuristic whale optimization algorithm for solving 0–1 knapsack problem. *Int. J. Mach. Learn. Cybern.* **2019**, *10*, 495–514. [CrossRef]
42. Chahar, V.; Kumar, D. Binary Whale Optimization Algorithm and its Application to Unit Commitment Problem. *Neural Comput. Appl.* **2020**, *32*, 2095–2123. [CrossRef]
43. Lee, H.; Tak, J.; Moon, W.; Lim, G. Effects of mutual impedance on the radiation characteristics of transducer arrays. *J. Acoust. Soc. Am.* **2004** *115*, 666–679. [CrossRef]
44. Li, D.; Chen, H. Directivity calculation for acoustic transducer arrays with considering mutual radiation impedance. In Proceedings of the International Conference on Graphic and Image Processing, Singapore, 14 March 2013; p. 876864. [CrossRef]

Article

Observing Meteorological Tides: Fifteen Years of Statistics in the Port of La Spezia (Italy)

Maurizio Soldani * and Osvaldo Faggioni

Istituto Nazionale di Geofisica e Vulcanologia, Via di Vigna Murata 605, 00143 Roma, Italy
* Correspondence: maurizio.soldani@ingv.it

Abstract: Sea level changes in coastal areas significantly influence port activities (e.g., the safety of navigation). Along Italian coastlines, sea level variations are mainly due to astronomical tides (well known, due to gravitational attraction between Earth, Moon and Sun); however, during the last fifteen years, a high number of "anomalous" tides has been observed: the study of the phenomenon has allowed to attribute its cause to variations in atmospheric pressure (the so-called meteorological tides: sea level drops when atmospheric pressure increases and vice versa); the statistical analysis of acquired data made it possible to evaluate the hydrobarometric transfer factor (a local parameter which represents the correlation between atmospheric pressure changes and consequent sea level variations): it was found that it is usually much larger within gulfs or port basins than offshore areas, where a pressure change of 1 hPa results in a sea level variation of about 1 cm; the statistical analysis described in the following, and aimed at correctly estimating the hydrobarometric transfer factor in harbors, can play a fundamental role in optimizing the management of port waters: its results allow to forecast meteorological tides and therefore future sea level (and depth) variations in a given port basin. The results of the study conducted in the port of La Spezia (North Western Italy) are presented here, together with possible applications on port activities and harbor water management.

Keywords: meteorological tides; marine environmental monitoring; sea level forecasting; harbor water management; port navigation safety

Citation: Soldani, M.; Faggioni, O. Observing Meteorological Tides: Fifteen Years of Statistics in the Port of La Spezia (Italy). *Appl. Sci.* **2022**, *12*, 12202. https://doi.org/10.3390/app122312202

Academic Editors: Enjin Zhao, Hao Qin and Lin Mu

Received: 24 October 2022
Accepted: 24 November 2022
Published: 29 November 2022

1. Introduction

The knowledge of sea level fluctuations in coastal areas is fundamental to increasing the safety of people who work in ports or onboard ships, to best managing port activities, and to reducing economic losses and environmental damage (improving the safety of navigation, optimizing ships' cargo and mooring, managing the refloating of stranded vessels, dimensioning maritime works, planning dredging activities, checking the chemical and physical parameters of water), as well as for civil protection purposes (mitigating the risk of flooding at the mouth of rivers) [1–7].

Sea level variations along Mediterranean coastlines are mostly due to astronomical tides (up and down motions caused by the gravitational attraction between the earth, moon and sun, then periodic and predictable deterministically), in particular to the diurnal and semi-diurnal components, shown in Table 1 [8,9].

However, a large number of anomalous tides have been observed over the last fifteen years inside many Italian harbors. The study of the phenomenon allowed us to associate these events to changes in atmospheric pressure above the sea basin under examination; in particular, sea level lowers/rises following an increase/decrease in atmospheric pressure (good/bad weather) and then in the weight of the overlying air column [10–17]: these low frequency oscillations, called meteorological tides, represent the geodetic adjustment (Newtonian compensation) of sea surface (induced effect), which compensates for atmospheric pressure perturbations (inducing cause), as shown in Figure 1 [18–23].

Table 1. The diurnal and semi-diurnal tide components and their periods expressed in hours (from [9]).

Name	Harmonic Constituent	Period/h
M2	Principal lunar semi-diurnal	12.4206
S2	Principal solar semi-diurnal	12.0000
N2	Larger lunar elliptic semi-diurnal	12.6583
K2	Luni-solar declinational semi-diurnal	11.9672
K1	Luni-solar declinational diurnal	23.9345
O1	Principal lunar declinational diurnal	25.8193
P1	Principal solar declinational diurnal	24.0659
Q1	Lunar flowelliptic diurnal	26.8680

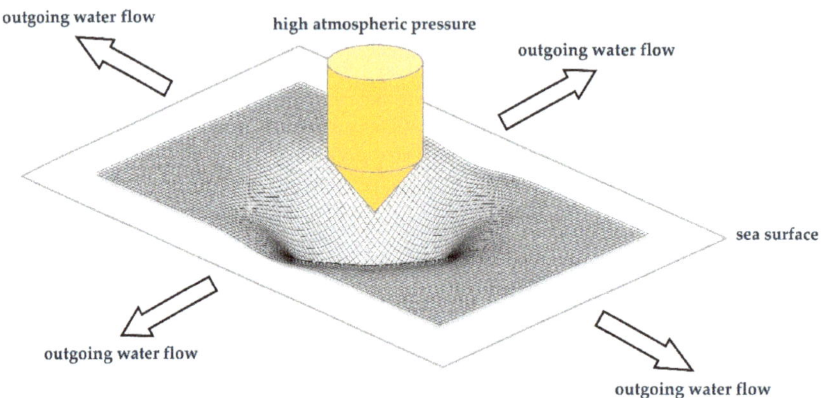

Figure 1. An increase in atmospheric pressure inducing a low meteorological tide (modified from [19]).

The evidence of the phenomenon was observed by means of ISPRA's (Italian Institute for Environmental Protection and Research) meteo-mareographic station located in La Spezia harbor (Italy); for example, on 13 November 2020 at 10:15 UTC (Universal Time Coordinates) and on 8 December 2020 at 09:25 UTC, the same astronomical tide was present (nearly 20 cm), but a pressure decrease of nearly 27.1 hPa resulted in an increase of about 34 cm in the sea level acquired.

In many Italian ports, we have verified that the hydrobarometric transfer factor, a parameter that represents the correlation between the atmospheric pressure variation and the consequent change in sea level, assumes often much larger values (even double) compared with offshore areas, where, as is well known, 1 hPa (about equal to 1 mBar) of atmospheric pressure variation corresponds to approximately 1 cm of change in sea level (the so-called inverted barometer effect); so, within harbors, a few hPa of decrease/increase in atmospheric pressure can cause several cm of sea level rise/fall; in fact, a sea basin behaves in the same way as a semi-constrained domain: by hindering the horizontal movement of the water mass, it amplifies its vertical displacement. Therefore, meteorological tides can cause exceptional changes in sea level within a port basin if they occur in-phase with astronomical ones.

To be able to forecast sea level in harbors, it is therefore necessary to estimate the correlation between atmospheric pressure and sea level, represented by the hydrobarometric transfer factor; it depends on a series of local parameters (first of all the morphology of the basin examined and the atmospheric dynamics over it), so it is not described by a deterministic law valid everywhere but obtained port by port through a statistical analysis of local data acquired [24–31].

In this work, we present the analysis carried out starting from data acquired since 2006 by ISPRA's monitoring station located inside the port of La Spezia (Eastern Ligurian Sea, Italy); based on the measurements of atmospheric pressure and sea level, the statistical analysis described below aims at estimating the hydrobarometric transfer factor in La Spezia harbor.

We also describe a prototype application developed to provide support to local authorities and port communities, in order to improve port navigation safety (obviously, the low tide hinders the port navigation, while the high tide facilitates it): based on the sea level measured or forecasted, the application updates water depths inside a port basin ("real" port bathymetric map) and detects, by means of a simple and intuitive graphic interface, hazardous areas for a certain ship moving inside the harbor at a given moment [32–38].

2. Materials and Methods

The starting point of this study is the monitoring of environmental parameters in the port of La Spezia, performed by means of the meteo-mareografic station working in the position 09°51′27.52″ E, 44°05′47.79″ N (see Figure 2) and belonging to the National Tidegauge Network managed by ISPRA [39–45].

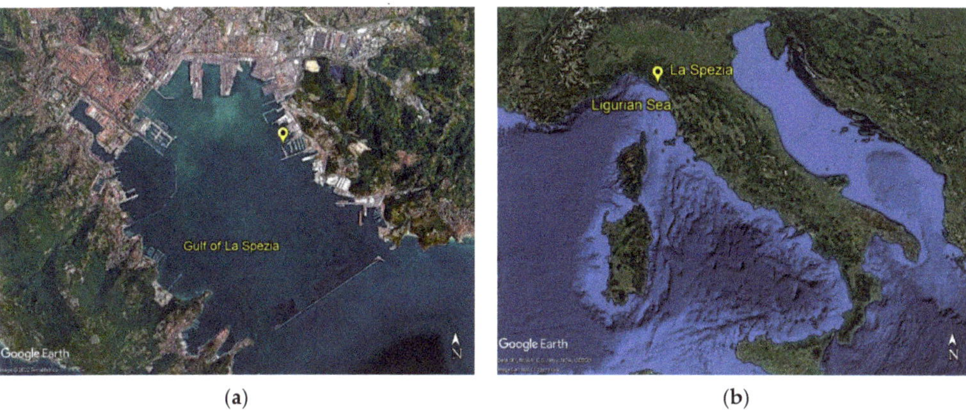

(a) (b)

Figure 2. The position of the monitoring station (pictures from Google Earth): (**a**) inside the port of La Spezia; (**b**) within the Eastern Ligurian Sea, Italy.

The instrumentation used consists of a hydrometer and a barometer; the first measures the sea level on the basis of the round trip time taken by a sequence of radar pulses sent from the air towards the sea surface; since the speed of propagation of electromagnetic waves in the air, the round trip time of the pulses, and the position of the radar transducer are known, sea levels are calculated; a typical sampling interval to acquire tide data is at least 10 min.

The barometer measures the atmospheric pressure by evaluating the deformations undergone by a silicon capacitive transducer; when the atmospheric pressure changes, the distance between the two plates and therefore the electrical capacitance also varies; measurements are typically made hourly (atmospheric pressure is characterized by very slow variations).

The date and time are expressed in UTC, while the sea level refers to IGM's (Italian Military Geographic Institute) 0 level. The data used in this work and further information about the ISPRA's monitoring station are available on the website www.mareografico.it (data from 2006 to 2009 are not available on the website; they have been kindly provided by ISPRA) (accessed on 28 April 2022).

Atmospheric pressure and sea level measurements (one sample every hour with resolutions equal to 10^{-1} hPa and 1 cm, respectively) were compared with each other in order to evaluate the hydrobarometric transfer factor, a local parameter that represents the correlation between the two quantities. Figure 3 shows the signals acquired between 28 May and 7 June 2009 inside the port of La Spezia.

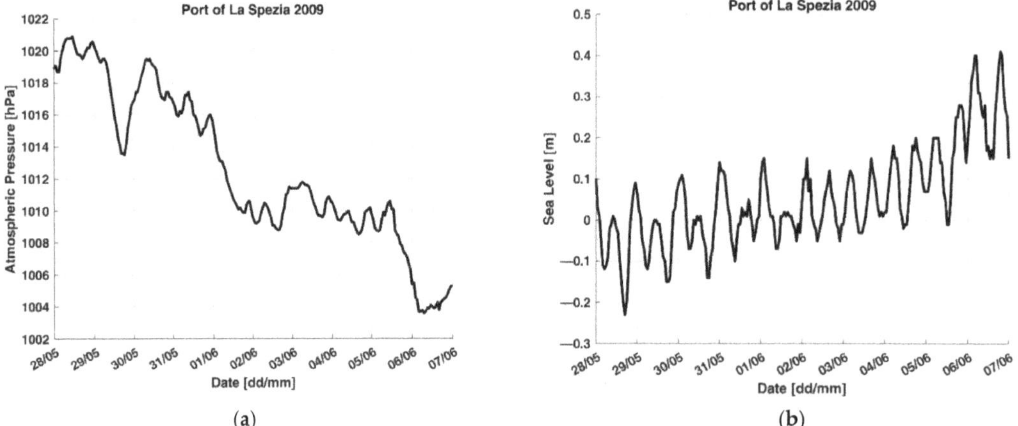

Figure 3. Measurements carried out between 28 May and 7 June 2009 inside the port of La Spezia: (**a**) atmospheric pressure; (**b**) sea level.

The sea level is the overlapping of different contributions, among which the main ones are astronomical and meteorological tides, as shown in the power spectral densities plotted in Figure 4, as regards the measurements performed in 2009: Frequency components due to diurnal and semidiurnal components are at approximately at 1.2 and 2.3 $\times$ 10^{-5} Hz respectively, while meteorological components are characterized by lower frequencies (slower oscillations) related to atmospheric pressure spectrum (DC components are not plotted because they do not carry useful information).

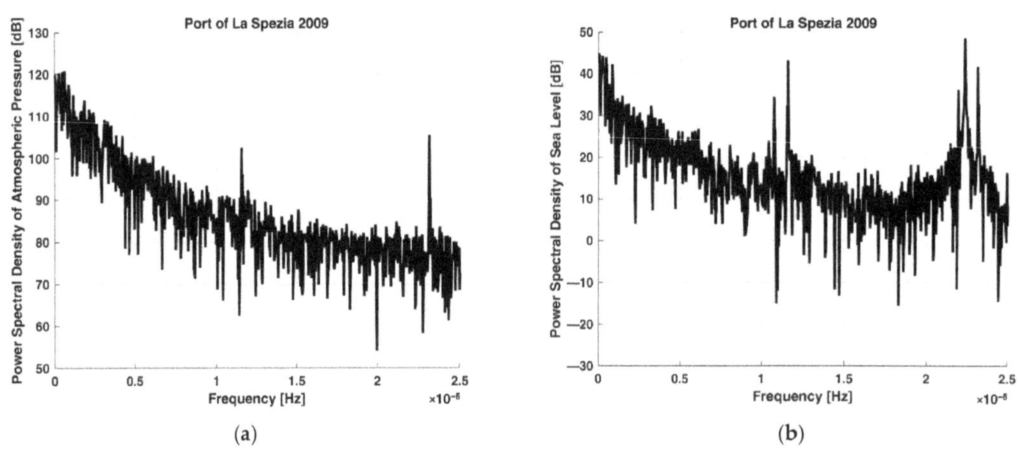

Figure 4. Power Spectral Densities of measurements carried out in 2009 within the port of La Spezia: (**a**) atmospheric pressure; (**b**) sea level.

First of all, it was therefore necessary to filter acquired data in order to remove high-frequency components due to causes different than meteorological ones (there could also be a residual of the disturbance due to sea waves, although the hydrometer works inside a still-pipe): atmospheric pressure and sea level signals are subjected to low-pass filtering with an appropriate cut frequency (10^{-5} Hz), so that only the contributions of meteorological origin survive. Figure 5 shows the result of the low-pass filtering applied to the data shown in Figure 3.

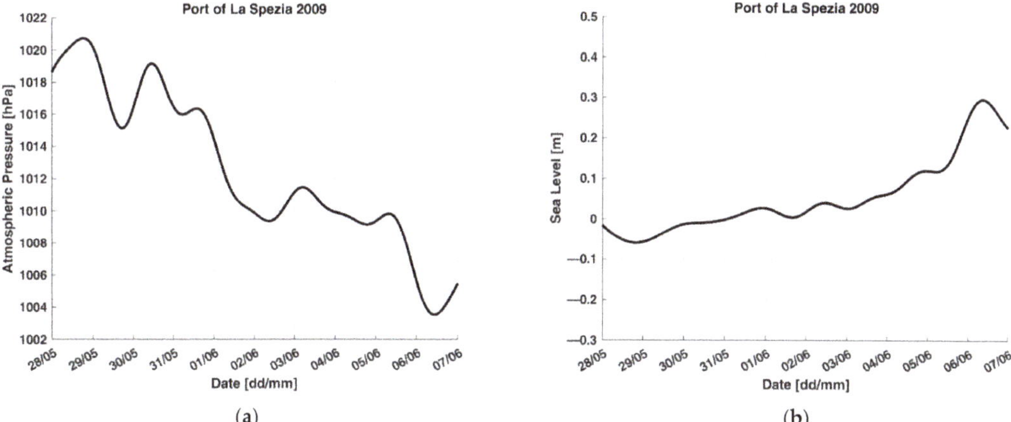

Figure 5. Measurements carried out between 28 May and 7 June 2009 inside the port of La Spezia, after Low-Pass filtering: (**a**) atmospheric pressure; (**b**) sea level.

By examining filtered signals, it is evident that a decrease Δp in atmospheric pressure equal to about 17.2 hPa causes an increase in low-frequency sea level Δh (low meteorological tide) equal to about 35.4 cm. So, for the event analyzed, the hydrobarometric transfer factor J_{ph} can be calculated as:

$$J_{ph} = \frac{\Delta h}{\Delta p} = \frac{35.4\ \text{cm}}{17.2\ \text{hPa}} = 2.1\ \text{cm} * \text{hPa}^{-1}, \quad (1)$$

which is more than double the offshore case; the gradients Δh and Δp are expressed in absolute values.

The analysis just described for a single case was repeated for all the events that occurred in the port of La Spezia since 14 March 2006 (the hydrometer of ISPRA's monitoring station has been working since 13 January 2006, but the barometer was not installed until two months later) in order to obtain an estimate of the hydrobarometric transfer factor for the water basin under examination, as described in the next paragraph.

3. Results

The analysis described in the previous paragraph was replicated for every significant hydrobarometric event occurred in the port of La Spezia from March 2006 (data from 24 January 2015 to 6 March 2019 are not available) to the end of 2021, to produce a fifteen-year statistics (this study was realized starting from the installation of ISPRA's monitoring station, in 2006).

The values of Δp, Δh and J_{ph} estimated event by event are listed in the Appendix A, in Table A2. Only events with Δp greater than about 5 hPa are taken into account. The highest observed value of meteorological tide in this period was 52.9 cm (between 28 February and 4 March 2020).

The acquisitions related to these events are shown, in Figures A1–A4 (in the Appendix A) and Figures S1–S45 (in the Supplementary Material); events occurring in the presence of wind were not considered in the statistical analysis, in order to exclude some phenomena due to causes such as storm surges, anyway not predominant in the site examined (the anemometer to acquire wind data has been working since 30 June 2010) and then not analyzed in this work; for example, the event shown in Figure 6 (from 28 September to

5 October 2020, Δp = 20.4 hPa, Δh = 43.5 cm) has been excluded from the statistics because there was a wind coming roughly from the South East (then from the mouth of the gulf), stronger than 10 m/s and persistent for more than one day.

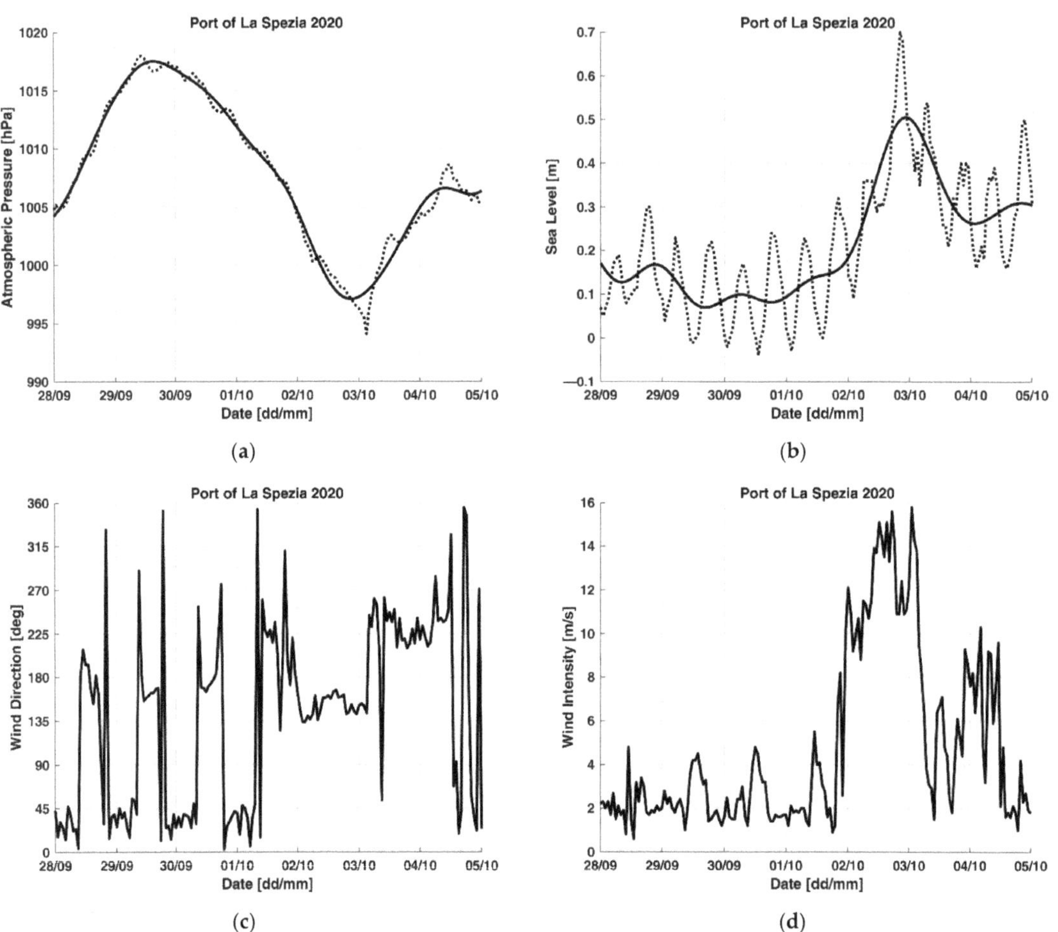

Figure 6. Measurements carried out between 28 September and 5 October 2020 inside the port of La Spezia: (**a**) atmospheric pressure; (**b**) sea level; (**c**) wind direction, clockwise from the North; (**d**) wind intensity; for (**a**,**b**) solid lines represent low-frequency components survived the Low-Pass filtering; dashed lines represent high-frequency components removed by the Low-Pass filtering.

First, the event shown in Figure S5 was removed from the statistics because its J_{ph} values differ from the mean value by more than 3 times the standard deviation; for this reason, it is considered outlier (rare event).

After doing that, the mean value and the standard deviation of the statistical distribution are 1.9 cm $*$ hPa^{-1} and 0.3 cm $*$ hPa^{-1}, respectively.

Then, a first estimate of the hydrobarometric transfer factor for the port of La Spezia can be represented by its average: in this case the mean value is about double the offshore; after this, the mean hydrobarometric transfer factor can be used to forecast a future sea level variation Δh starting from the measured atmospheric pressure change Δp: it is sufficient to multiply Δp by J_{ph} (taking into account that when the pressure goes up the level falls and vice versa); this corresponds to suppose a linear dependence:

$$\Delta h = J_{ph} * \Delta p \,, \tag{2}$$

as represented by the red straight line in Figure 7, whereas the black dots correspond to the measured pairs (Δp, Δh) listed in Table A2.

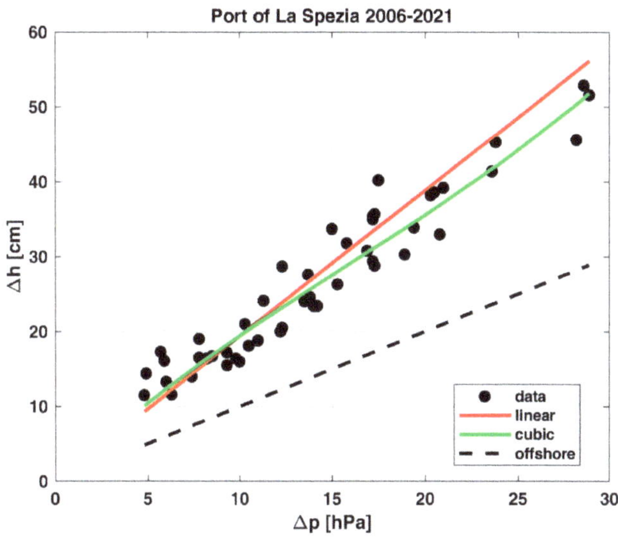

Figure 7. Measured pairs (Δp, Δh) from 2006 to 2021 in the port of La Spezia (black dots), linear (red) and cubic (green) trends and comparison with the offshore case (black dashed line).

A better estimate of the relationship between atmospheric pressure and sea level gradients can be obtained by considering a non-linear law $\Delta h = f(\Delta p)$; using the least squares method applied to the pairs (Δp, Δh) the fixed degree polynomial can be extrapolated that best fits the data cloud (the hydrobarometric transfer factor becomes a sequence of coefficients); for example, assuming a cubic dependence of Δh on Δp and imposing the passage to the origin (because $\Delta p = 0$ implies $\Delta h = 0$), this estimate of Δh is obtained:

$$\Delta h = J_{ph3} * \Delta p^3 + J_{ph2} * \Delta p^2 + J_{ph1} * \Delta p \,, \tag{3}$$

where: $J_{ph3} = 0.001$ cm $*$ hPa^{-3}, $J_{ph2} = -0.05$ cm $*$ hPa^{-2}, $J_{ph1} = 2.3$ cm $*$ hPa^{-1}, as plotted in Figure 7 (green line).

The couples (Δp, Δh) measured from 2006 to 2021 in the port of La Spezia, linear and cubic approximations are shown in Figure 7, together with the comparison with the "inverted barometer effect" of the offshore case.

Once the change in atmospheric pressure is measured or predicted, the expected sea level variation can be derived from the Equation (3).

Obviously, the two different estimates (linear and cubic) lead to two different errors between the measured value and the estimated trends: starting from data in Table A2, an average error on the expected sea level (difference between forecast and measurement) of 11.3% was obtained in the case of linear approximation, while in the case of cubic function the mean error is 9.6% (against a greater computational load), which is satisfactory for our purposes.

Therefore, the fundamental role of the hydrobarometric transfer factor consists in converting a variation of atmospheric pressure into a forecasted sea level change (meteorological component) for a given port; finally, meteorological tides will have to be added (or subctracted) to the astronomical tides (predictable by means of tide charts) to obtain the forecasted sea level in that basin.

Contributions to changes in sea level due to other causes such as, for example, seiches, storm surges or wind effects, which are not predominant in the site examined, are not taken into account in this work.

It should be emphasized that the hydrobarometric transfer factor must be updated year after year and periodically recalculated, e.g., following events that modify the topography of the basin examined (dredging operations, coastal erosion, bottom subsidence, sediment deposition).

Moreover, a multi-decade statistics is necessary to examine any changes in the hydrobarometric transfer factor due to climate change, as well as the variation over the years of occurrence frequency of events observed [46,47].

4. Discussion

As seen in the previous paragraph, the hydrobarometric transfer factor is usually much greater within port basins than in offshore areas: A few hPa of atmospheric pressure variation can cause several cm of astronomical tide that, if in phase with the astronomical one, can generate anomalous sea level variations.

The knowledge of the hydrobarometric transfer factor allows for correctly estimating expected meteorological tides in harbors and, together with the joint prediction of astronomical components from tide charts, forecasting sea level within port basins, an aspect of fundamental importance to better managing port operations.

In fact, the monitoring/forecasting of sea level (and then of water depth) in coastal areas is extremely important for managing, planning, and optimizing:

- Maritime transport and port navigation safety (e.g., to reduce the risk of accidents or to plan the refloating of a ship and minimize the risk of environmental damages and economic losses) [48–51];
- Ship loading (how much to load a ship in the departure port based on the expected tide in the port of arrival) [52–55];
- Dock performances and vessel moorings;
- Dimensioning of maritime works based on the maximum sea level expected;
- Dredging activities;
- The control of chemical and physical parameters of water;

as well as preventing the risk of flooding at the mouths of rivers (civil protection purposes) by providing early warning to the population involved.

For this reason, the results of this study can have important applications in coastal areas: A software tool has been developed by the research group to which the authors belong with the aim of providing useful operational support to port communities, local authorities, and decision makers; it dynamically updates the initial port bathymetric map (usually acquired through multibeam surveys and updated after changes in the harbor topography, e.g., following dredging operations) based on sea level measured or expected and, if the draught of a certain ship is known, identifies the permitted/alert/prohibited areas for that same ship at a given time. An intuitive graphical interface implements what are called "virtual traffic lights" by coloring the forbidden areas red, the alert zones yellow, and the allowed ones green, based on two thresholds that in their turn depend on the vessel draught. Red areas are those with depths less than the lower threshold (usually equal to the vessel's draught), green areas those with depths greater than the upper threshold (much greater than the vessel's draught); finally, yellow areas are the intermediate ones.

The application continuously recalculates the "real" bathymetry (water depth variable over time) of a harbor using sea level data acquired in real time (by downloading them from the monitoring station via an Internet connection), measured in the past and saved in a dataset (to analyze a posteriori past events of particular importance such as the stranding of ships) or forecasted in the future by means of the hydrobarometric transfer factor and tide charts, in order to signal in advance potentially dangerous areas and avoid critical situations induced by sea level changes for a given ship. It can be very useful, e.g., for supporting a certain ship in choosing the best route to follow or the best time to enter or leave a port or the most suitable quay to moor.

For example, in Figure 8, virtual traffic lights in the middle of the port of La Spezia on 30 June 2010 at 12:00 UTC are shown, when the tide gauge was measuring −0.54 m. Thresholds were chosen equal to 12 and 13.5 m, e.g., for a Panamax cargo ship, whose draught is about 12 m (areas with depth less than 12 m are prohibited, those with depth greater than 13.5 m are permitted, the others are warning areas).

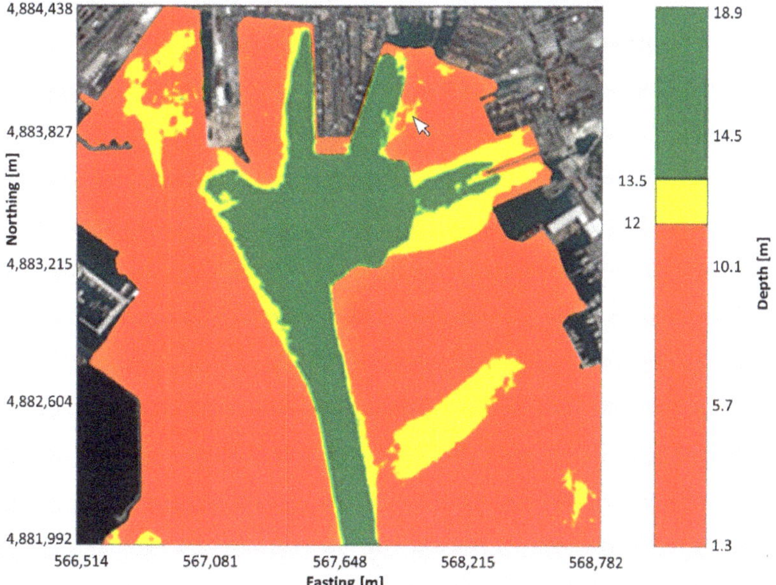

Figure 8. Virtual traffic lights in the port of La Spezia on 30 June 2010 at 12:00 UTC (threshold levels 12 and 13.5 m).

Easting and northing are expressed in UTM (Universal Transverse Mercator) coordinates, zone 32T; grid spacing is 2 m; depth refers to IGM's 0 level, depth resolution 1 cm; bathymetric data are courtesy of the Port System Authority of the Eastern Ligurian Sea, La Spezia, Italy.

Instead, Figure 9 refers to 24 November 2010 at 18:00 UTC (sea level = 0.83 m), for the same ship (and then the same thresholds).

Note how, following the rise of 1.37 m in sea level, the forbidden area narrows and many warning positions become allowed, while a yellow waterway appears in the west side of the port.

In particular, the state of the position indicated by the mouse pointer (coordinates 567,956, 4,883,930 m) in the middle top of the map switches from alert to allowed, as indicated by the color of the traffic light in Figures 8 and 9, since its depth increases from 12.45 to 13.82 m.

It is worth highlighting that the increase in sea level is partly due to a 9.5 hPa fall in atmospheric pressure between 30 June 2010 at 12:00 UTC and 24 November 2010 at 18:00 UTC (from 1016.9 to 1007.4 hPa).

Obviously at the same instant, for a vessel with greater draught (for example 14 m for a container ship), thresholds would be higher, and forbidden areas would expand, as shown in Figure 10; the traffic light for the position indicated by the mouse pointer becomes red.

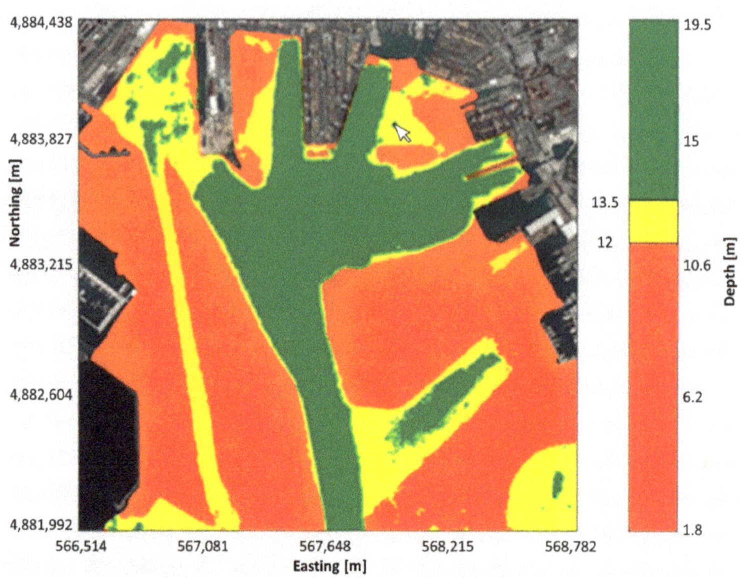

Figure 9. Virtual traffic lights in the port of La Spezia on 24 November 2010 at 18:00 UTC (threshold levels 12 and 13.5 m).

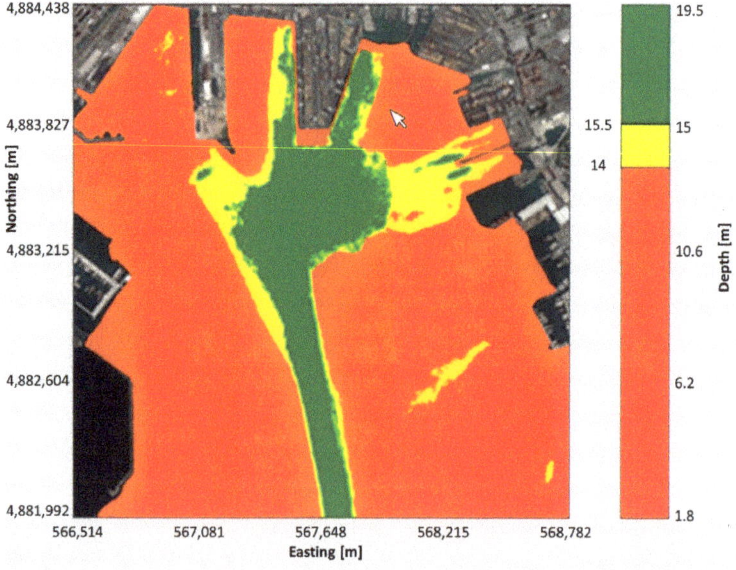

Figure 10. Virtual traffic lights in the port of La Spezia on 24 November 2010 at 18:00 UTC (threshold levels 14 and 15.5 m).

Therefore, this interface represents a useful tool for detecting potentially dangerous areas for a given ship at a certain moment.

Additionally other contributions due to different phenomena can be passed as input to the application, such as seiches, storm surges, or wind effects, which particularly in certain locations must be taken into account.

5. Conclusions

In Mediterranean harbors, tides are mainly due to astronomical and meteorological components; while the first ones are well known and predictable everywhere by means of a deterministic law (tide charts), the second ones (due to atmospheric disturbances) need more study: starting from monitoring environmental parameters in La Spezia harbor (Ligurian Sea, North-Western Italy), we performed a statistical analysis over the last fifteen years (starting from the installation of ISPRA's monitoring station in 2006) to evaluate the hydrobarometric transfer factor, which represents the correlation between atmospheric pressure (cause) and sea level variations (effect) and therefore can play a fundamental role in forecasting meteorological tides in harbors.

We found that the hydrobarometric transfer factor in La Spezia harbor is larger (sometimes more than double) than in offshore areas (where 1 hPa of atmospheric pressure gradient induces nearly 1 cm of sea level rise/fall); thus, some hPa of atmospheric gradient can induce several cm of sea level rise/fall.

In particular, we found that using J_{ph} = 1.9 cm ∗ hPa^{-1} results in an estimate of the average error of about 11.3%, while approximating the dependence of sea level on atmospheric pressure by means of a nonlinear (cubic) law reduces the mean error to about 9.6%. This error very rarely induces confusion between different colors along the edges of adjacent areas in the map representing virtual traffic lights described above.

Furthermore, meteorological tides can cause exceptional changes in sea level if they occur in conjunction with astronomical components: the observation of the phenomenon allowed us to highlight anomalou" tides, sea level changes that are very different from the expected astronomical tides.

The hydrobarometric transfer factor allows to forecast meteorological tides in the sea basin examined and then, simply by adding the contribution of the astronomical tides, to know in advance the sea level (and therefore water depth) expected in the near future; this can represent a useful tool for optimizing port activities and logistic operations by planning them in advance.

Since sea level changes in coastal areas affect in a relevant way the safety of port communities, a prototype application has been developed that updates the port bathymetric map based on sea level acquired in real time or forecasted, with the aim of planning and optimizing port activities or managing emergencies (it is also able to load old measurements stored in a dataset, e.g., to analyze accidents occurred in the past); the software tool classifies the port bathymetric map in forbidden (red), warning (yellow), or permitted (green) areas for a certain ship at a given moment, based on its draught (virtual traffic lights); the graphical interface is able to detect and easily signal hazardous situations in harbors in order to provide a useful support to decision makers (port authorities, coast guards), with the aim of increasing safety for people working in ports.

Supplementary Materials: The following supporting information can be downloaded at: https://www.mdpi.com/article/10.3390/app122312202/s1, Figure S1. Measurements carried out between 21 and 27 February 2007 inside the port of La Spezia: (**a**) atmospheric pressure; (**b**) sea level (Δp = 10 hPa, Δh = 16 cm, J_{ph} = 1.6 cm ∗ hPa^{-1}). Figure S2. Measurements carried out between 25 February and 1 March 2007 inside the port of La Spezia: (**a**) atmospheric pressure; (**b**) sea level (Δp = 9.8 hPa, Δh = 16.4 cm, J_{ph} = 1.7 cm ∗ hPa^{-1}). Figure S3. Measurements carried out between 9 and 16 May 2007 inside the port of La Spezia: (**a**) atmospheric pressure; (**b**) sea level (Δp = 5.9 hPa, Δh = 16.1 cm, J_{ph} = 2.7 cm ∗ hPa^{-1}). Figure S4. Measurements carried out between 31 May and 2 June 2007 inside the port of La Spezia: (**a**) atmospheric pressure; (**b**) sea level (Δp = 4.9 hPa, Δh = 14.4 cm, J_{ph} = 2.9 cm ∗ hPa^{-1}). Figure S5. Measurements carried out between 1 and 4 June 2007 inside the port of La Spezia: (**a**) atmospheric pressure; (**b**) sea level (Δp = 4.9 hPa, Δh = 16.8 cm, J_{ph} = 3.4 cm ∗ hPa^{-1}). Figure S6. Measurements carried out between 3 and 10 August 2007 inside the port of La Spezia: (**a**) atmospheric pressure; (**b**) sea level (Δp = 12.2 hPa, Δh = 20 cm, J_{ph} = 1.6 cm ∗ hPa^{-1}). Figure S7. Measurements carried out between 20 and 27 October 2007 inside the port of La Spezia: (**a**) atmospheric pressure; (**b**) sea level (Δp = 13.7 hPa, Δh = 27.6 cm, J_{ph} = 2 cm ∗ hPa^{-1}). Figure S8. Measurements carried out between 3 and 13 April 2008 inside the port of La Spezia: (**a**) atmospheric pressure; (**b**) sea level (Δp = 17.5 hPa, Δh = 40.2 cm, J_{ph} = 2.3 cm ∗ hPa^{-1}). Figure S9. Measurements carried out between 10 and 17 April 2008 inside the port of La Spezia: (**a**) atmospheric pressure; (**b**) sea level (Δp = 16.9 hPa, Δh = 30.8 cm, J_{ph} = 1.8 cm ∗ hPa^{-1}). Figure S10. Measurements carried out between 26 August and 9 September 2008 inside the port of La Spezia: (**a**) atmospheric pressure; (**b**) sea level (Δp = 7.8 hPa, Δh = 19 cm, J_{ph} = 2.4 cm ∗ hPa^{-1}). Figure S11. Measurements carried out between 26 November and 2 December 2008 inside the port of La Spezia: (**a**) atmospheric pressure; (**b**) sea level (Δp = 28.9 hPa, Δh = 51.6 cm, J_{ph} = 1.8 cm ∗ hPa^{-1}). Figure S12. Measurements carried out between 29 January and 9 February 2009 inside the port of La Spezia: (**a**) atmospheric pressure; (**b**) sea level (Δp = 28.2 hPa, Δh = 45.6 cm, J_{ph} = 1.6 cm ∗ hPa^{-1}). Figure S13. Measurements carried out between 28 May and 7 June 2009 inside the port of La Spezia: (**a**) atmospheric pressure; (**b**) sea level (Δp = 17.2 hPa, Δh = 35.4 cm, J_{ph} = 2.1 cm ∗ hPa^{-1}). Figure S14. Measurements carried out between 6 and 18 September 2009 inside the port of La Spezia: (**a**) atmospheric pressure; (**b**) sea level (Δp = 13.8 hPa, Δh = 24.6 cm, J_{ph} = 1.8 cm ∗ hPa^{-1}). Figure S15. Measurements carried out between 28 November and 1 December 2009 inside the port of La Spezia: (**a**) atmospheric pressure; (**b**) sea level (Δp = 20.5 hPa, Δh = 38.6 cm, J_{ph} = 1.9 cm ∗ hPa^{-1}). Figure S16. Measurements carried out between 18 and 20 February 2010 inside the port of La Spezia: (**a**) atmospheric pressure; (**b**) sea level (Δp = 19.4 hPa, Δh = 33.9 cm, J_{ph} = 1.7 cm ∗ hPa^{-1}). Figure S17. Measurements carried out between 4 and 15 June 2010 inside the port of La Spezia: (**a**) atmospheric pressure; (**b**) sea level (Δp = 8.2 hPa, Δh = 16.4 cm, J_{ph} = 2 cm ∗ hPa^{-1}). Figure S18. Measurements carried out between 7 and 16 August 2010 inside the port of La Spezia: (**a**) atmospheric pressure; (**b**) sea level (Δp = 7.8 hPa, Δh = 16.5 cm, J_{ph} = 2.1 cm ∗ hPa^{-1}). Figure S19. Measurements carried out between 27 October and 02 November 2010 inside the port of La Spezia: (**a**) atmospheric pressure; (**b**) sea level (Δp = 17.3 hPa, Δh = 35.7 cm, J_{ph} = 2.1 cm ∗ hPa^{-1}). Figure S20. Measurements carried out between 26 January and 8 February 2011 inside the port of La Spezia: (**a**) atmospheric pressure; (**b**) sea level (Δp = 18.9 hPa, Δh = 30.3 cm, J_{ph} = 1.6 cm ∗ hPa^{-1}). Figure S21. Measurements carried out between 18 and 28 June 2011 inside the port of La Spezia: (**a**) atmospheric pressure; (**b**) sea level (Δp = 11.3 hPa, Δh = 24.1 cm, J_{ph} = 2.1 cm ∗ hPa^{-1}). Figure S22. Measurements carried out between 18 September and 2 October 2011 inside the port of La Spezia: (**a**) atmospheric pressure; (**b**) sea level (Δp = 20.8 hPa, Δh = 33 cm, J_{ph} = 1.6 cm ∗ hPa^{-1}). Figure S23. Measurements carried out between 21 and 27 October 2011 inside the port of La Spezia: (**a**) atmospheric pressure; (**b**) sea level (Δp = 15 hPa, Δh = 33.7 cm, J_{ph} = 2.2 cm ∗ hPa^{-1}). Figure S24. Measurements carried out between 25 October and 3 November 2011 inside the port of La Spezia: (**a**) atmospheric pressure; (**b**) sea level (Δp = 17.3 hPa, Δh = 28.8 cm, J_{ph} = 1.7 cm ∗ hPa^{-1}). Figure S25. Measurements carried out between 28 October and 7 November 2011 inside the port of La Spezia: (**a**) atmospheric pressure; (**b**) sea level (Δp = 17.2 hPa, Δh = 29.4 cm, J_{ph} = 1.7 cm ∗ hPa^{-1}). Figure S26. Measurements carried out between 12 and 20 July 2012 inside the port of La Spezia: (**a**) atmospheric pressure; (**b**) sea level (Δp = 10.5 hPa, Δh = 18.1 cm, J_{ph} = 1.7 cm ∗ hPa^{-1}). Figure S27. Measurements carried out between 8 and 16 October 2012 inside the port of La Spezia: (**a**) atmospheric pressure; (**b**) sea level (Δp = 14.2 hPa, Δh = 23.4 cm, J_{ph} = 1.6 cm ∗ hPa^{-1}). Figure S28. Measurements carried out between 5 and 12 May 2013 inside the port of La Spezia: (**a**) atmospheric pressure; (**b**) sea level (Δp = 5.7 hPa, Δh = 17.3 cm, J_{ph} = 3 cm ∗ hPa^{-1}). Figure S29. Measurements carried out between 23 and 28 June 2013 inside the port of La Spezia: (**a**) atmospheric pressure; (**b**) sea level (Δp = 9.3 hPa, Δh = 17.2 cm, J_{ph} = 1.8 cm ∗ hPa^{-1}). Figure S30. Measurements carried out between 19 and 26 January 2014 inside the port of La Spezia: (**a**) atmospheric pressure; (**b**) sea level (Δp = 17.2 hPa, Δh = 35 cm, J_{ph} = 2 cm ∗ hPa^{-1}). Figure S31. Measurements carried out between 10 and 26 February 2014 inside the port of La Spezia: (**a**) atmospheric pressure; (**b**) sea level (Δp = 23.6 hPa, Δh = 41.4 cm, J_{ph} = 1.8 cm ∗ hPa^{-1}). Figure S32. Measurements carried out between 12 and 16 October 2019 inside the port of La Spezia: (**a**) atmospheric pressure; (**b**) sea level (Δp = 12.3 hPa, Δh = 20.5 cm, J_{ph} = 1.7 cm ∗ hPa^{-1}). Figure S33. Measurements carried out between 15 and 18 October 2019 inside the port of La Spezia: (**a**) atmospheric pressure; (**b**) sea level (Δp = 9.3 hPa, Δh = 15.5 cm, J_{ph} = 1.7 cm ∗ hPa^{-1}). Figure S34. Measurements carried out between 17 and 21 October 2019 inside the port of La Spezia: (**a**) atmospheric pressure; (**b**) sea level (Δp = 4.8 hPa, Δh = 11.5 cm, J_{ph} = 2.4 cm $*$ hPa^{-1}). Figure S35. Measurements carried out between 19 and 23 October 2019 inside the port of La Spezia: (**a**) atmospheric pressure; (**b**) sea level (Δp = 6 hPa, Δh = 13.3 cm, J_{ph} = 2.2 cm $*$ hPa^{-1}). Figure S36. Measurements carried out between 19 and 25 November 2019 inside the port of La Spezia: (**a**) atmospheric pressure; (**b**) sea level (Δp = 12.3 hPa, Δh = 28.7 cm, J_{ph} = 2.3 cm $*$ hPa^{-1}). Figure S37. Measurements carried out between 27 November and 9 December 2019 inside the port of La Spezia: (**a**) atmospheric pressure; (**b**) sea level (Δp = 21 hPa, Δh = 39.2 cm, J_{ph} = 1.9 cm $*$ hPa^{-1}). Figure S38. Measurements carried out between 28 February and 4 March 2020 inside the port of La Spezia: (**a**) atmospheric pressure; (**b**) sea level (Δp = 28.6 hPa, Δh = 52.9 cm, J_{ph} = 1.8 cm $*$ hPa^{-1}). Figure S39. Measurements carried out between 31 May and 8 June 2020 inside the port of La Spezia: (**a**) atmospheric pressure; (**b**) sea level (Δp = 20.3 hPa, Δh = 38.2 cm, J_{ph} = 1.9 cm $*$ hPa^{-1}). Figure S40. Measurements carried out between 7 and 17 June 2020 inside the port of La Spezia: (**a**) atmospheric pressure; (**b**) sea level (Δp = 6.3 hPa, Δh = 11.6 cm, J_{ph} = 1.8 cm $*$ hPa^{-1}). Figure S41. Measurements carried out between 21 and 31 August 2020 inside the port of La Spezia: (**a**) atmospheric pressure; (**b**) sea level (Δp = 15.3 hPa, Δh = 26.3 cm, J_{ph} = 1.7 cm $*$ hPa^{-1}). Figure S42. Measurements carried out between 21 and 29 June 2021 inside the port of La Spezia: (**a**) atmospheric pressure; (**b**) sea level (Δp = 7.4 hPa, Δh = 1.4 cm, J_{ph} = 1.9 cm $*$ hPa^{-1}). Figure S43. Measurements carried out between 21 and 27 July 2021 inside the port of La Spezia: (**a**) atmospheric pressure; (**b**) sea level (Δp = 8.5 hPa, Δh = 16.7 cm, J_{ph} = 2 cm $*$ hPa^{-1}). Figure S44. Measurements carried out between 31 July and 24 August 2021 inside the port of La Spezia: (**a**) atmospheric pressure; (**b**) sea level (Δp = 14 hPa, Δh = 23.4 cm, J_{ph} = 1.7 cm $*$ hPa^{-1}). Figure S45. Measurements carried out between 18 and 25 September 2021 inside the port of La Spezia: (**a**) atmospheric pressure; (**b**) sea level (Δp = 11 hPa, Δh = 18.8 cm, J_{ph} = 1.7 cm $*$ hPa^{-1}).

Author Contributions: Conceptualization, M.S. and O.F.; methodology, M.S. and O.F.; software, M.S. and O.F.; validation, M.S. and O.F.; formal analysis, M.S. and O.F.; investigation, M.S. and O.F.; resources, M.S. and O.F.; data curation, M.S.; writing—original draft preparation, M.S.; writing—review and editing, M.S. and O.F.; visualization, M.S. and O.F.; supervision, O.F.; project administration, M.S. and O.F.; funding acquisition, M.S. and O.F. All authors have read and agreed to the published version of the manuscript.

Funding: This research was funded within the framework of the MENFOR Project (MEteo-tide Newtonian FORecasting) by several Italian port authorities, in particular the Port Authority of La Spezia (now named the Port System Authority of the Eastern Ligurian Sea) for what we described in this article and from the European Union and Regional Government of Liguria (Italy) by means of the Regional Plan of Innovative Actions—European Funds for the Regional Development.

Institutional Review Board Statement: Not applicable.

Informed Consent Statement: Not applicable.

Data Availability Statement: Meteo-mareographic data used in this article, acquired by means of the monitoring station in La Spezia belonging to ISPRA's National Tidegauge Network, are mostly available on the website www.mareografico.it (accessed on 28 April 2022); data from 2006 to 2009 are not available on the website; they have been kindly provided by ISPRA; additionally, archives of environmental data available on weather websites were consulted.

Acknowledgments: The authors wish to thank the Port Authority of La Spezia (now named the Port System Authority of the Eastern Ligurian Sea) for having kindly provided bathymetric data (a special thanks to D. Vetrala and I. Roncarolo) and D.A. Leoncini for his past contribution in the development of the software tool. Part of this research was carried out when the first author was at OGS—National Institute of Oceanography and Applied Geophysics (Trieste, Italy). Finally, the authors thank the anonymous reviewers whose comments and suggestions helped improve this work.

Conflicts of Interest: The authors declare no conflict of interest.

Appendix A

Table A1. Estimated values for representative events occurred from 2006 to 2021 in the port of La Spezia.

Start	End	Δp hPa	Δh cm	J_{ph} cm * hPa^{-1}	Figure
11/06/2006	01/07/2006	13.5	24	1.8	A1
16/08/2006	23/08/2006	10.3	21	2	A2
12/10/2006	26/10/2006	15.8	31.8	2	A3
23/10/2006	12/11/2006	23.8	45.3	1.9	A4
21/02/2007	27/02/2007	10	16	1.6	S1
25/02/2007	01/03/2007	9.8	16.4	1.7	S2
09/05/2007	16/05/2007	5.9	16.1	2.7	S3
31/05/2007	02/06/2007	4.9	14.4	2.9	S4
01/06/2007	04/06/2007	4.9	16.8	3.4	S5
03/08/2007	10/08/2007	12.2	20	1.6	S6
20/10/2007	27/10/2007	13.7	27.6	2	S7
03/04/2008	13/04/2008	17.5	40.2	2.3	S8
10/04/2008	17/04/2008	16.9	30.8	1.8	S9
26/08/2008	09/09/2008	7.8	19	2.4	S10
26/11/2008	02/12/2008	28.9	51.6	1.8	S11
29/01/2009	09/02/2009	28.2	45.6	1.6	S12
28/05/2009	07/06/2009	17.2	35.4	2.1	S13
06/09/2009	18/09/2009	13.8	24.6	1.8	S14
28/11/2009	01/12/2009	20.5	38.6	1.9	S15
18/02/2010	20/02/2010	19.4	33.9	1.7	S16
04/06/2010	15/06/2010	8.2	16.4	2	S17
07/08/2010	16/08/2010	7.8	16.5	2.1	S18
27/10/2010	02/11/2010	17.3	35.7	2.1	S19
26/01/2011	08/02/2011	18.9	30.3	1.6	S20
18/06/2011	28/06/2011	11.3	24.1	2.1	S21
18/09/2011	02/10/2011	20.8	33	1.6	S22
21/10/2011	27/10/2011	15	33.7	2.2	S23
25/10/2011	03/11/2011	17.3	28.8	1.7	S24
28/10/2011	07/11/2011	17.2	29.4	1.7	S25
12/07/2012	20/07/2012	10.5	18.1	1.7	S26
08/10/2012	16/10/2012	14.2	23.4	1.6	S27
05/05/2013	12/05/2013	5.7	17.3	3	S28
23/06/2013	28/06/2013	9.3	17.2	1.8	S29
19/01/2014	26/01/2014	17.2	35	2	S30
10/02/2014	26/02/2014	23.6	41.4	1.8	S31
12/10/2019	16/10/2019	12.3	20.5	1.7	S32
15/10/2019	18/10/2019	9.3	15.5	1.7	S33

Table A2. Estimated values for representative events occurred from 2006 to 2021 in the port of La Spezia.

Start	End	Δp hPa	Δh cm	J_{ph} cm ∗ hPa^{-1}	Figure
17/10/2019	21/10/2019	4.8	11.5	2.4	S34
19/10/2019	23/10/2019	6	13.3	2.2	S35
19/11/2019	25/11/2019	12.3	28.7	2.3	S36
27/11/2019	09/12/2019	21	39.2	1.9	S37
28/02/2020	04/03/2020	28.6	52.9	1.8	S38
31/05/2020	08/06/2020	20.3	38.2	1.9	S39
07/06/2020	17/06/2020	6.3	11.6	1.8	S40
21/08/2020	31/08/2020	15.3	26.3	1.7	S41
21/06/2021	29/06/2021	7.4	14	1.9	S42
21/07/2021	27/07/2021	8.5	16.7	2	S43
31/07/2021	24/08/2021	14	23.4	1.7	S44
18/09/2021	25/09/2021	11	18.8	1.7	S45

Figures A1–A4 (in this Appendix A) and Figures S1–S45 (in the Supplementary Material) show data acquired during the events listed in Table A2; solid lines represent low-frequency components survived the Low-Pass filtering; dashed lines represent high-frequency components removed by the Low-Pass filtering.

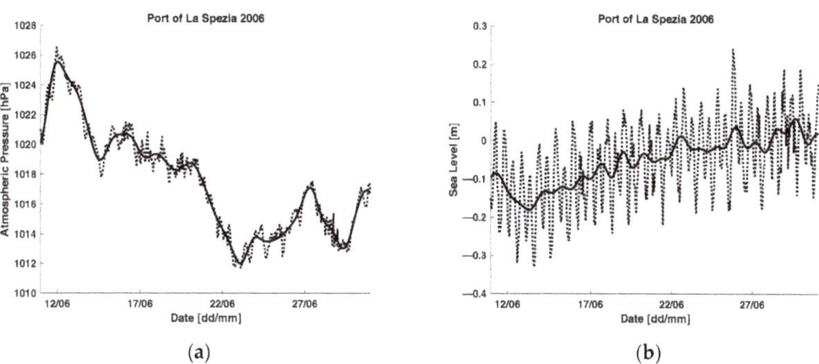

(a) (b)

Figure A1. Measurements carried out between 11 June and 01 July 2006 inside the port of La Spezia: (**a**) atmospheric pressure; (**b**) sea level (Δp = 13.5 hPa, Δh = 24 cm, J_{ph} = 1.8 cm ∗ hPa^{-1}).

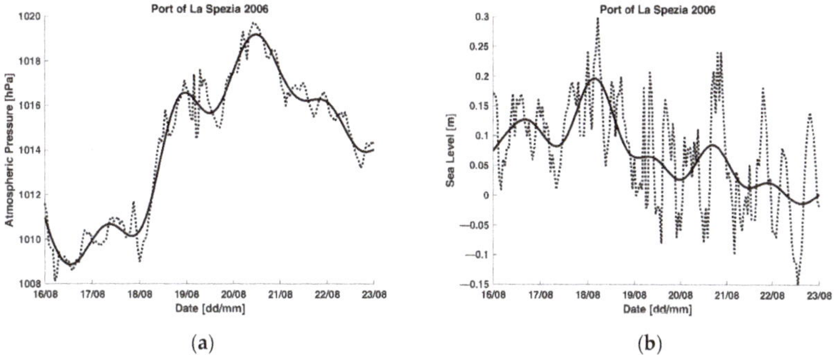

(a) (b)

Figure A2. Measurements carried out between 16 and 23 August 2006 inside the port of La Spezia: (**a**) atmospheric pressure; (**b**) sea level (Δp = 10.3 hPa, Δh = 21 cm, J_{ph} = 2 cm ∗ hPa^{-1}).

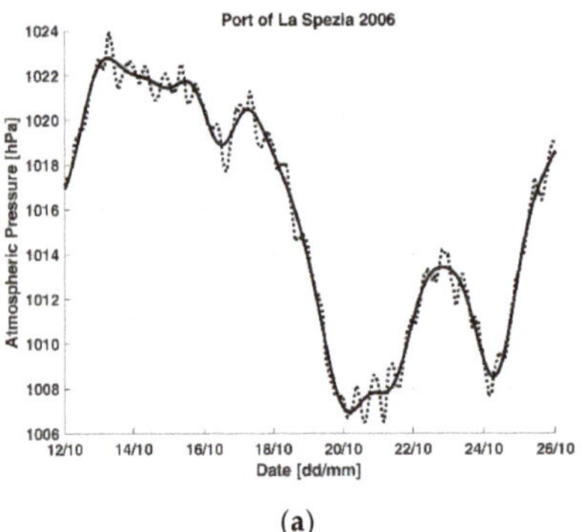

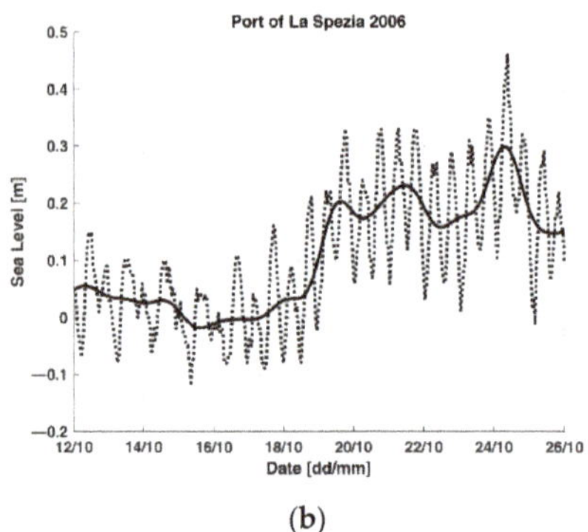

(a) (b)

Figure A3. Measurements carried out between 12 and 26 October 2006 inside the port of La Spezia: (**a**) atmospheric pressure; (**b**) sea level (Δp = 15.8 hPa, Δh = 31.8 cm, J_{ph} = 2 cm $*$ hPa^{-1}).

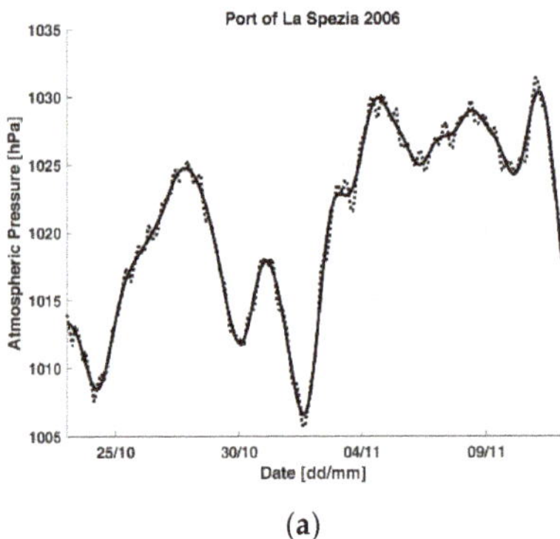

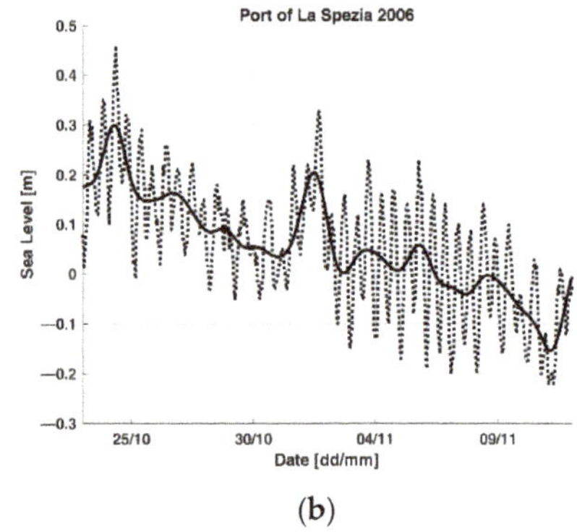

(a) (b)

Figure A4. Measurements carried out between 23 October and 12 November 2006 inside the port of La Spezia: (**a**) atmospheric pressure; (**b**) sea level (Δp = 23.8 hPa, Δh = 45.3 cm, J_{ph} = 1.9 cm $*$ hPa^{-1}).

References

1. Giuffrida, N.; Stojaković, M.; Twrdy, E.; Ignaccolo, M. The Importance of Environmental Factors in the Planning of Container Terminals: The Case Study of the Port of Augusta. *Appl. Sci.* **2021**, *11*, 2153. [CrossRef]
2. Meyers, S.D.; Luther, M.E. The impact of sea level rise on maritime navigation within a large, channelized estuary. *Marit. Pol. Manag.* **2020**, *47*, 920–936. [CrossRef]
3. Nguyen, T.-H.; Garrè, L.; Amdahl, J.; Leira, B.J. Benchmark study on the assessment of ship damage conditions during stranding. *Ships Offshore Struct.* **2012**, *7*, 197–213. [CrossRef]
4. Ogura, T.; Inoue, T.; Uchihira, N. Prediction of Arrival Time of Vessels Considering Future Weather Conditions. *Appl. Sci.* **2021**, *11*, 4410. [CrossRef]
5. Petraška, A.; Čižiūnienė, K.; Jarašūnienė, A.; Maruschak, P.; Prentkovskis, O. Algorithm for the assessment of heavyweight and oversize cargo transportation routes. *J. Bus. Econ. Manag.* **2017**, *18*, 1098–1114. [CrossRef]
6. Titz, M.A. Port state control versus marine environmental pollution. *Marit. Pol. Manag.* **1989**, *16*, 189–211. [CrossRef]
7. Vandermeulen, J.H. Environmental trends of ports and harbours: Implications for planning and management. *Marit. Pol. Manag.* **1996**, *23*, 55–66. [CrossRef]
8. Istituto Idrografico della Marina. *Tavole di Marea 2022*; Istituto Idrografico della Marina: Genoa, Italy, 2021.
9. Wei, G.; Wang, Q.; Peng, W. Accurate Evaluation of Vertical Tidal Displacement Determined by GPS Kinematic Precise Point Positioning: A Case Study of Hong Kong. *Sensors* **2019**, *19*, 2559. [CrossRef]
10. Allen, J.S.; Denbo, D.W. Statistical Characteristics of the Large-Scale Response of Coastal Sea Level to Atmospheric Forcing. *J. Phys. Oceanogr.* **1984**, *14*, 1079–1094. [CrossRef]
11. Chelton, D.B.; Enfield, D.B. Ocean signals in tide gauge records. *J. Geophys. Res. Solid Earth* **1986**, *91*, 9081–9098. [CrossRef]
12. Dickman, S.R. Theoretical investigation of the oceanic inverted barometer response. *J. Geophys. Res. Solid Earth* **1988**, *93*, 14941–14946. [CrossRef]

13. Dobslaw, H.; Thomas, M. Atmospheric induced oceanic tides from ECMWF forecasts. *Geophys. Res. Lett.* **2005**, *32*, L10615. [CrossRef]
14. Fu, L.-L.; Pihos, G. Determining the response of sea level to atmospheric pressure forcing using TOPEX/POSEIDON data. *J. Geosphys. Res.* **1994**, *99*, 24633–24642. [CrossRef]
15. Moon, I.-J. Impact of a coupled ocean wave-tide-circulation system on coastal modeling. *Ocean Model* **2005**, *8*, 203–236. [CrossRef]
16. Willebrand, J.; Philander, S.G.H.; Pacanowski, R.C. The Oceanic Response to Large-Scale Atmospheric Disturbances. *J. Phys. Oceanogr.* **1980**, *10*, 411–429. [CrossRef]
17. Wunsch, C.; Stammer, D. Atmospheric loading and the oceanic "inverted barometer" effect. *Rev. Geophys.* **1997**, *35*, 79–107. [CrossRef]
18. Dong, D.; Gross, R.S.; Dickey, J.O. Seasonal variations of the Earth's gravitational field: An analysis of atmospheric pressure, ocean tidal, surface water excitation. *Geophys. Res. Lett.* **1996**, *23*, 725–728. [CrossRef]
19. Faggioni, O.; Arena, G.; Bencivenga, M.; Bianco, G.; Bozzano, R.; Canepa, G.; Lusiani, P.; Nardone, G.; Piangiamore, G.L.; Soldani, M.; et al. The Newtonian approach in meteorological tide waves forecasting: Preliminary observations in the East Ligurian harbours. *Ann. Geophys.* **2006**, *49*, 1177–1187. [CrossRef]
20. Garrett, C.; Majaess, F. Nonisostatic Response of Sea Level to Atmospheric Pressure in the Eastern Mediterranean. *J. Phys. Oceanogr.* **1984**, *14*, 656–665. [CrossRef]
21. Merriam, J.B. Atmospheric pressure and gravity. *Geophys. J. Int.* **1992**, *109*, 488–500. [CrossRef]
22. Spratt, R.S. Modelling the effect of atmospheric pressure variations on gravity. *Geophys. J. Int.* **1982**, *71*, 173–186. [CrossRef]
23. Trenberth, K.E. Seasonal variations in global sea level pressure and the total mass of the atmosphere. *J. Geophys. Res.* **1981**, *86*, 5238–5246. [CrossRef]
24. Bâki Iz, H. The effect of regional sea level atmospheric pressure on sea level variations at globally distributed tide gauge stations with long records. *J. Geod. Sci.* **2018**, *8*, 55–71. [CrossRef]
25. Crépon, M. Influence de la pression atmosphérique sur le niveau moyen de la Méditerranée Occidentale et sur le flux à travers le détroit de Gibraltar. *Cah. Oceanogr.* **1965**, *1*, 15–32.
26. Deser, C.; Tomas, R.A.; Sun, L. The Role of Ocean-Atmosphere Coupling in the Zonal-Mean Atmospheric Response to Arctic Sea Ice Loss. *J. Clim.* **2015**, *28*, 2168–2186. [CrossRef]
27. Garrett, C.; Toulany, B. Sea level variability due to meteorological forcing in the northeast Gulf of St. Lawrence. *J. Geophys. Res.* **1982**, *87*, 1968–1978. [CrossRef]
28. Le Traon, P.-Y.; Gauzelin, P. Response of the Mediterranean mean sea level to atmospheric pressure forcing. *J. Geophys. Res.* **1997**, *102*, 973–984. [CrossRef]
29. Ponte, R.M.; Gaspar, P. Regional analysis of the inverted barometer effect over the global ocean using TOPEX/POSEIDON data and model results. *J. Geosphys. Res.* **1999**, *104*, 15587–15601. [CrossRef]
30. Tsimplis, M.N. The Response of Sea Level to Atmospheric Forcing in the Mediterranean. *J. Coast. Res.* **1995**, *11*, 1309–1321.
31. Tsimplis, M.N.; Vlahakis, G.N. Meteorological forcing and sea level variability in the Aegean Sea. *J. Geophys. Res.* **1994**, *99*, 9879–9890. [CrossRef]
32. Faggioni, O.; Soldani, M. Geomatics for underwater electromagnetic harbour protection systems and Newtonian systems for coastal navigation safety—Theory. In Proceedings of the Giornate INGV sull'Ambiente Marino—INGV Workshop on Marine Environment, Rome, Italy, 26–27 June 2019; Abstract Volume. Miscellanea INGV. Sagnotti, L., Beranzoli, L., Caruso, C., Guardato, S., Simoncelli, S., Eds.; Istituto Nazionale di Geofisica e Vulcanologia: Rome, Italy, 2019; Volume 51, pp. 111–114. [CrossRef]
33. Faggioni, O.; Soldani, M.; Piangiamore, G.L.; Ferrante, A.; Bencivenga, M.; Arena, G.; Nardone, G. harbour Water Management for port structures and sea bottom design, coast proximity navigation management, water quality control. In Proceedings of the 1st Mediterranean Days of Coastal and Port Engineering, Palermo, Italy, 7–9 October 2008; PIANC: Brussels, Belgium, 2008.
34. Soldani, M. Geomatics for Port Safety and Security. In *R3 in Geomatics: Research, Results and Review. Communications in Computer and Information Science*; Parente, C., Troisi, S., Vettore, A., Eds.; Springer: Cham, Switzerland, 2020; Volume 1246, pp. 91–102. [CrossRef]
35. Soldani, M. The contribution of Geomatics to increase safety and security in ports. *Appl. Geomat.* **2021**, *12*. [CrossRef]
36. Soldani, M.; Faggioni, O. Geomatics for underwater electromagnetic harbour protection systems and Newtonian systems for coastal navigation safety—Applications. In Proceedings of the Giornate INGV sull'Ambiente Marino—INGV Workshop on Marine Environment, Rome, Italy, 26–27 June 2019; Abstract Volume. Miscellanea INGV. Sagnotti, L., Beranzoli, L., Caruso, C., Guardato, S., Simoncelli, S., Eds.; Istituto Nazionale di Geofisica e Vulcanologia: Rome, Italy, 2019; Volume 51, pp. 115–118. [CrossRef]
37. Soldani, M.; Faggioni, O. A System to Improve Port Navigation Safety and Its Use in Italian harbours. *Appl. Sci.* **2021**, *11*, 10265. [CrossRef]
38. Soldani, M.; Faggioni, O. A tool to aid the navigation in La Spezia harbour (Italy). In *Geomatics for Green and Digital Transition*; ASITA2022; Communications in Computer and Information Science; Borgogno-Mondino, E., Zamperlin, P., Eds.; Springer: Cham, Switzerland, 2022; Volume 1651, pp. 89–101. [CrossRef]
39. Faggioni, O. The Information Protection in Automatic Reconstruction of Not Continuous Geophysical Data Series. *J. Data Anal. Inform. Process.* **2019**, *7*, 208–227. [CrossRef]

40. Faggioni, O. Measurement and Forecasting of Port Tide Hydrostatic Component in North Tyrrhenian Sea (Italy). *Open J. Mar. Sci.* **2020**, *10*, 52–77. [CrossRef]
41. Faggioni, O.; Soldani, M.; Leoncini, D.A. Metrological Analysis of Geopotential Gravity Field for Harbor Waterside Management and Water Quality Control. *Int. J. Geophys.* **2013**, *2013*, 398956. [CrossRef]
42. Kanasewich, E.R. *Time Sequence Analysis in Geophysics*, 3rd ed.; The University of Alberta Press: Edmonton, AB, Canada, 1981.
43. Krishnamurti, T.N. Numerical Weather Prediction. *Annu. Rev. Fluid Mech.* **1995**, *27*, 195–224. [CrossRef]
44. Lynch, P. The origins of computer weather prediction and climate modeling. *J. Comp. Phys.* **2008**, *227*, 3431–3444. [CrossRef]
45. Telford, W.M.; Geldart, L.P.; Sheriff, R.E. *Applied Geophysics*, 2nd ed.; Cambridge University Press: Cambridge, UK, 1990. [CrossRef]
46. Cabos, W.; de la Vara, A.; Álvarez-García, F.J.; Sánchez, E.; Sieck, K.; Pérez-Sanz, J.-I.; Limareva, N.; Sein, D.V. Impact of ocean-atmosphere coupling on regional climate: The Iberian Peninsula case. *Clim. Dyn.* **2020**, *54*, 4441–4467. [CrossRef]
47. El-Gindy, A.A.H.; Eid, F.M. Long-term variations of monthly mean sea level and its relation to atmospheric pressure in the Mediterranean Sea. *Int. Hydrogr. Rev.* **1990**, *67*, 147–159.
48. Bartlett, D.; Celliers, L. (Eds.) *Geoinformatics for Marine and Coastal Management*, 1st ed.; CRC Press: Boca Raton, FL, USA, 2016. [CrossRef]
49. Lam, S.Y.-W.; Yip, T.L. The role of geomatics engineering in establishing the marine information system for maritime management. *Marit. Pol. Manag.* **2008**, *35*, 53–60. [CrossRef]
50. Palikaris, A.; Mavraeidopoulos, A.K. Electronic Navigational Charts: International Standards and Map Projections. *J. Mar. Sci. Eng.* **2020**, *8*, 248. [CrossRef]
51. Weintrit, A. Geoinformatics in Shipping and Marine Transport. In *Challenge of Transport Telematics: TST 2016. Communications in Computer and Information Science*; Mikulski, J., Ed.; Springer: Cham, Switzerland, 2016; Volume 640, pp. 13–25. [CrossRef]
52. Bąk, A.; Zalewski, P. Determination of the Waterway Parameters as a Component of Safety Management System. *Appl. Sci.* **2021**, *11*, 4456. [CrossRef]
53. Ducruet, C.; Berli, J.; Bunel, M. Geography versus topology in the evolution of the global container shipping network (1977–2016). In *Geographies of Maritime Transport: Transport, Mobilities and Spatial Change*; Wilmsmeier, G., Monios, J., Eds.; Edward Elgar Publishing: Cheltenham, UK, 2020; pp. 49–70. [CrossRef]
54. Nohheman, W. Benefits of dredging through reduced tidal waiting. *Marit. Pol. Manag.* **1981**, *8*, 17–20. [CrossRef]
55. Vidmar, P.; Perkovič, M.; Gucma, L.; Łazuga, K. Risk Assessment of Moored and Passing Ships. *Appl. Sci.* **2020**, *10*, 6825. [CrossRef]

Article

High-Precision Numerical Research on Flow and Structure Noise of Underwater Vehicle

Hao Cao * **and Lihua Wen**

School of Astronautics, Northwestern Polytechnical University, Xi'an 710072, China
* Correspondence: caohao@163.com

Abstract: This paper presents the results of research on the noise generated by an underwater vehicle in the operational state. The study combines the large eddy simulated turbulence model and Lighthill's acoustic analogy theory and extracts the transient flow field data as the excitation conditions for acoustic calculations. The results of the numerical calculations of the external acoustic field were obtained under vehicle wall pressure pulsation condition, Lighthill volume excitation condition, and the vibration excitation condition of the underwater vehicle. It is found that the noise is concentrated at the front and tail of underwater vehicle, and its level is closely related to the form of vortex shedding. The peak frequency of structural radiation noise of underwater vehicle is consistent with its peak frequency of mean square vibration velocity. The basis for selecting the boundary conditions of the sound field according to the incoming flow conditions is also evaluated. The research results provide a reference for the noise reduction design of underwater vehicles, thus improving their concealment in combat.

Keywords: underwater vehicle; large eddy simulation; Lighthill's acoustic analogy; noise

1. Introduction

Torpedoes, submarines, and other underwater vehicles are the main combat tools of national navies and play an important role in assault operations, striking military targets, and reconnaissance owing to their high navigation speed, long range, and stealth. However, in the context of rapid development of soft and hard kill technologies such as sonar technology to combat underwater high-speed vehicles, how to improve the survivability of underwater vehicles has become a key technology in the military field. Underwater vehicles inevitably generate noise during their missions, and for every 5 dB increase in noise, the probability of hitting a target is reduced by 25%, thus directly affecting their stealth and deterrence. Therefore, underwater high-speed vehicles with silent characteristics have received much attention from various navies. The underwater noise is mainly generated by the pulsating pressure on the surface of vehicle and the disturbance within the turbulent boundary layer around the surface, and it is difficult to reduce the noise using general methods. Therefore, it is important to study the mechanism of underwater vehicle noise generation to improve its concealment.

The study of noise begins with a focus on turbulence and pulsating pressure, which was usually studied by theoretical analysis and experimental verification. However, theoretical analysis is only applicable to a simplified model, which cannot describe the phenomenon of a complex flow; experimental verification is also not widely used because of its high cost, long cycle time, and small measurement range. With the development of computer technology and improvement of computer performance, computational fluid dynamics (CFD) has become an important tool for solving flow problems. After the SUBOFF submarine model was proposed in the United States in 1989 and a series of experimental data were obtained, many numerical simulation studies based on this model and data were also carried out [1]. Alin analyzed a fully attached SUBOFF flow field variation using

Citation: Cao, H.; Wen, L. High-Precision Numerical Research on Flow and Structure Noise of Underwater Vehicle. *Appl. Sci.* **2022**, *12*, 12723. https://doi.org/10.3390/app122412723

Academic Editors: Lin Mu, Enjin Zhao and Hao Qin

Received: 14 November 2022
Accepted: 5 December 2022
Published: 12 December 2022

Unsteady Reynolds Averaged Navier Stokes (URANS) [2], Detached Eddy Simulation (DES), and Large Eddy Simulation (LES) [3], discussed the accuracy of different numerical computational models, and finally found the advantages of DES and LES for solving the pulsation-related terms. Bensow [4] and Holloway [5] analyzed the viscous flow field and flow-separation phenomena in SUBOFF submarines using LES and verified the accuracy of LES in flow field simulations when compared with experimental values. Kim and Sung [6] directly used Direct Numerical Simulation (DNS) to calculate the pulsation pressure within the boundary layer of a two-dimensional model flow field. However, DNS is poorly applicable because of the high requirements of computer hardware. Therefore, to balance the computational resources and solution accuracy, it is gradually becoming a research trend to use LES to obtain turbulent flow field data.

After separating the acoustic terms from the equations using the acoustic analogy theory proposed by Lighthill [7,8], the flow field calculations were successfully combined with the acoustic field calculations and numerical calculations on developing noise. Wang et al. [9] simulated the complex flow field of a submarine with a propeller based on LES and analyzed the sound pressure spectrum and noise-pointing characteristics of the submarine using ACTRAN software. It was found that the noise of the submarine with a propeller was concentrated in the low-frequency band, and the high-pressure region of the propeller contributed the most to the noise. Piomelli et al. [10] studied the effect of small-scale eddies on noise and found that the low-frequency part of the noise was better solved using the LES turbulence model. Both Bailly [11] and Moon [12] studied flat plate noise. Bailly [11] used the acoustic analogy method to study flat plates subjected to pulsating pressure and verified the feasibility of the acoustic analogy method by comparing the data with direct acoustic field calculations. Moon [12] studied the sound pressures of flat plates under different operating conditions based on a combination of LES and linearized perturbed compressible equations (LPCE), and found that the numerical calculation results are consistent with the experimental data. Lighthill's acoustic analogy theory can guarantee the accuracy of calculation, and a combination of high-precision numerical simulation with the theory is widely used in the field of noise solution.

In this study, we first obtained the data of velocity pulsation and pressure pulsation of a type of underwater vehicle by conducting a numerical simulation of flow field with high accuracy using LES. These pulsation data are then used as the excitation for the subsequent noise calculation, and finally the noise radiation characteristics and acoustic vibration coupling phenomenon of this underwater vehicle are determined. The first part of the article presents the theory involved in the numerical solution and the modeling of flow field and acoustic field. The Section 1 of the second part shows the results of numerical calculations of external flow field of underwater vehicle. The second part of the Section 2 analyzes the results of numerical calculations of external sound field of underwater vehicle under two boundary conditions: vehicle wall pressure pulsation condition and Lighthill volume excitation condition. The Section 3 of the second part considers the material and structural properties of underwater vehicle and analyzes its structural radiation noise. Finally, the study is summarized, and future research directions are given.

2. Theory And Model Building

2.1. Large Eddy Simulation

LES uses the N-S equation to directly model large-scale vortices. The equations for large-scale vortex control are obtained by filtering vortices that are smaller than the filter width or grid size using the filtering process.

The filtering function can be expressed as follows:

$$G(x, x') = \begin{cases} 1/V, x' \in V \\ 0, x' \notin V \end{cases} \tag{1}$$

where V is the control volume.

The filtered equation is given by:

$$\frac{\partial \rho}{\partial t} + \frac{\partial}{\partial x_i}(\rho \overline{u_i}) = 0 \quad (2)$$

$$\frac{\partial}{\partial t}(\rho \overline{u_i}) + \frac{\partial}{\partial x_j}(\rho \overline{u_i u_j}) = \frac{\partial}{\partial x_j}(\sigma_{ij}) - \frac{\partial \overline{p}}{\partial x_i} - \frac{\partial \tau_{ij}}{\partial x_j} \quad (3)$$

where ρ is the fluid density; t is the time; p is the pressure; σ_{ij} is the stress tensor due to molecular viscosity; τ_{ij} is the sublattice scale stress.

2.2. Lighthill's Acoustic Analogy Theory

The Lighthill acoustic analogy theory approximates the complex sound source formed by fluid motion as a static medium equivalent source to solve difficult problems to accurately describe the sound source. The theory assumes that the coupling between the fluid and sound field exists only in the near field, and considers the far field as the acoustic radiation region, whose flow has almost no effect on the sound field, where $\rho_a = \rho - \rho_0$, $P_a = P - P_0$, ρ_0 and P_0 represents the density and sound pressure. Then, the equation can be expressed as follows:

$$\frac{\partial^2 \rho_a}{\partial t^2} - c_0^2 \frac{\partial^2 \rho_a}{\partial x_i x_j} = \frac{\partial^2 T_{ij}}{\partial x_i x_j} \quad (4)$$

$$T_{ij} = \rho v_i v_j + \delta_{ij}\left[(P - P_0) - c^2\left(\rho - \rho^2\right)\right] - \tau_{ij} \quad (5)$$

where c_0 is the velocity of sound under isentropic conditions; T_{ij} is the Lighthill stress tensor; δ_{ij} is the elastic constant; τ_{ij} is the viscous stress.

As the viscous stress is a small amount of Reynolds stress and can be neglected, the final equation can be simplified as follows:

$$\frac{\partial^2 \rho_a}{\partial t^2} - c_0^2 \frac{\partial^2 \rho_a}{\partial x_i x_j} = \frac{\partial^2 T_{ij}}{\partial x_i x_j} \quad (6)$$

$$T_{ij} = \rho_0 v_i v_j \quad (7)$$

This equation is the basic equation of Lighthill's acoustic analog theory, which can be solved for turbulent pulsating pressure direct radiation noise.

2.3. Model Building

Figure 1a shows the underwater vehicle studied in this paper, which is a standard rotary body of 3.29 m in length. Figure 1b shows the computational domain of cylindrical flow field created from the model shape. The length of the front end part is taken as the length of the underwater vehicle. To accurately capture the trailing vortex region at the trailing edge of underwater vehicle, the trailing end is taken as twice the length of underwater vehicle. The diameter of computational domain is taken as 4.45 m. FLUENT MESHING was used for meshing, and the y^+ value was taken as 1. To accurately capture the vortex region that may be generated by the flow and the region with large parameter variations, the cylindrical region and tail section that wraps the whole underwater vehicle are locally encrypted. The final total number of generated meshes is about 27 million.

To facilitate the observation and analysis of sound field radiation characteristics and to avoid the effect of geometric features such as the sharp corners of calculation domain on sound field radiation, the acoustic field calculation domain was selected as the spherical region shown in Figure 1c. The diameter of the sphere was taken as three times the length of underwater vehicle. The noise in the low-frequency band (below 2000 Hz) was calculated using ACTRAN. According to the required calculation frequency, the accuracy of acoustic calculation results can be ensured when each wavelength contains at least six acoustic grids, so the maximum size of acoustic grid is set to 0.12 m. ICEM was used to divide the grid, and the total number of grids is about 6 million.

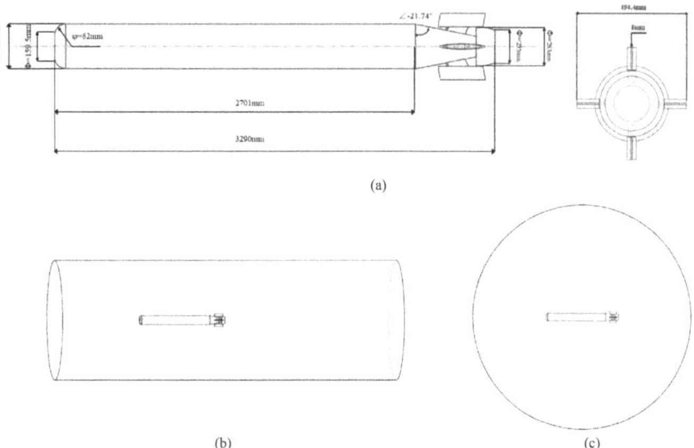

Figure 1. Diagram of the model. (**a**) Underwater vehicle model (**b**) Flow field calculation domain (**c**) Acoustic computational domain.

To detect the flow field and acoustic field of underwater vehicle, 14 monitoring points were set up at the front and tail of underwater vehicle. The distance of each monitoring point from the wall is 1 m, as shown in Figure 2.

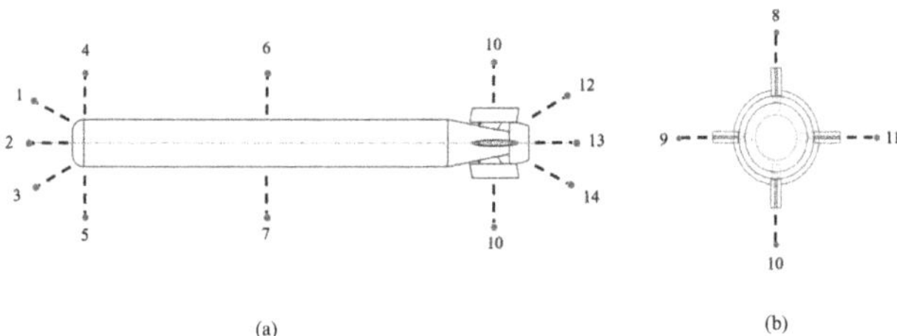

Figure 2. Monitoring point location diagram. (**a**) Side view (**b**) Rear view.

3. Results And Discussion

3.1. Calaulation Results of Flow Field

The study set the entrance boundary condition as the velocity entrance. The velocity was taken as a torpedo navigation speed of 19 m/s. The torpedo wall was set as no-slip condition, and the outer field boundary was set as no-shear condition. To provide a better initial field for the transient calculation to converge and stabilize faster, the standard k–e model was first selected for steady-state calculation. In the steady-state calculation section, the broadband noise source model was turned on to perform a preliminary estimation and analysis of the entire fluid domain sound field. The results are shown in Figure 3a.

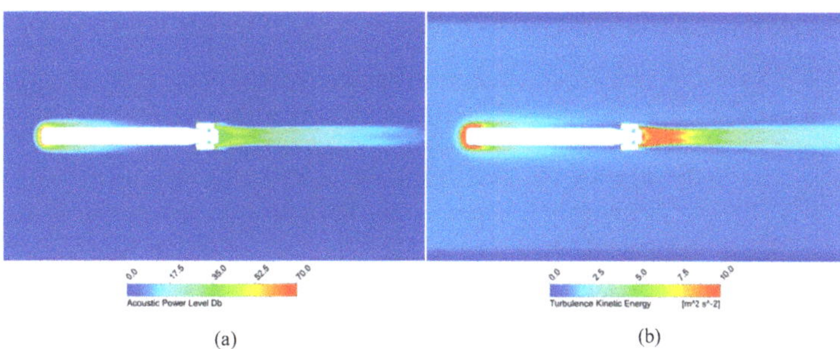

(a) (b)

Figure 3. Steady-state calculation results. (**a**) Distribution of sound power level (**b**) Distribution of turbulent kinetic energy.

The accuracy of noise data obtained from the broadband noise source model simulation is not high because it uses a constant Reynolds averaging model to calculate the noise. However, this model has a natural advantage in quickly determining the location of noise source. The sound power level diagram shows that the noise source is mainly concentrated in the torpedo head and wake vortex area behind the torpedo tail. A comparison with the turbulent kinetic energy shows that the noise generation region highly overlaps with the turbulence concentration region, and it can be assumed that the noise mainly originates from the turbulence region.

Based on the results of steady-state calculations, an LES with a Smagorinsky–Lilly subgrid scale model was used to carry out the transient solution process, and the calculation time step was set to 1×10^{-5} s according to the grid size and acoustic conditions. The Q criterion was used to describe the vortex structure of transient flow field, as shown in Figure 4. It was found that the vortex structure is concentrated at the rear of the torpedo blade, which corresponds to the abovementioned sound power level distribution and turbulent kinetic energy distribution, indicating that the vortex structure is closely related to the generation of flow noise and is the cause of flow noise.

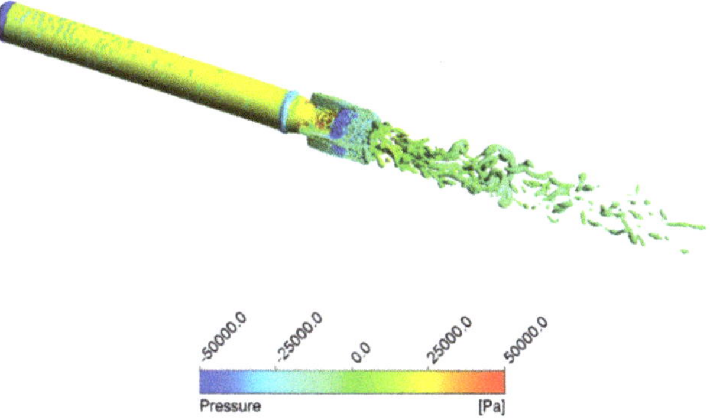

Figure 4. Vortex structure.

3.2. Calculation Results of Flow Noise

According to the acoustic solution theory, sound sources can be divided into three types: monopole, dipole, and quadrupole sources. In the flow field radiation noise, the noise caused by turbulent phenomena such as boundary layer and vortex development in the flow field is a quadrupole noise source, and the vehicle wall pressure pulsation caused by turbulent activity is a dipole noise source. Monopole and dipole noises are surface sources, whereas quadrupole noise is a volume source. Some studies reported that when the flow velocity is low, the surface–source noise can be mainly considered, and the volume–source noise is neglected. Therefore, most researchers do not consider the quadrupole source when solving and directly use the vehicle wall pulsation pressure as the source excitation for calculation. However, the underwater vehicle is faster. The quadrupole noise ratio may increase in this case, and the flow field results calculated in the previous section may improve. The underwater vehicle has a large wake vortex area, which is filled with large and small vortices, and its generation, development, and breaking inevitably generate noise. Therefore, it can be inferred that the noise calculation in this region cannot ignore the influence of quadrupole noise source generated by the vortex motion. A comparative analysis of the study was performed using wall pulsation pressure and Lighthill volume sound source as the excitation for acoustic calculations. For all the results reported in this paper, the reference sound pressure is taken as 1×10^{-6} Pa.

3.2.1. Calculation Results for Pulsating Pressure Excition

In the case where the pulsating pressure on wall is used as the excitation, the vehicle wall pressure pulsation is extracted from the flow field data and used as a boundary condition for further calculation. The extracted wall pressure pulsation at some frequencies is shown in Figure 5, and the unit was set to dB.

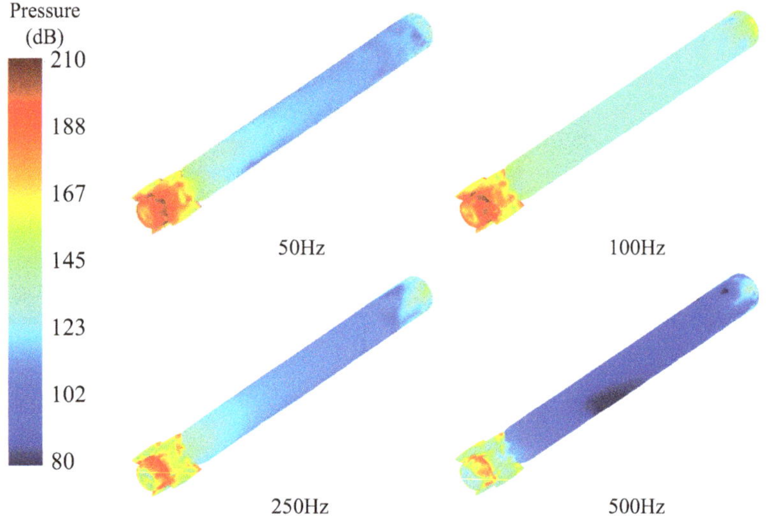

Figure 5. Pulsating pressure on the vehicle wall.

Figure 5 shows that the pulsating pressure distribution on the wall of underwater vehicle is inextricably related to the distribution of vortex region of flow field, and there is a high pulsating pressure distribution in the vortex-dense area of both the leading and trailing edges of underwater vehicle, more obvious at the trailing edge. It is easy to observe that the pulsating pressure on the vehicle wall is higher in the low-frequency part and lower in the high-frequency part. For the same frequency, the underwater vehicle has a large pulsating pressure area in the wake region, which is caused by the wake vortex area. Below 500 Hz, as the frequency increases, the large pulsating pressure region in the tail gradually decreases, and by around 500 Hz, it decreases to a very small portion and does not change much until higher frequencies are applied. Large vortices have a greater impact on the low-frequency noise situation, while small vortices affect the high-frequency noise results. A comparison of the vortex structure diagram of transient flow field can lead to the conclusion that large vortices play a major role in generating the pulsating pressure on the vehicle wall.

Figure 6a shows the calculation results of sound pressure at monitoring points 1, 2, and 3. It can be observed that the spectral characteristics of sound pressure levels at these monitoring points are very similar, although monitoring point 2 is located above the axis, while monitoring points 1 and 3 are symmetrically distributed on both sides of the axis. This indicates that the acoustic radiation distribution at the front of the underwater vehicle is approximately the same in all the directions. This can be explained by the absence of substantial vortex and turbulence properties in the front, which would not produce a large source of stream noise. In addition, the amplitude and trend of the frequency domain distribution curves of sound pressure at symmetric monitoring points (e.g., receivers 6 and 7 and receivers 8–11) are almost the same, although there is a certain directionality at the corresponding frequency values.

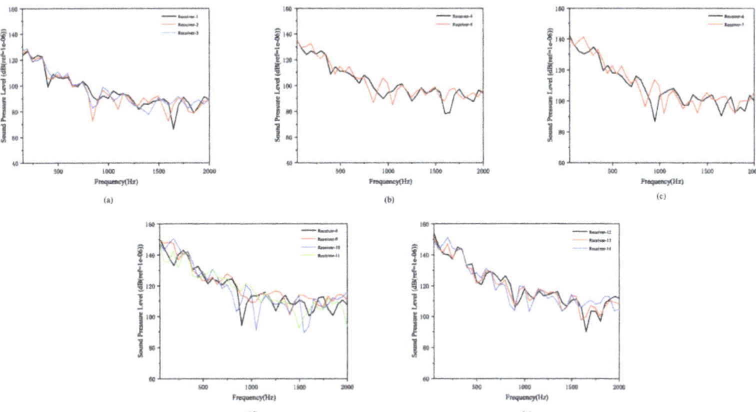

Figure 6. Spectral characteristics of sound pressure at different monitoring points along the flow direction considering pressure pulsation excitation. (**a**) Receiver 1–3 (**b**) Receiver 4–6 (**c**) Receiver 6–7 (**d**) Receiver 8–11 (**e**) Receiver 12–14.

Several representative monitoring points were selected to compare the changes in sound pressure levels along the flow direction, as shown in Figure 6. In the case where the pulsating pressure is the excitation, the sound pressure level substantially varies along the flow direction, with the highest value of about 135 dB in the low-frequency part and also around 90 dB in the high-frequency part at several monitoring points (receivers 1–5) at the front of underwater vehicle. By the middle of the underwater vehicle (receivers 6 and 7), the sound pressure level improved considerably, probably by about 6 dB, and in the whole tail section of underwater vehicle and the subsequent part of the area (receivers 8–14), the sound pressure level increased integrally by about 8 dB. The highest value of low-frequency part in this region is about 150 dB, whereas the high-frequency part is also about 110 dB.

The sound pressure distribution diagram of the middle section of underwater vehicle can show the sound field radiation more clearly, as shown in Figure 7. The sound pressure level shows a general trend of gradually decreasing with increasing frequency, but a small increase in sound pressure level might have occurred at some frequencies. The highest values of sound pressure levels appear around the blades in the wake of torpedo, which is due to the complex shape of blades and the tendency to obtain stronger disturbances in the flow. The change in sound pressure along the radius direction does not always decrease, indicating the oscillatory nature of sound waves.

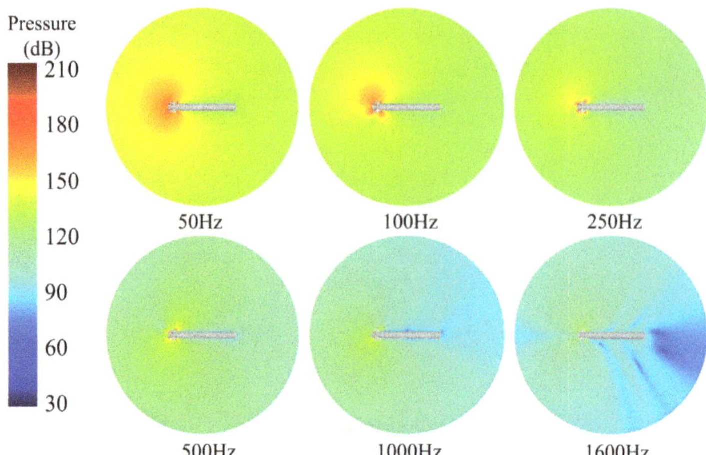

Figure 7. Sound pressure level distribution for pressure pulsation excitation.

3.2.2. Calculation Results for Lighthill Volume Excition

The calculation results for the Lighthill volume as an acoustic excitation are shown in Figure 8, and a comparison of the calculation results regarding Lighthill and pulsating pressure is shown in Figure 9. Except for monitoring point 13 in the tail vortex area, the comparisons of the rest (monitoring points 1–12 and 14) are not very different, and they all show a greater sound pressure level of the calculation results from the pulsating pressure as the excitation. Therefore, only the cases of monitoring points 1 and 13 are shown in Figure 9. When the difference in sound pressure level is 20 dB, the difference in sound pressure is at least one order of magnitude (as Equation (8)), so it is reasonable for researchers to ignore quadrupole sources in fluid-radiated noise calculations and consider pulsating pressure alone as a dipole source, slightly affecting the calculation results.

$$L = 20\lg\frac{p}{p_{ref}} \tag{8}$$

where L is the sound pressure level; p is the sound pressure; p_{ref} is the reference sound pressure, taken as 1×10^{-6} Pa.

In one case at monitoring point 13, the calculated sound pressure level for Lighthill volume excitation within 200 Hz is substantially higher than the calculation results for pulsating pressure excitation, verifying our assumption that the quadrupole source has non-negligibility in some regions due to very intense vortex activity at a higher velocity flow. Of course, for most applications, appropriate neglect can simplify the calculation process and provide acceptable results.

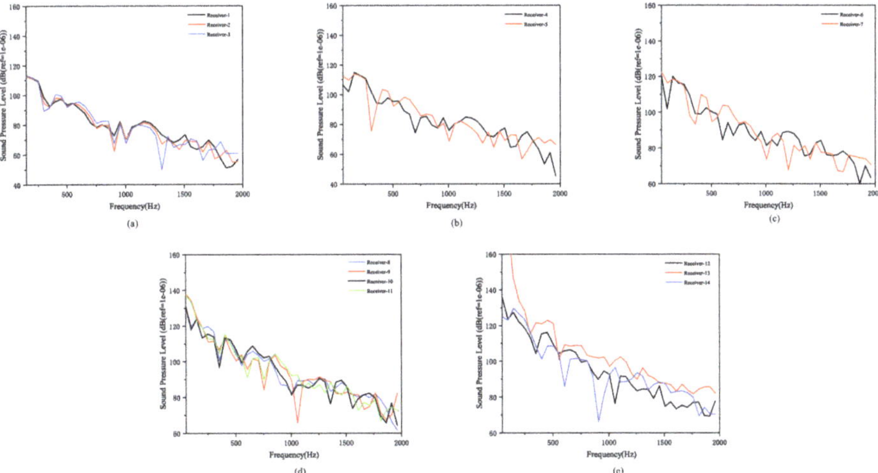

Figure 8. Spectral characteristics of sound pressure at different monitoring points along the flow direction considering Lighthill volume excition. (**a**) Receiver 1–3 (**b**) Receiver 4–6 (**c**) Receiver 6–7 (**d**) Receiver 8–11 (**e**) Receiver 12–14.

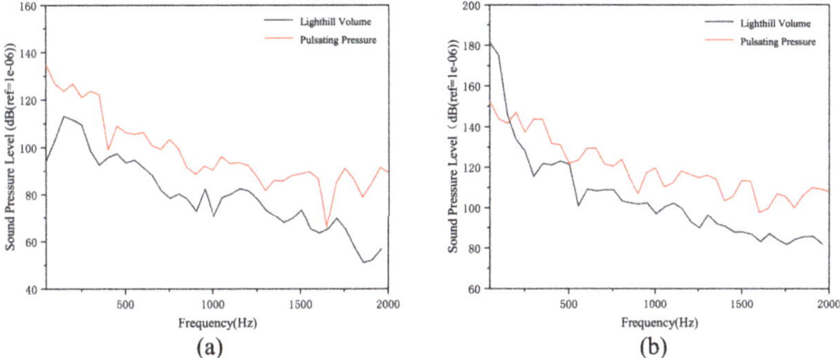

Figure 9. Comparison results of spectral characteristics. (**a**) Receiver 1 (**b**) Receiver 13.

3.3. Coupling Calculation and Structural Radiation Noise Analysis

To fully characterize the radiated noise of underwater vehicle, the relevant structure-radiated noise calculations were also carried out. In structure-radiated noise calculations, the structure is considered as an elastomer, which deforms and vibrates under the pulsating pressure at the vehicle wall and acts as a radiated noise source. Therefore, the structural mesh is imported, and the pulsating pressure is applied to the solid surface for coupled calculations. When the deformation of the surface of structure is small, the changes in flow field caused by the deformation of structure are negligible. Therefore, only the action of the fluid on the solid can be considered to simplify the calculation process, while ensuring the accuracy and precision of the calculation.

In this section, the results of flow field calculations are used as input boundary conditions for subsequent related calculations, and the computational domain is still taken as a spherical region of the same size. To simulate the shell vibration of a realistic underwater vehicle, the interior of underwater vehicle shown in Figure 1a was selectively hollowed out with a shell thickness of 20 mm. The material is set to Aluminum alloy with a Young's modulus of 8.2×10^{10} N/m^2, Poisson's ratio of 0.33, and density of 2820 kg/m^3, and its intersection was set to the coupling surface.

Using the vehicle wall pressure pulsation data as a boundary condition, coupled calculations were carried out to obtain the vibration displacement of underwater vehicle, as shown in Figure 10. The results show that under the effect of wall pulsation pressure, the vibration displacements throughout the solid domain are similar to the wall pressure pulsation distribution, both showing a trend towards larger values in the low-frequency part and smaller values in the high-frequency part. At 50 Hz, the underwater vehicle's stern section reached a maximum vibration displacement of the order of 10^{-5} m, while the high-frequency section is much less than this. For example, the maximum vibration displacement of a 750 Hz torpedo is only in the order of 10^{-9} m. It is precise because the low-frequency part of vibration is more intense, and the large eddies can be considered to play a crucial role while generating the vibration.

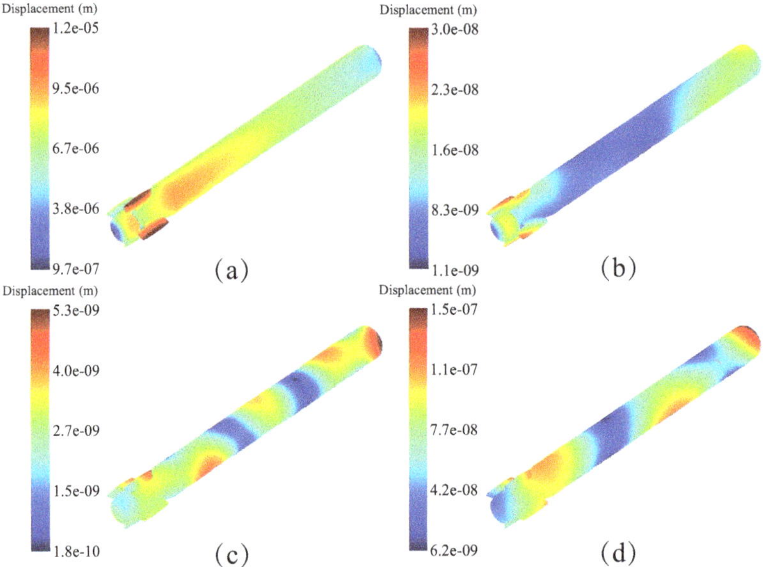

Figure 10. (**a**–**d**) Vibration displacement distribution.

The mean square vibration speed characteristics of underwater vehicle were obtained, as shown in Figure 11. In general, the mean square speed decreases as the frequency increases, i.e., the mean square speed is generally higher in the low-frequency part than in the high-frequency band. The results of other researchers show that the location of peak frequency point is only related to the structural model itself and is not related to the strength of flow excitation vibration at different flow velocities. Therefore, it can be assumed that the larger peak vibration velocity points for this structural underwater vehicle are located roughly at the frequency points of 300 Hz, 850 Hz, 1400 Hz, and 1850 Hz.

The frequency domain distribution curve of the structure's radiated noise shows distinctly different characteristics from the flow field radiated noise, showing a more pronounced vibration profile.

Several representative monitoring points were selected to show the frequency domain curves of sound pressure levels at their locations, as shown in Figure 12. It can be observed that whether the monitoring point is located at the leading, middle, or trailing edge of underwater vehicle, the structure-radiated noise has several corresponding peaks, all corresponding to the same frequency and peak frequency corresponding to the mean square vibration speed characteristic of torpedo. It can be inferred that this characteristic is related to the structure of underwater vehicle, and it is an inherent characteristic. From the sound pressure level curves of each monitoring point, it can be observed that the structure-radiated noise is not very much related to distribution along the flow direction. Most of the monitoring points have the same trend and amplitude. In the low-frequency part, most of them are located around 125 dB, and the main peak points are located around 130 dB. Compared to the central part of underwater vehicle (receivers 4–7), the sound pressure levels at the leading edge of underwater vehicle (receivers 1–3) are in contrast larger than those at the trailing edge, which is an important feature that clearly differs from the fluid-radiated noise. In the high-frequency part, the sound pressure levels are also around 100 dB. In addition, monitoring point 13 has consistently higher sound pressure levels in fluid-radiated noise and lower values in structure-radiated noise, indicating that structure-radiated noise is very different from fluid-radiated noise in terms of propagation direction.

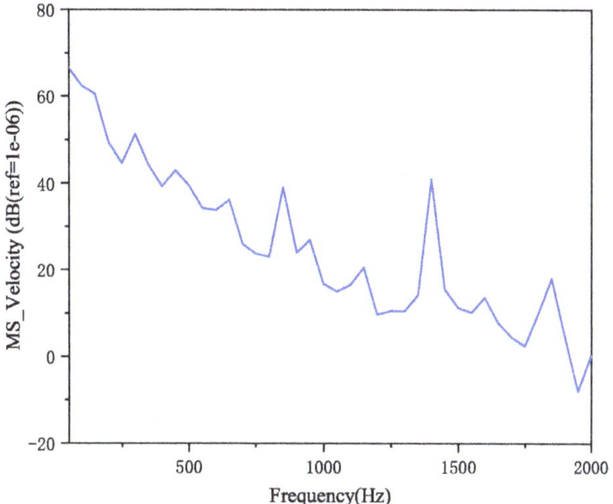

Figure 11. Mean square vibration characteristics.

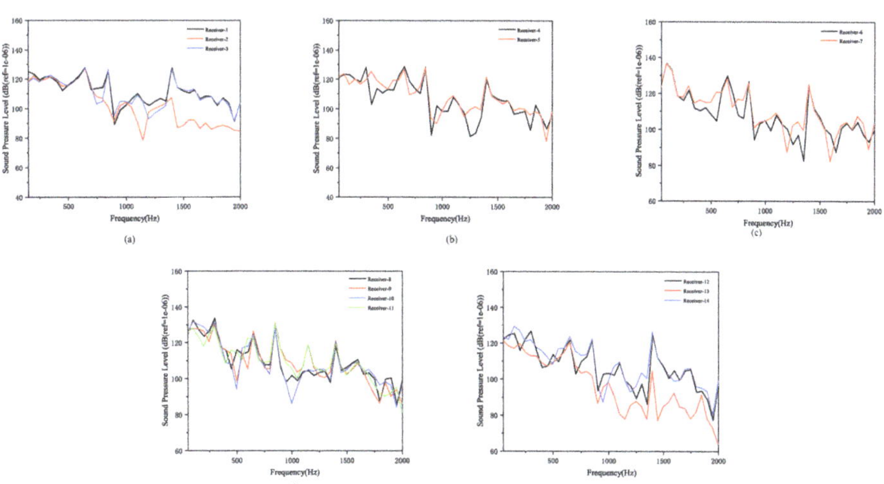

Figure 12. Spectral characteristics of sound pressure at different monitoring points along the flow direction considering vibration excitation. (**a**) Receiver 1–3 (**b**) Receiver 4–6 (**c**) Receiver 6–7 (**d**) Receiver 8–11 (**e**) Receiver 12–14.

For Monitoring Points 1, 2, and 3, the sound pressure level frequency domain distribution curves are shown in Figure 12a. Apart from the obvious peak in the curve, the sound pressure levels at monitoring point 2 are substantially different from the other two monitoring points. After the peak sound pressure level point at 1400 Hz, the sound pressure level in the high-frequency part of monitoring point 2 decreased below 90 dB, while the sound pressure level of the other two monitoring points remained at around 100 dB. After comparing the curves of monitoring points in the trailing edge section (shown in Figure 12e), it is demonstrated that the structure radiates less noise in the direction of axis of the underwater vehicle, which is particularly noticeable at high frequencies.

To facilitate further analysis of radiation characteristics of sound field, a cloud of sound pressure levels for structure-radiated noise generated by structural vibration due to pulsating pressure at the vehicle wall is given in Figure 13. It can be observed that the pulsating sound pressure cloud at 100 Hz is a left-right symmetrical pattern, which is due to the axisymmetrical shape of underwater vehicle, i.e., the low-frequency band has a stronger acoustic directivity at the head and tail, but the blade section radiates more widely, indicating that the tail is more directive than the head of underwater vehicle. Above 100 Hz, the sound field gradually becomes more "flap-like", and the flap-like features become more pronounced as the frequency increases. When the frequency reaches 2000 Hz, the "flap-like" distribution is already very dense. As the frequency increases further, the "flap-like" features gradually develop to the left and right, and this flap-like feature, which characterizes the fluctuation and oscillation of sound pressure, often leads to a lower sound pressure level region in the direction of underwater vehicle axis.

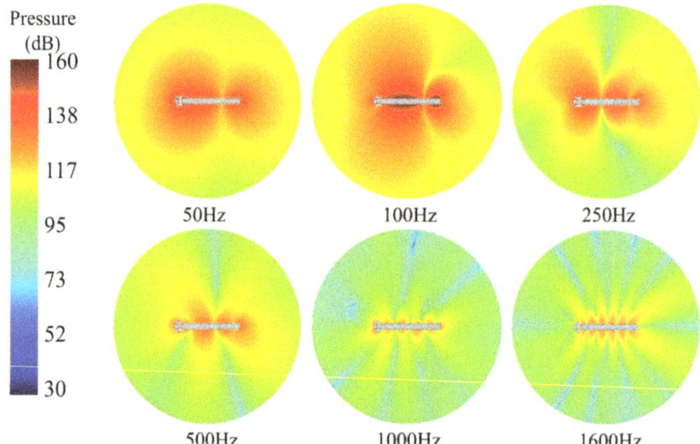

Figure 13. Sound pressure level distribution of structure-radiated noise at specific frequencies.

4. Conclusions

To improve the survivability and effective strike capability of underwater vehicles, this study used various approaches to simulate the radiation characteristics of their sound fields. This study mainly focuses on fluid-radiated noise and structure-radiated noise, and some necessary comparisons and analyses were made.

The head and tail of an underwater vehicle are the main sources of fluid-radiated noise, especially in the trailing vortex region of trailing edge section. Pressure pulsations generated by transient characteristics as well as the formation, development, and breaking of the vortex itself are relatively important sources of noise. In most cases, it is reasonable to ignore the quadrupole source such as volume excitation and consider only the wall pulsation pressure as the dipole source for excitation. However, in the case under the conditions of this study, ignoring the quadrupole source may have a bad effect, which needs to be considered and traded off in some practical applications.

The frequency domain distribution curve of structure-radiated noise is very different from that of fluid-radiated noise. In the high frequency part, peaks in the sound pressure levels occur, and these peaks are not lower than those in the low-frequency part. The peaks appear at approximately the same frequency at each monitoring point, which is related to the structure of underwater vehicle and is an inherent characteristic of the underwater vehicle. The sound pressure level diagram of structure-radiated noise is also very different from that of fluid-radiated noise, which shows obvious flap-like distribution characteristics at higher frequencies, and the number of flaps increases with frequency and gradually develops to both sides of the torpedo. The sound pressure level of the high-frequency part in the direction of torpedo axis is relatively low, corresponding to the frequency domain distribution curve. For the conditions studied in this paper, the structure-radiated noise and fluid-radiated noise have the same order of magnitude and do not differ much.

In the future, we will carry out multiconditions and multimorphology underwater vehicle flow field and sound field calculation, and explore a more universal trend from changing the flow velocity and changing the model, to provide guidance to general underwater vehicle researchers for model design and noise reduction design.

Author Contributions: Conceptualization, H.C.; Software, H.C.; Writing—original draft, H.C.; Writing—review & editing, H.C.; Supervision, L.W. All authors have read and agreed to the published version of the manuscript.

Funding: This research received no external funding.

Institutional Review Board Statement: Not applicable.

Informed Consent Statement: Not applicable.

Data Availability Statement: The data that support the findings of this study are available from the corresponding author upon reasonable request.

Conflicts of Interest: The authors have no conflicts to disclose.

References

1. Groves, N.C.; Huang, T.T.; Chang, M.S. *Geometric Characteristics of DARPA (Defense Advanced Research Projects Agency) SUBOFF Models (DTRC Model Numbers 5470 and 5471)*; Technical Report; David Taylor Research Center Bethesda MD Ship Hydromechanics Dept: Belhesda, MD, USA , 1989.
2. Alin, N.; Fureby, C.; Svennberg, S.; Sandberg, W.; Ramamurti, R.; Wikstrom, N.; Bensow, R.; Persson, T. 3D unsteady computations for submarine-like bodies. In Proceedings of the 43rd AIAA Aerospace Sciences Meeting and Exhibit, Reno, NV, USA, 10–13 January 2005; p. 1104.
3. Alin, N.; Bensow, R.; Fureby, C.; Huuva, T.; Svennberg, U. Current capabilities of DES and LES for submarines at straight course. *J. Ship Res.* **2010**, *54*, 184–196. [CrossRef]
4. Bensow, R.; Persson, T.; Fureby, C.; Svennberg, U.; Alin, N. Large eddy simulation of the viscous flow around submarine hulls. In Proceedings of the 25th ONR Symposium on Naval Hydrodynamics, St. Johns, NL, Canada, 8–13 August 2004.
5. Holloway, A.; Jeans, T.; Watt, G. Flow separation from submarine shaped bodies of revolution in steady turning. *Ocean Eng.* **2015**, *108*, 426–438. [CrossRef]
6. Kim, J.; Sung, H.J. Wall pressure fluctuations and flow-induced noise in a turbulent boundary layer over a bump. *J. Fluid Mech.* **2006**, *558*, 79–102. [CrossRef]
7. Lighthill, M.J. On sound generated aerodynamically I. General theory. *Proc. R. Soc. Lond. Ser. A Math. Phys. Sci.* **1952**, *211*, 564–587.
8. Lighthill, M.J. On sound generated aerodynamically II. Turbulence as a source of sound. *Proc. R. Soc. Lond. Ser. A Math. Phys. Sci.* **1954**, *222*, 1–32.
9. Wang, S.; Tang, J.; Wang, W.; Zhang, X. Numerical prediction of propeller excitation force and hydrodynamic noise of submarine with propeller. *Chin. Ship Res.* **2019**, *14*, 43–51.
10. Piomelli, U.; Streett, C.L.; Sarkar, S. On the computation of sound by large-eddy simulations. *J. Eng. Math.* **1997**, *32*, 217–236. [CrossRef]
11. Bailly, C.; Bogey, C.; Gloerfelt, X. Some useful hybrid approaches for predicting aerodynamic noise. *Comptes Rendus Mec.* **2005**, *333*, 666–675. [CrossRef]
12. Moon, Y.; Seo, J.; Bae, Y.; Roger, M.; Becker, S. A hybrid prediction method for low-subsonic turbulent flow noise. *Comput. Fluids* **2010**, *39*, 1125–1135. [CrossRef]

Article

A New 3D Marine Controlled-Source Electromagnetic Modeling Algorithm Based on Two New Types of Quadratic Edge Elements

Bin Xiong [1,2], Yuguo Lu [1,2], Hanbo Chen [1,2,*], Ziyu Cheng [1], Hao Liu [1] and Yu Han [1]

[1] School of Earth Science, Guilin University of Technology, Guilin 541004, China
[2] Institute of Urban Underground Space and Energy Studies, Shenzhen 518116, China
* Correspondence: chenhanbo@glut.edu.cn

Abstract: This study proposes a new algorithm for a higher-order vector finite element method based on two new types of second-order edge elements to solve the electromagnetic field diffusion problem in a 3D anisotropic medium. To avoid source singularity in the quasistatic variant of Maxwell's function, a secondary field formulation was adopted. The modeling domain was discretized using two types of quadratic edge hexahedral elements, which were obtained using the edge unification method to reduce variables on each side of two conventional quadratic edge elements. Compared with the traditional quadratic element, the number of unknowns that needed to be solved was significantly reduced. The sparse linear equation of the finite element system was solved using an open-source direct solver called MUMPS. The numerical results demonstrated that the proposed algorithm has the same level of accuracy as the conventional vector finite element method and has a significant advantage over it in terms of computational cost.

Keywords: vector finite element method; marine controlled-source electromagnetic method; quadratic edge element; hexahedral element; anisotropic

Citation: Xiong, B.; Lu, Y.; Chen, H.; Cheng, Z.; Liu, H.; Han, Y. A New 3D Marine Controlled-Source Electromagnetic Modeling Algorithm Based on Two New Types of Quadratic Edge Elements. *Appl. Sci.* **2023**, *13*, 2821. https://doi.org/10.3390/app13052821

Academic Editors: Victor Franco Correia and Marek Krawczuk

Received: 9 December 2022
Revised: 29 January 2023
Accepted: 17 February 2023
Published: 22 February 2023

1. Introduction

The marine controlled-source electromagnetic method (MCSEM) uses controllable artificial field sources to transmit electromagnetic signals in the ocean and measures the electric or magnetic fields at a location far away from the field source to detect the electrical distribution of the media below the seabed. The resistance characteristics of oil and gas reservoirs are the most important properties, which can generate characteristic surface electromagnetic signals. In other words, MCSEM technology can distinguish underground oil and gas from other fluids. In theory, CSEM measurement is to record data with multiple power-receiver offsets, several different frequencies, and a certain power-receiver arrangement. The marine controlled-source electromagnetic (MCSEM) method has become a popular geophysical exploration tool for offshore hydrocarbon (HC) exploration [1–3]. During the last decade, marine CSEM has been widely used to reduce ambiguities in data interpretation and reduce the risk of exploration. With the rapid development of MCSEM exploration instruments, large-scale and high-quality data have been obtained, resulting in the need to develop high-precision MCSEM three-dimensional (3D) data interpretation tools. The 3D inversion algorithm technology of MCSEM is commonly used for data interpretation. Forward modeling is the basis of inversion, and high-precision forward modeling technology is key to achieving high-precision inversion.

The integral equation (IE) [4–6], finite difference (FD) [7–12], and finite element (FE) methods [13–17] are popular numerical techniques for electromagnetic (EM) field modeling. In the FE method, it is easier to use an unstructured mesh to create an irregular body, locally refine the interest region, and coarsen the boundary area of the model domain [15]. With this processing, numerical modeling becomes efficient and effective. Therefore, these advantages make the FE method more suitable for simulating the complex EM response of irregular models than FD and IE methods.

Compared with the conventional finite element method, which is based on a node, the edge-based finite element method has the advantage that divergence-free conditions are automatically satisfied by appropriately selected basic functions [18]. In recent years, the vector finite element method based on edge elements has been used to solve the electromagnetic field problem, and is known as Maxwell's equations [19,20]. It should be noted that to date, most formulations for the 3D CSEM problem have implemented the lowest order. In addition, some studies have used higher-order elements to solve three-dimensional forward modeling of CSEM, but these were all aimed at tetrahedral elements [14,21]. To the best of our knowledge, no high-order hexahedron element is currently used for three-dimensional forward modeling of the controlled-source electromagnetic method. As we know, high-order edge elements can provide accurate solutions but result in an increase in unknown variables and lead to larger computing costs. Fortunately, we can improve the accuracy of the solution at a small computational cost by eliminating some variables when using higher-order elements [22,23].

In this study, we applied the unification method to the edge to reduce one edge variable on each side of the conventional serendipity and the Lagrange-type quadratic hexahedral element, resulting in two new types of second order edge elements. We also introduced a new vector FE method based on the proposed elements to solve the 3D marine CSEM modeling problem in an anisotropic medium. For a complex bathymetry simulation, we used a general hexahedral element to discretize the modeling domain. We validated our code by using several models in isotropic and anisotropic media.

2. Theory

In geophysical applications, the displacement current is typically ignored when the frequency of the electromagnetic field is low. Maxwell's equations can then be simplified as follows [24]:

$$\nabla \times \boldsymbol{E} = i\omega\mu_0\boldsymbol{H} \tag{1}$$

$$\nabla \times \boldsymbol{H} = \sigma\boldsymbol{E} + \boldsymbol{J}_\mathrm{s} \tag{2}$$

where the space magnetic permeability is denoted by μ_0, angular frequency is denoted by ω, source current is denoted by $\boldsymbol{J}_\mathrm{s}$, and conductivity tensor is denoted by σ:

$$\sigma = \begin{pmatrix} \sigma_x & 0 & 0 \\ 0 & \sigma_y & 0 \\ 0 & 0 & \sigma_z \end{pmatrix} \tag{3}$$

where σ_x, σ_y and σ_z are principal conductivities. In this study, we discuss only the principal axis anisotropy case. However, our formulation also works in the case of general anisotropy.

Calculate the curl at the left and right ends of Equation (1), and use Equation (2) to eliminate the magnetic field:

$$\nabla \times \nabla \times \boldsymbol{E} - i\omega\mu_0\sigma\boldsymbol{E} = i\omega\mu_0\boldsymbol{J}_\mathrm{s} \tag{4}$$

The total and secondary field formulations are two types of formulations that are typically used for solving the 3D MCSEM modeling problem. In contrast to the total field formulation, the secondary field formulation's source term comprises a primary electric field. This type of source term is smoother than the current source used in the total-field formulation. Therefore, in this study, we use a secondary field formulation to address the marine CSEM modeling problem in an anisotropic medium. In this formulation, the total field can be decomposed into two parts: the background (primary field, $\boldsymbol{E}_\mathrm{p}$) and the anomaly fields (secondary field, $\boldsymbol{E}_\mathrm{s}$):

$$\boldsymbol{E} = \boldsymbol{E}_\mathrm{s} + \boldsymbol{E}_\mathrm{p} \tag{5}$$

Similarly, the conductivity tensor can be decomposed into background (σ_b) and anomalous conductivities ($\Delta\sigma$):

$$\sigma = \sigma_b + \Delta\sigma \tag{6}$$

With substitution of Equations (5) and (6) into Equation (4), the Helmholtz equation with the electric component of the secondary field can be written as:

$$\nabla \times \nabla \times \boldsymbol{E}_s - i\omega\mu_0\sigma\boldsymbol{E}_s = i\omega\mu_0\Delta\sigma\boldsymbol{E}_p \tag{7}$$

The secondary magnetic component can easily be calculated using Equation (8) once the secondary electric component in Equation (7) can be solved:

$$\boldsymbol{H}_s = \frac{1}{i\omega\mu_0}\nabla \times \boldsymbol{E}_s \tag{8}$$

3. Edge-Based Finite Element Analysis

In this study, we used the edge-based FE method with quadratic elements, as discussed in the previous section. We used the edge shape function $\boldsymbol{N}_i$ to approximate the secondary electric field within an edge element:

$$\boldsymbol{E}_S = \sum_{i}^{N_{edge}} \boldsymbol{N}_i \boldsymbol{E}_{S,j} \tag{9}$$

where N_{edge} denotes the total number of edges in the edge element. For the first-order edge element, the $N_{edge} = 12$; for conventional serendipity and Lagrange-type second-order hexahedral elements, $N_{edge} = 36$ and 54, respectively.

By substituting (9) into (7) and applying the Galerkin method, we can determine that the weak form of the original differential equation is:

$$\mathrm{R}_i = \int_\Omega \boldsymbol{N}_i \cdot \left[\nabla \times \nabla \times \boldsymbol{E}_s - i\omega\mu_0\sigma\boldsymbol{E}_s - i\omega\mu_0\Delta\sigma\boldsymbol{E}_p\right] dv \tag{10}$$

where Ω denotes the modeling area.

Applying first vector Green's theorem to Equation (10), the secondary electric field is given, and it is continuous across the inner boundary; the surface term in Equation (10) vanishes. Then, the discretized form of Equation (10) for each element can be expressed as follows:

$$\mathrm{R}_i^{Ne} = \sum_{1}^{Ne}\left[\boldsymbol{K}^e\boldsymbol{E}_s^e - i\omega\mu_0\boldsymbol{M}_1^e\boldsymbol{E}_s^e - i\omega\mu_0\boldsymbol{M}_2^e\boldsymbol{E}_p^e\right] \tag{11}$$

where $\boldsymbol{K}^e$ and $\boldsymbol{M}_1^e$ and $\boldsymbol{M}_2^e$ are the local stiffness matrices defined as follows:

$$\boldsymbol{K}_{i,j}^e = \int_{\Omega_e} \left(\nabla \times \boldsymbol{N}_i^e\right) \cdot \left(\nabla \times \boldsymbol{N}_j^e\right) dv \tag{12}$$

$$\boldsymbol{M}_{1i,j}^e = \int_{\Omega_e} \boldsymbol{N}_i^e \cdot \sigma \cdot \boldsymbol{N}_j^e dv \tag{13}$$

$$\boldsymbol{M}_{2i,j}^e = \int_{\Omega_e} \boldsymbol{N}_i^e \cdot \Delta\sigma \cdot \boldsymbol{N}_j^e dv \tag{14}$$

where Ω_e denotes the domain of an element.

By assembling a global matrix with local matrices, a linear equation for the elements is obtained. The equation is typically sparse:

$$\boldsymbol{Ae} = \boldsymbol{b}, \tag{15}$$

Here $\boldsymbol{A}$ denotes the global stiffness matrix, and $\boldsymbol{b}$ denotes the RHSs of the equations.

To solve Equation (14), a proper condition should be added to the equation. For simplicity, let us consider the homogeneous Dirichlet boundary condition:

$$e|_{\Gamma} = 0 \tag{16}$$

Here, Γ is the boundary area.

IDR(s) iterative solvers with an ILU preconditioner were used to solve the linear equations in Equation (15).

4. First and Conventional Quadratic Edge Elements

In the edge-based FE method, we can use regular rectangular, general tetrahedral, general hexahedral, or other complex elements to discretize the modeling domain. To simplify the problem and simulate bathymetry, hexahedral elements were used to discretize the modeling domain in this study.

To calculate the stiffness matrix of the hexahedral element, a general hexahedral element should be transformed into an origin-centered cubic element. A general hexahedron in Cartesian coordinates xyz is shown in Figure 1a, and a transformed reference domain (cubic element) in Cartesian coordinates $\xi\zeta\eta$ is shown in Figure 1b.

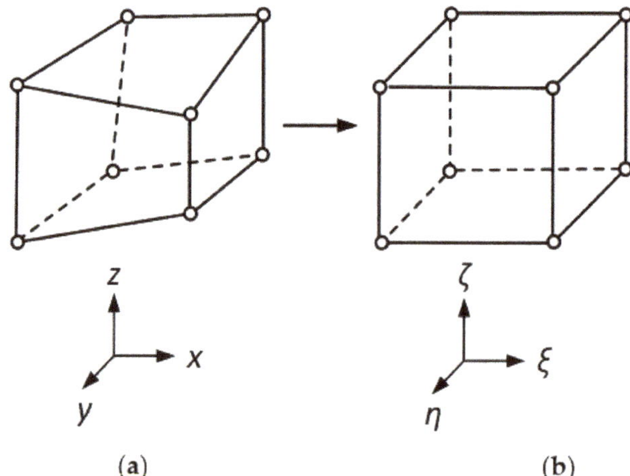

Figure 1. (**a**) Hexahedron element in Cartesian coordinates xyz; (**b**) the transformed cubic element in Cartesian coordinates $\xi\zeta\eta$.

The transformation can be described by the following formulas [25]:

$$x = \sum_{i=1}^{8} N_i^e(\xi, \eta, \zeta) x_i^e \tag{17}$$

$$y = \sum_{i=1}^{8} N_i^e(\xi, \eta, \zeta) y_i^e \tag{18}$$

$$z = \sum_{i=1}^{8} N_i^e(\xi, \eta, \zeta) z_i^e \tag{19}$$

where N_i^e is the scalar node-based shape function, which can be defined as follows:

$$N_i^e(\xi, \eta, \zeta) = \frac{1}{8}(1 + \xi\xi_i)(1 + \eta\eta_i)(1 + \zeta\zeta_i) \tag{20}$$

where i is a local node index of the element.

In the analysis of the vector finite element, two volume integrals (12)–(14) must be evaluated. For a hexahedron element, it is easy to transform with two integrals in the $\xi\eta\zeta$-coordinate using the Jacobian transform:

$$K_{i,j}^e = \int_{-1}^{1}\int_{-1}^{1}\int_{-1}^{1}(\nabla\times \mathbf{N}_i^e(\xi,\eta,\zeta))\cdot\left(\nabla\times \mathbf{N}_j^e(\xi,\eta,\zeta)\right)|\mathbf{J}(\xi,\eta,\zeta)|\mathrm{d}\xi\mathrm{d}\eta\mathrm{d}\zeta, \tag{21}$$

$$M_{1i,j}^e = \int_{-1}^{1}\int_{-1}^{1}\int_{-1}^{1}\mathbf{N}_i^e(\xi,\eta,\zeta)\cdot\sigma\cdot\mathbf{N}_j^e(\xi,\eta,\zeta)|\mathbf{J}(\xi,\eta,\zeta)|\mathrm{d}\xi\mathrm{d}\eta\mathrm{d}\zeta, \tag{22}$$

$$M_{2i,j}^e = \int_{-1}^{1}\int_{-1}^{1}\int_{-1}^{1}\mathbf{N}_i^e(\xi,\eta,\zeta)\cdot\Delta\sigma\cdot\mathbf{N}_j^e(\xi,\eta,\zeta)|\mathbf{J}(\xi,\eta,\zeta)|\mathrm{d}\xi\mathrm{d}\eta\mathrm{d}\zeta, \tag{23}$$

$\mathbf{J}(\xi,\eta,\zeta)$ and $|\mathbf{J}(\xi,\eta,\zeta)|$ denote the Jacobian matrix and determinant of $\mathbf{J}(\xi,\eta,\zeta)$, respectively.

$$\mathbf{J}(\xi,\eta,\zeta) = \begin{bmatrix} \frac{\partial x}{\partial \xi} & \frac{\partial y}{\partial \xi} & \frac{\partial z}{\partial \xi} \\ \frac{\partial x}{\partial \eta} & \frac{\partial y}{\partial \eta} & \frac{\partial z}{\partial \eta} \\ \frac{\partial x}{\partial \zeta} & \frac{\partial y}{\partial \zeta} & \frac{\partial z}{\partial \zeta} \end{bmatrix} \tag{24}$$

In the next section, we introduce several conventional first-order and quadratic hexahedral elements and two quadratic edge hexahedral elements of the new type.

4.1. First-Order Edge Hexahedral Element

A conventional first-order hexahedral element is shown in Figure 2, which has eight nodes and twelve edges. The shape function of the edge on the sides along the ξ-axis direction are accordingly defined as [25]:

$$\mathbf{N}_i = \frac{1}{8}(1+\eta_i\eta)(1+\zeta_i\zeta)\Delta\xi \tag{25}$$

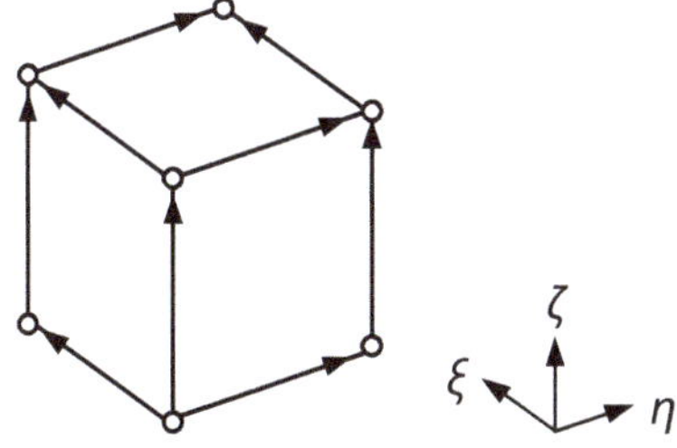

Figure 2. Conventional first-order edge hexahedral element.

Similarly, the shape function of the edge on the sides along the η-axis or ζ-axis direction is defined as:

$$\mathbf{N}_j = \frac{1}{8}(1+\xi_j\xi)(1+\zeta_j\zeta)\Delta\eta \tag{26}$$

and

$$\mathbf{N}_k = \frac{1}{8}(1+\xi_k\xi)(1+\zeta_k\zeta)\Delta\zeta \tag{27}$$

where ξ, η and ζ denote the local coordinate in the elements.

4.2. A Quadratic Hexahedron Element of a Conventional Serendipity Type

Figure 3 shows a conventional serendipity quadratic hexahedron element with 12 edges on its surface, 20 nodes in the element, and 24 edges on the sides. The shape function for the serendipitous quadratic hexahedron element can be defined as follows: the shape function of the edge on sides which along ξ-axis direction should be:

$$N_i = \frac{1}{8}(1 + \eta_i\eta)(1 + \zeta_i\zeta)(4\xi_i\xi + \eta_i\eta + \zeta_i\zeta - 1)\Delta\xi \tag{28}$$

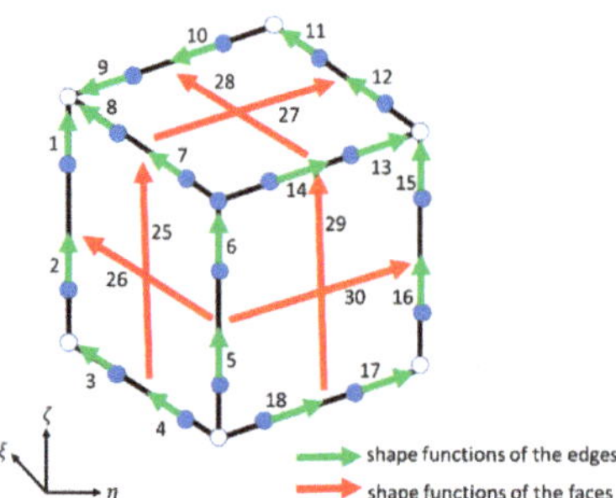

Figure 3. The location and number of shape functions of a conventional serendipity quadratic hexahedron element (SC). The green arrows and the red arrows represent the shape functions on the edge and on the face, respectively.

The edge shape function of the edges along the ξ-axis direction on the $\zeta = \pm 1$ surface:

$$N_i = \frac{1}{4}\left(1 - \eta^2\right)(1 + \zeta_i\zeta)\Delta\xi \tag{29}$$

The edge shape function of the edges along the ξ-axis direction on the $\eta = \pm 1$ surface:

$$N_i = \frac{1}{4}\left(1 - \zeta^2\right)(1 + \eta_i\eta)\Delta\xi \tag{30}$$

The sides along the η-axis direction on the sides:

$$N_j = \frac{1}{8}(1 + \xi_j\xi)(1 + \zeta_j\zeta)(4\eta_j\eta + \xi_j\xi + \zeta_j\zeta - 1)\Delta\eta \tag{31}$$

The edges along the η-axis direction on the $\xi = \pm 1$ surface:

$$N_j = \frac{1}{4}\left(1 - \zeta^2\right)(1 + \xi_j\xi)\Delta\eta \tag{32}$$

The edges along the η-axis direction on the $\zeta = \pm 1$ surface:

$$N_j = \frac{1}{4}\left(1 - \xi^2\right)(1 + \zeta_j\zeta)\Delta\eta \tag{33}$$

The edges along the ζ-axis direction on the $\eta = \pm 1$ surface:

$$N_k = \frac{1}{4}\left(1 - \xi^2\right)(1 + \eta_k\eta)\Delta\zeta \tag{34}$$

4.3. Lagrange Quadratic Hexahedron Element

Figure 4 shows a Lagrange quadratic hexahedron element (LC) with 27 nodes and 24 edges on the sides, 24 edges on the surface, and 8 edges within the element. The shape function of the Lagrange quadratic hexahedron element is defined as the edges on the sides along the ξ-axis direction:

$$N_i = \frac{1}{8}(1 + 4\xi_i\xi)\eta\zeta(\eta_i + \eta)(\zeta_i + \zeta)\Delta\xi \tag{35}$$

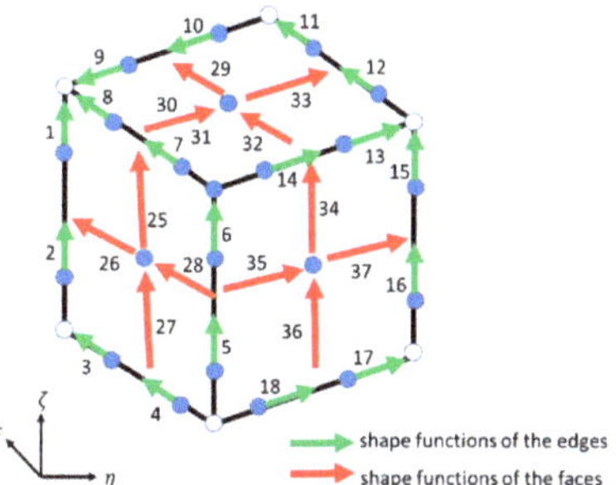

Figure 4. The location and number of shape functions of a Lagrange quadratic hexahedron element (LC). The green arrows and the red arrows represent the shape functions on the edges and on the faces, respectively.

The edges along the ξ-axis direction on the $\zeta = \pm 1$ surface:

$$\mathbf{N}_i = \frac{1}{4}(1 + 4\xi_i\xi)\zeta\left(1 - \eta^2\right)(\zeta_i + \zeta)\Delta\xi \tag{36}$$

The edges along the ξ-axis direction on the $\eta = \pm 1$ surface:

$$\mathbf{N}_i = \frac{1}{4}(1 + 4\xi_i\xi)\eta\left(1 - \zeta^2\right)(\eta_i + \eta)\Delta\xi \tag{37}$$

The edges along the ξ-axis direction within the element:

$$\mathbf{N}_i = \frac{1}{2}(1 + 4\xi_i\xi)\left(1 - \eta^2\right)\left(1 - \zeta^2\right)\Delta\xi \tag{38}$$

The edges along the η-axis direction on the sides:

$$\mathbf{N}_j = \frac{1}{8}(1 + 4\eta_j\eta)\xi\zeta(\xi_j + \xi)(\zeta_j + \zeta)\Delta\eta \tag{39}$$

The edges along the η-axis direction on the $\zeta = \pm 1$ surface:

$$\mathbf{N}_j = \frac{1}{4}(1 + 4\eta_j\eta)\zeta\left(1 - \xi^2\right)(\zeta_j + \zeta)\Delta\eta \tag{40}$$

The edges along the η-axis direction on the $\xi = \pm 1$ surface:

$$\mathbf{N}_j = \frac{1}{4}(1 + 4\eta_j\eta)\xi\left(1 - \zeta^2\right)(\xi_j + \xi)\Delta\eta \tag{41}$$

The edges along the η-axis direction within the element:

$$\mathbf{N}_j = \frac{1}{2}(1 + 4\eta_j\eta)\left(1 - \xi^2\right)\left(1 - \zeta^2\right)\Delta\eta \tag{42}$$

The edges along the ζ-axis direction on the sides:

$$\mathbf{N}_k = \frac{1}{8}(1 + 4\zeta_k\zeta)\xi\eta(\xi_k + \xi)(\eta_k + \eta)\Delta\zeta \tag{43}$$

The edges along the ζ-axis direction on the $\xi = \pm 1$ surface:

$$\mathbf{N}_k = \frac{1}{4}(1 + 4\zeta_k\zeta)\xi\left(1 - \eta^2\right)(\xi_k + \xi)\Delta\zeta \tag{44}$$

The edges along the ζ-axis direction on the $\eta = \pm 1$ surface:

$$\mathbf{N}_k = \frac{1}{4}(1 + 4\zeta_k\zeta)\eta\left(1 - \xi^2\right)(\eta_k + \eta)\Delta\zeta \tag{45}$$

The edges along the ζ-axis direction within the element:

$$\mathbf{N}_k = \frac{1}{2}(1 + 4\zeta_k\zeta)\left(1 - \eta^2\right)\left(1 - \xi^2\right)\Delta\zeta \tag{46}$$

4.4. A New Type of Quadratic Edge Element

There are two edges on the side of a conventional quadratic hexahedron element. Within an edge element, because the curls of the edge shape functions are linearly dependent, many edge variables are redundant. The computation time and storage requirements of these elements has increased significantly. Therefore, eliminating these redundant variables helps improve efficiency at no cost to computational accuracy. In this section, we adopt a new method to eliminate redundant variables from the conventional quadratic element.

Let us first consider a one-sided conventional quadratic edge element, as shown in Figures 3 and 4. We named the two edges on this side edge1 and edge2. $\mathbf{N}_1$ and $\mathbf{N}_2$ are the shape functions of edge1 and edge2, and their corresponding edge variables are A_1 and A_2. Because one of these is redundant, we decided to eliminate A_2 and consider $A_1 = A_2$. Assume A_{side} is the vector field interpolation contribution of the two edges:

$$\begin{aligned} A_{\text{side}} &= \mathbf{N}_1 A_1 + \mathbf{N}_2 A_2 \\ &= (\mathbf{N}_1 + \mathbf{N}_2) A_1 \end{aligned} \tag{47}$$

Make a line integration of A along this side and name it as A_{unif}. Note that the following integration is orthogonal:

$$\int_{eKL} \mathbf{N}_{IL} \mathrm{d}s = \delta_{(IL)(KL)} \tag{48}$$

Here, ds is the differential vector of the line element along edge eKL, and $\delta_{(IL)(KL)}$ represents the Kronecker delta. Thus, the integration A_{unif} is written as:

$$A_{\text{unif}} = \int_{\text{edge1}} A \cdot \mathrm{d}s + \int_{\text{edge2}} A \cdot \mathrm{d}s = A_1 + A_2 = 2A_1 \tag{49}$$

Then Equation (47) is rewritten as:

$$A_{\text{side}} = (\mathbf{N}_1 + \mathbf{N}_2)A_1 = (\mathbf{N}_1 + \mathbf{N}_2)\frac{A_{\text{unif}}}{2} = \mathbf{N}_{\text{unif}} A_{\text{unif}} \tag{50}$$

In Equation (50), we obtain a new shape function $N_{\text{unif}} = (N_1 + N_2)/2$ for the edge, as shown in Figure 5b. According to this, N_{unif} can be written in detail as follows:

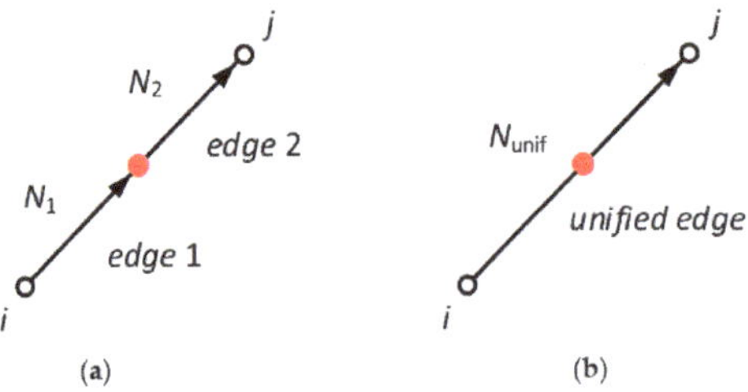

Figure 5. Two-edge unification: (**a**) edge before unification; (**b**) edge after unification.

For the serendipity quadratic hexahedron element, the edges on the sides along the ξ-axis direction:

$$N_k = \frac{1}{2}(1 + 4\zeta_k\zeta)\left(1 - \eta^2\right)\left(1 - \xi^2\right)\Delta\zeta \quad (51)$$

The edges on the sides along the η-axis direction:

$$\mathbf{N}_j = \frac{1}{8}(1 + \xi_j\xi)(1 + \zeta_j\zeta)(\xi_j\xi + \zeta_j\zeta - 1)\Delta\eta \quad (52)$$

The edges on the sides along the ζ-axis direction:

$$\mathbf{N}_k = \frac{1}{8}(1 + \xi_k\xi)(1 + \eta_k\eta)(\xi_k\xi + \eta_k\eta - 1)\Delta\zeta \quad (53)$$

For the Lagrange quadratic hexahedron element, the edges on the sides along the ξ-axis direction:

$$\mathbf{N}_i = \frac{1}{8}\eta\zeta(\eta_i + \eta)(\zeta_i + \zeta)\Delta\xi \quad (54)$$

The edges on the sides along the η-axis direction:

$$\mathbf{N}_j = \frac{1}{8}\xi\zeta(\xi_j + \xi)(\zeta_j + \zeta)\Delta\eta \quad (55)$$

The edges on the sides along the ζ-axis direction:

$$\mathbf{N}_k = \frac{1}{8}\xi\eta(\xi_k + \xi)(\eta_k + \eta)\Delta\zeta \quad (56)$$

The shape functions of the edges in the element and the faces remained intact. Therefore, we obtained two new quadratic edge elements (Figure 6). There is only one edge on the sides of the two new quadratic edge elements (Figure 6); therefore, the redundant variable can be effectively eliminated. Table 1 lists various types of elements and the distribution of their basis functions.

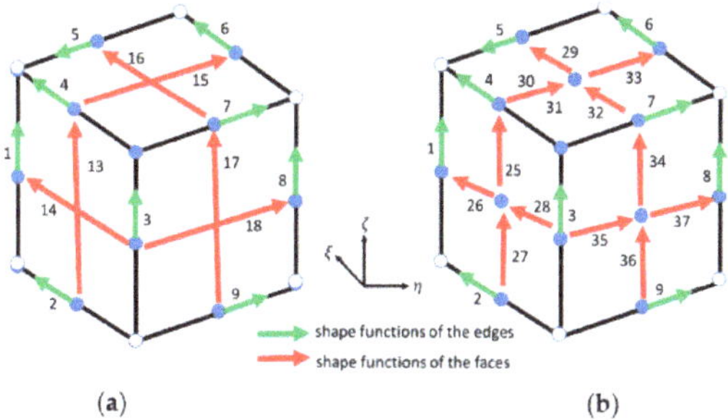

Figure 6. The location and number of shape functions of two new types quadratic elements: (**a**) new serendipity quadratic element (SN); (**b**) new Lagrange quadratic element (LN). The green arrows and the red arrows represent the shape functions on the edges and on the faces, respectively.

Table 1. Elements and the distribution of their basic functions.

Element	Edges	Faces	Interior Bubbles	Total
1st	12	0	0	12
SC	24	12	0	36
SN	12	12	0	24
LC	24	24	6	54
LN	12	24	6	42

4.5. Interpolation and Convergence

Interpolation capacity and convergence tests are performed using two new types of second-order element. Those functions are approximated in the domain defined by $\Omega = [-1\ 1]^3$. Figure 7 shows the error of interpolation of the tested functions and their respective convergence rates. The two basis functions (SC, LC) have similar convergence rates, both close to one.

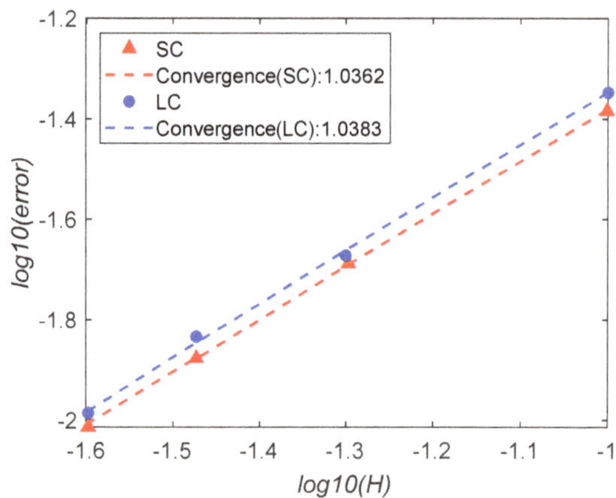

Figure 7. Interpolation errors and convergence rates for the two basis functions approximated by EFEM. H is the length of edge of a element.

5. Model Studies

All MCSEM response simulations were performed on a platform with hardware and operation, as follows. The CPU was a double Intel processor, each processor having 6 cores and 12 threads, and RAM memory was 64 GB at a frequency of approximately 1333 MHz. The operating system was a 64-bit Windows 7 professional version.

5.1. Verification of the New Algorithm

To verify the new algorithm, an isotropic horizontal layered geoelectrical model, as shown in Figure 8, was considered first. In this model, air and seawater were 1000 m thick. A 100 m thin-layer reservoir was embedded in the sediment at 1500 m to 1600 m depth. The EM field response was based on the model presented in Table 2. An x-oriented 1 Hz frequency current source was placed at the location $(0,\ 0,\ 950\ \text{m})$. The receiving stations were located on the seafloor $(z = 1000\ \text{m})$. The modeling domain was 3 km $\times$ 3 km square and 4 km in the vertical direction $(-1\ \text{km} \leq z \leq 3\ \text{km})$. A non-uniform rectangular grid was used to discretize Ω. We refined the grid near the current source, receiving stations, and target layer. The analytical response was calculated using the numerical integration of the Hankel transform [25–29].

Table 2. The conductivities of an isotropic horizontally layered geoelectric model.

Layer	Conductivity (s/m)
Air	10^{-6}
Sea Water	3.30
Sediment	1.00
Reservoir	0.01

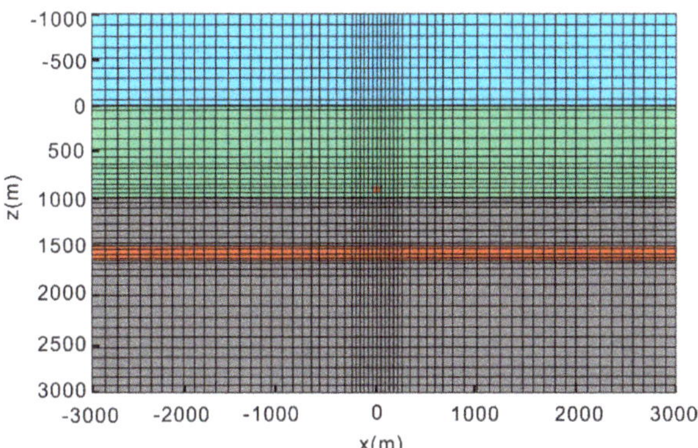

Figure 8. Horizontally layered geoelectric model with rectangular mesh; the red star denotes the electric dipole source. The blue, green, red and gray areas represent the air layer, sea water, oil and gas, and bedrock, respectively.

Figures 9 and 10 show the difference in the x-component of the electric field by applying conventional serendipity (SC)-and Lagrange (LC)-type second-order hexahedral elements, and the proposed elements (new serendipity (SN) and Lagrange element (LN)). These new quadratic edge elements were at the same level of accuracy as conventional quadratic elements, both higher than that of linear edge elements.

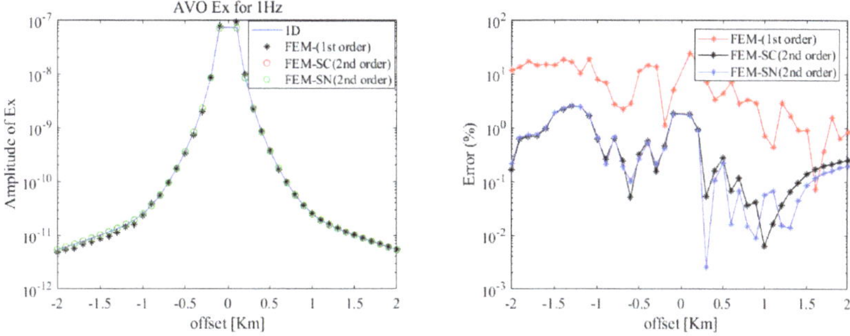

Figure 9. A comparison of the x-component of the electric field using different FEM methods (serendipity-type second-order element) with the analytical solution.

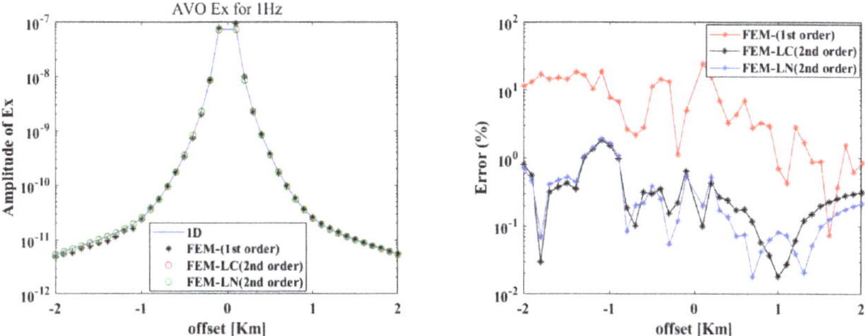

Figure 10. A comparison of the x-component of the secondary electric field by using different FEM methods (Lagrange-type second-order element) with the analytical solution.

Table 3 lists the CPU times for different elements. Less time was required when using new quadratic elements than when using conventional quadratic elements. Therefore, one can say that the computation is reduced at no cost to accuracy.

Table 3. The comparison of CPU time by using different methods.

Element	Number of Elements	DOF	CPU Time (s)
1st order	332,112	867,730	30.5
SC	332,112	3,952,952	51.7
SN	332,112	2,960,408	76.2
LC	332,112	5,920,816	111.4
LN	332,112	4,924,272	221.9

5.2. *Off-Shore Hydrocarbon ReservoirMmodel*

Consider a geoelectric model with a hydrocarbon reservoir embedded in sediment, as shown in Figures 11 and 12. In this model, air and seawater were 1000 m thick, similar to the previous model. The conductivity values of each component are listed in Table 4. The modeling domain was a 4 km $\times$ 4 km $\times$ 4 km cube with 1 km $\leq z \leq$ 3 km vertically . The hydrocarbon reservoir was buried 2.0–2.1 km deep and occupied a 1 km $\times$ 1 km area horizontally. An x-oriented 1 Hz frequency current source was placed at the location $(-3000,\ 0,\ 950)$ m. Receiving stations were placed on the seafloor ($z = 1000$ m). We used the same type of rectangular grid used in the previous model to discretize the model domain. We also refined the local grid for the current source, receiving stations, and the target layer area.

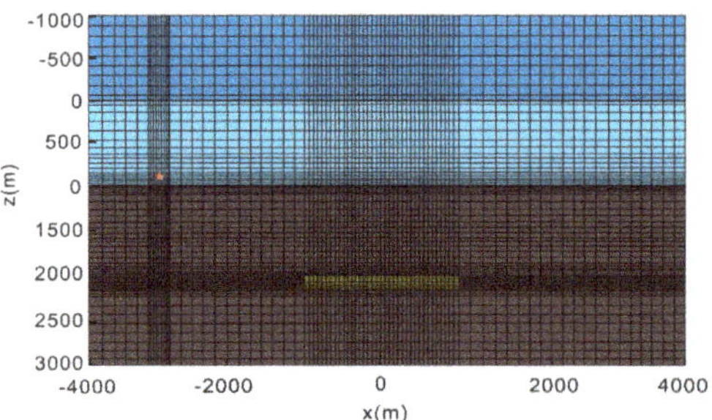

Figure 11. Horizontally layered geoelectrical model with rectangular mesh. The red star identifies the location of the current source. The dark blue, light blue, yellow, and brown areas represent the air layer, sea water, reservoir, and sediment, respectively.

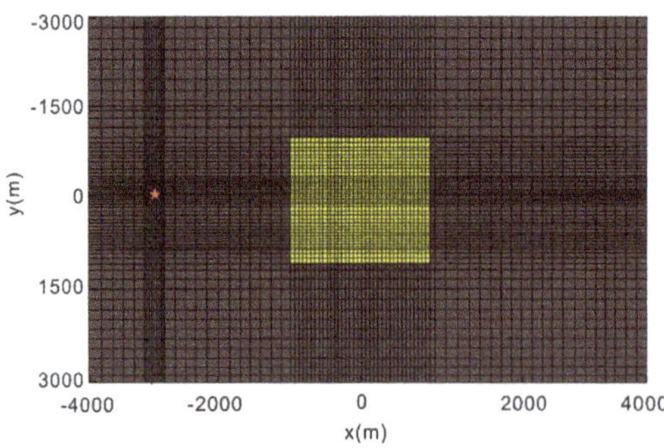

Figure 12. Plane view of the grid. The yellow area is the area occupied with reservoir, and the red star identifies the location of the current source.

Table 4. The conductivities of a hydrocarbon reservoir geoelectrical model.

Layer	Conductivity in Horizontal (s/m)	Conductivity in Vertical (s/m)
Air	10^{-6}	10^{-6}
Sea Water	3.30	3.30
Sediment	1.00	0.80
Reservoir	0.05	0.05

Figures 13 and 14 show the x-component of the secondary electric field calculated with different edge elements. The results of the four types of second-order elements were in agreement with each other. Furthermore, a strong anomalous field distortion caused by the background conductivity anisotropy can also be observed. Table 5 shows the cost-time of calculation with four types of quadratic edge elements. Similar to the previous model, the efficiencies of the two types of quadratic edge elements (SN and LN) were higher than those of conventional serendipity or Lagrange-type quadratic hexahedron elements (SC, LC).

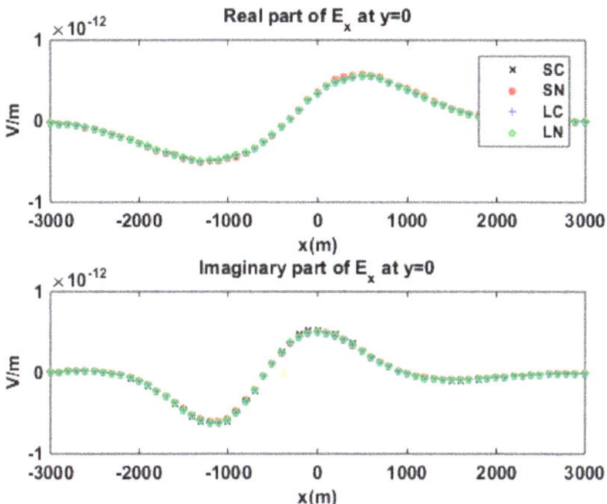

Figure 13. Secondary electric component in the x-direction, calculated by using different elements for isotropic sediment.

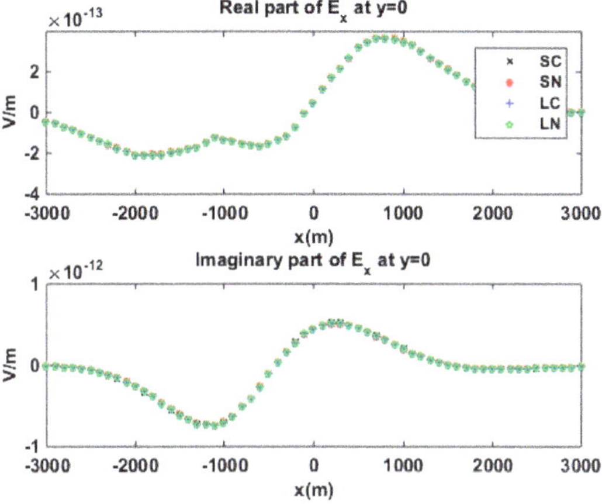

Figure 14. Secondary electric component in the x-direction, calculated by using different elements for anisotropic sediment.

Table 5. The cost-time of CPU when using different methods.

Element	Number of Elements	DOF	CPU Time (s)
SC	560,000	6,679,300	141.7
SN	560,000	4,999,050	156.2
LC	560,000	9,998,100	311.4
LN	560,000	8,317,850	471.9

5.3. *Pyramid Type Bathymetry Model*

Further verification of the new algorithm was performed by considering a pyramid-type bathymetry model (Figure 15) without a reservoir. The top area of the pyramid-type bathymetry model was 0.5 km × 0.5 km in square, at 990 m depth. The bottom area was 1.0 km × 1.0 km square, at 1000 m in depth. Air and sea water were 1000 m thick. In the domain of pyramid-type bathymetry, the depth of seawater ranges from 900 m to 1000 m. Figure 16 shows the pyramid-type bathymetry model with a reservoir. The domain of the reservoir was $\Omega = \{-1000\text{ m} \leq x, y \leq 1000\text{ m}, 1500\text{ m} \leq z \leq 1600\text{ m}\}$. The modeling domain was $\Omega = \{-4\text{ km} \leq x, y \leq 4\text{ km}, -1\text{ km} \leq z \leq 3\text{ km}\}$. In these models, the conductivities of air, seawater, and sediment were the same as previously, whereas the reservoir's conductivity was 0.05 s/m. The electric dipole source was x-oriented at $(-3000,\ 0,\ 800)$ m at 1 Hz frequency, and the receiving station was located at $z = 800$ m.

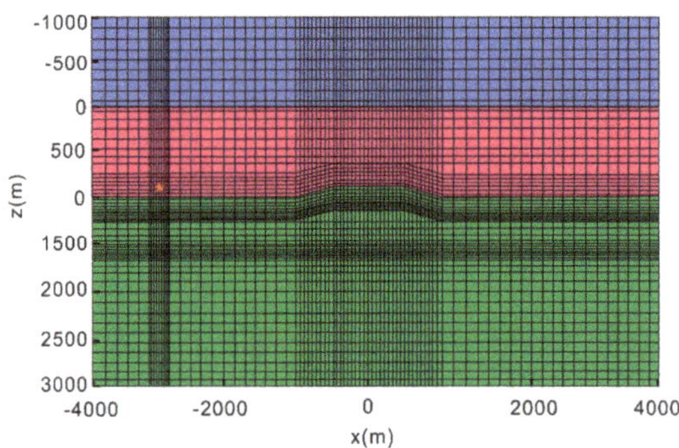

Figure 15. Section view of the grid used for the bathymetry model without reservoir. The red star locates the dipole source. The dark blue, pink, blue, and green areas represent the air layer, sea water, oil and gas, and bedrock, respectively.

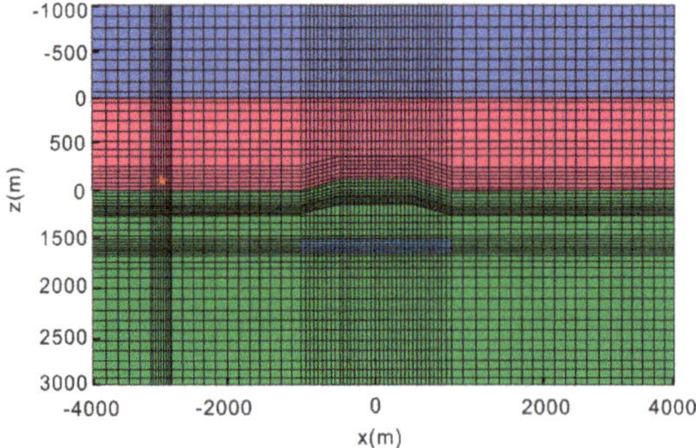

Figure 16. Section view of the grid used for the bathymetry model with reservoir. The reservoir occupies the blue area, and the current source is located at the red star. The dark blue, pink, blue, and green areas represent the air layer, sea water, oil and gas, and bedrock, respectively.

Figures 17–20 show the secondary electric component in the x-direction of the bathymetry model, which has no reservoir or one reservoir in the isotropic sediment. Figures 21–24 show a comparison of the x-component of the secondary electric field of the bathymetry model with and without a reservoir in anisotropic sediment. Bathymetry and anisotropic background clearly affected the distribution of the marine controlled-source electromagnetic field (Figures 18–24), which should be considered in the processing of MCSEM data. Figures 25 and 26 show a comparison of the degrees of freedom (DOF) and CPU time using different elements, respectively. Clearly, these two new elements contributed significantly to reducing the CPU time in computing.

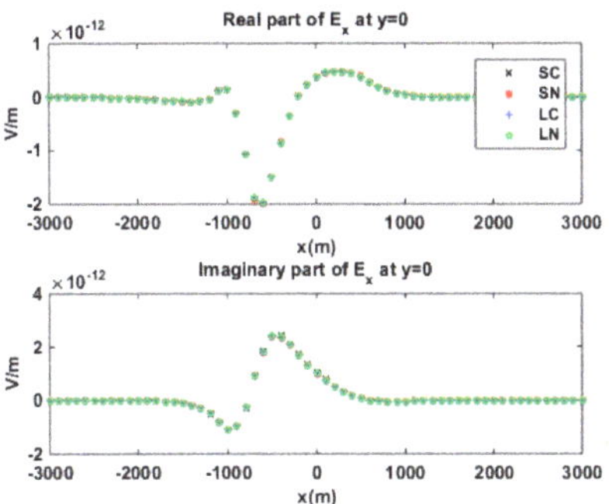

Figure 17. No-reservoir bathymetry model in isotropic sediment. The secondary electric field was calculated by using different second-order edge elements; here is the comparison.

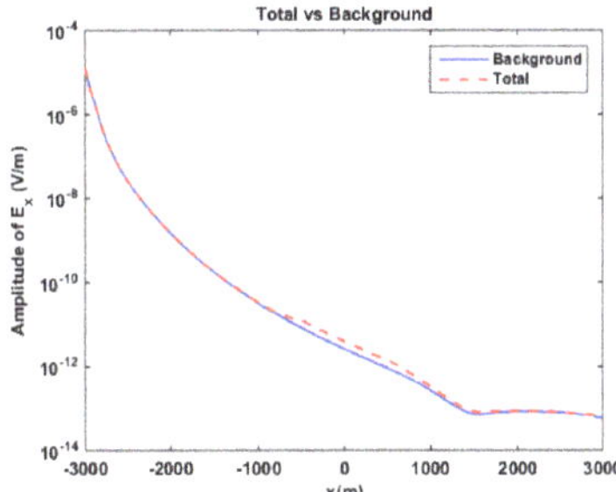

Figure 18. No-reservoir bathymetry model in isotropic sediment. Amplitude comparison between background field (blue line) and total field (dashed red line) for E_x component, at $y = 0$ m.

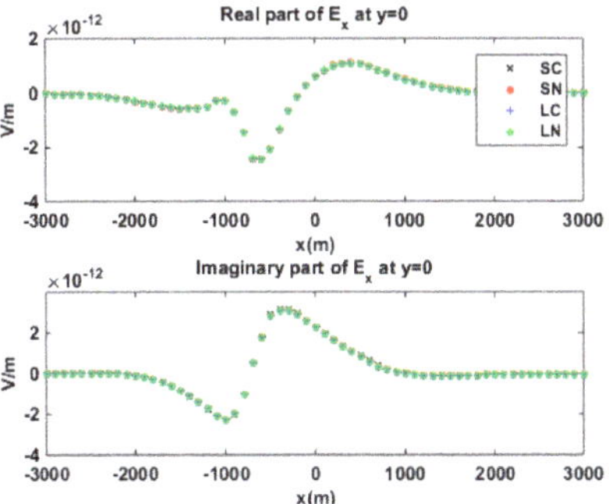

Figure 19. Reservoir-embedded bathymetry model in isotropic sediment. Second electric component in the x-direction, which was calculated by using different quadratic edge elements.

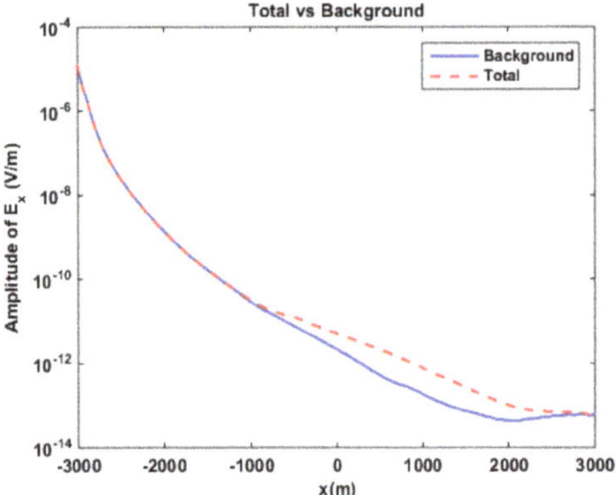

Figure 20. Reservoir-embedded bathymetry model in isotropic sediment. Amplitude comparison between background field (blue line) and total field (dashed red line) for E_x component, at $y = 0$ m.

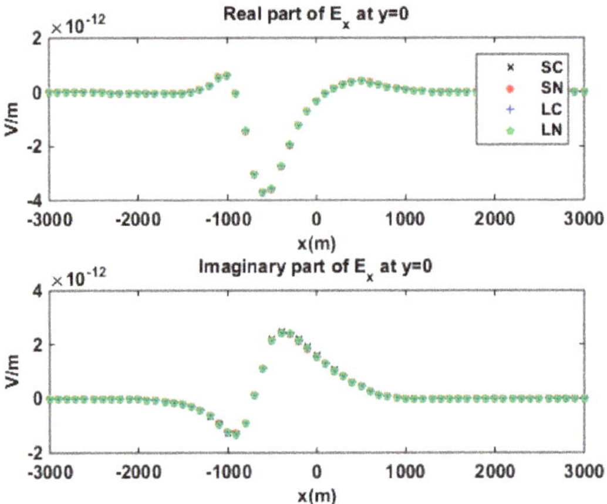

Figure 21. No-reservoir bathymetry model in anisotropic sediment. Second electric component in the x-direction, which was calculated by using different quadratic edge elements.

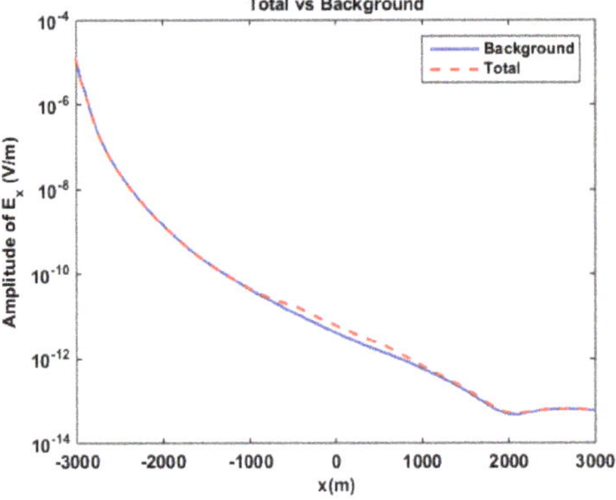

Figure 22. No-reservoir bathymetry model in anisotropic medium. Amplitude comparison between background field (blue line) and total field (dashed red line) for E_x component, at $y = 0$ m.

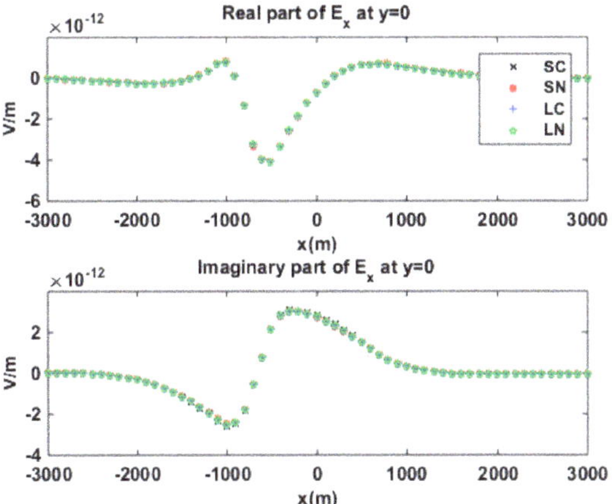

Figure 23. Reservoir-embedded bathymetry model in anisotropic sediment. Second electric component in the x-direction, which was calculated by using different quadratic edge elements.

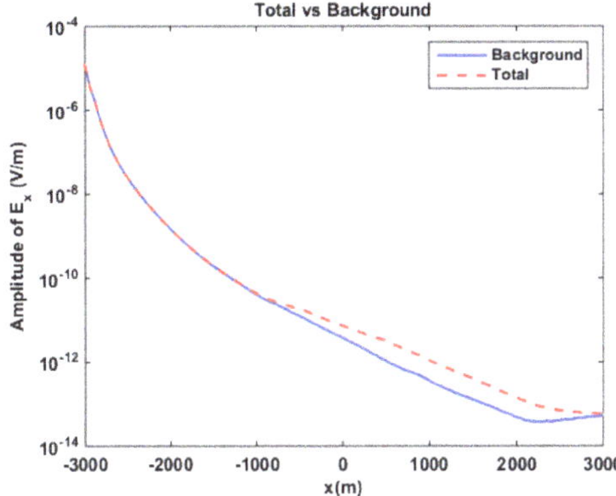

Figure 24. Bathymetry model with a reservoir in anisotropic medium. Amplitude comparison between background field (blue line) and total field (dashed red line) for E_x component, at $y = 0$ m.

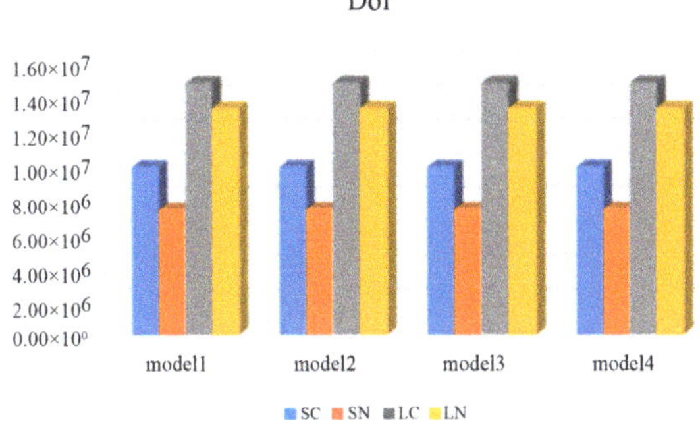

Figure 25. A comparison of DOF using different elements. Model 1 = bathymetry + isotropic sediment; model 2 = bathymetry + isotropic sediment + reservoir; model 3 = bathymetry + anisotropic sediment; model 4 = bathymetry + anisotropic sediment + reservoir.

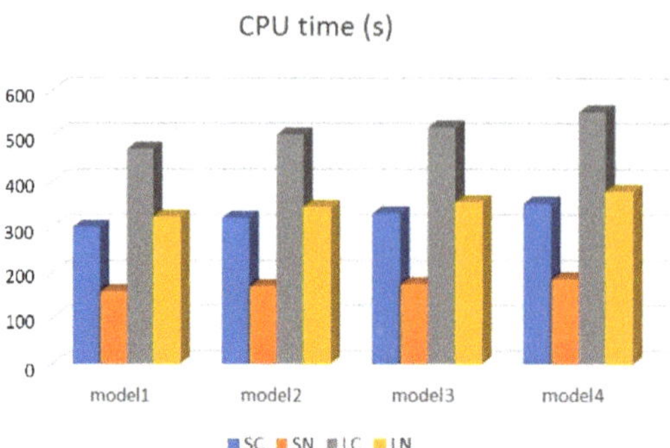

Figure 26. Comparison of CPU time using different elements. Model 1 = bathymetry + isotropic sediment; model 2 = bathymetry + isotropic sediment + reservoir; model 3 = bathymetry + anisotropic sediment; model 4 = bathymetry + anisotropic sediment + reservoir.

6. Conclusions

We developed a new vector finite element algorithm to solve the 3D marine CSEM modeling problem in an anisotropic medium. An algorithm based on a secondary field formulation and two new quadratic edge elements was proposed. Compared with the vector FEM algorithm based on the conventional quadratic edge element, the proposed algorithm significantly reduced the computational cost without losing accuracy. The algorithm in this paper has good generality and is suitable for solving other electromagnetic induction problems in isotropic and anisotropic media, including airborne and borehole electromagnetic methods.

Author Contributions: Conceptualization, B.X. and H.C.; methodology, H.C.; software, H.C. and Y.L.; validation, Z.C. and Y.H.; formal analysis, Y.L.; investigation, H.L.; resources, H.C.; data curation, H.C.; writing—original draft preparation, H.C.; writing—review and editing, Y.L.; visualization, B.X.; supervision, B.X. and H.C.; project administration, B.X.; funding acquisition, H.C. and B.X. All authors have read and agreed to the published version of the manuscript.

Funding: This research was supported by the National Natural Science Foundation of China No. 42174080. It was also supported by the Natural Science Foundation of Guangxi (No. 2020GXNSFAA297079) and the China Postdoctoral Science Foundation funded project (No. 2021MD703820) and the Research Start-up Foundation of Guilin University of Technology (No. RD2100002165).

Institutional Review Board Statement: The study did not require ethical approval.

Informed Consent Statement: The study did not involve humans.

Data Availability Statement: The data is unavailable due to privacy or ethical restrictions.

Conflicts of Interest: The authors declare no conflict of interest.

References

1. Constable, S.; Srnka, L.J. An introduction to marine controlled source electromagnetic methods for hydrocarbon exploration. *Geophysics* **2007**, *72*, WA3–WA12. [CrossRef]
2. Constable, S. Ten years of marine CSEM for hydrocarbon exploration. *Geophysics* **2010**, *75*, A67–A81. [CrossRef]
3. Weitemeyer, K.; Gao, G.; Constable, S.; Alumbaugh, D. The practical application of 2D inversion to marine controlled-source electromagnetic data. *Geophysics* **2010**, *75*, F199–F211. [CrossRef]
4. Wannamaker, P.; Hohmann, G.; San Filipo, W. Electromagnetic modeling of 3-dimensional bodies in layered earths using integral equations. *Geophysics* **1984**, *49*, 60–74. [CrossRef]
5. Avdeev, D.B.; Kuvshinov, A.V.; Pankratov, O.V.; Newman, G.A. High performance three-dimensional electromagnetic modeling using modified Neumann series: Wide band numerical solution and examples. *J. Geomag. Geoelectr.* **1997**, *49*, 1519–1539. [CrossRef]
6. Zhdanov, M.S.; Fang, S. Quasi-linear series in three-dimensional electromagnetic modeling. *Radio Sci.* **1997**, *32*, 2167–2188. [CrossRef]

7. Newman, G.A.; Alumbaugh, D.L. Frequency-domain modelling of airborne electromagnetic responses using staggered finite differences. *Geophys. Prospect.* **1995**, *43*, 1021–1042. [CrossRef]
8. Davydycheva, S.; Druskin, V.; Habashy, T. An efficient finite difference scheme for electromagnetic logging in 3D anisotropic inhomogeneous media. *Geophysics* **2003**, *68*, 1525–1536. [CrossRef]
9. Abubakar, A.; Habashy, T.M.; Druskin, V.L.; Knizhnerman, L.; Alumbaugh, D. 2.5D forward and inverse modeling for interpreting low-frequency electromagnetic measurements. *Geophysics* **2008**, *73*, F165–F177. [CrossRef]
10. Commer, M.; Newman, G.A. New advances in three-dimensional controlled-source electromagnetic inversion. *Geophys. J. Int.* **2008**, *172*, 513–535. [CrossRef]
11. Sasaki, Y.; Meju, M.A. Useful characteristics of shallow and deep marine CSEM responses inferred from 3D finite-difference modeling. *Geophysics* **2009**, *74*, F67–F76. [CrossRef]
12. Streich, R.; Becken, M. Electromagnetic fields generated by finite length wire sources: Comparison with point dipole solutions. *Geophys. Prospect.* **2011**, *59*, 361–374. [CrossRef]
13. Badea, E.A.; Everett, M.E.; Newman, G.A.; Biro, O. Finite-element analysis of controlled-source electromagnetic induction using Coulomb gauged potentials. *Geophysics* **2001**, *66*, 786–799. [CrossRef]
14. Schwarzbach, C.; Borner, R.-U.; Spitzer, K. Three-dimensional adaptive higher order finite element simulation for geo-electromagnetics—A marine CSEM example. *Geophys. J. Int.* **2011**, *187*, 63–74. [CrossRef]
15. Ansari, S.; Farquharson, C.G. 3D finite-element forward modeling of electromagnetic data using vector and scalar potentials and unstructured grids. *Geophysics* **2014**, *79*, E149–E165. [CrossRef]
16. Cai, H.; Xiong, B.; Han, M.; Zhdanov, M. 3D controlled-source electromagnetic modeling in anisotropic medium using edge-based finite element method. *Comput. Geosci.* **2014**, *73*, 164–176. [CrossRef]
17. Cai, H.; Xiong, B.; Zhdanov, M. Three-dimensional marine controlled-source electromagnetic modelling in anisotropic medium using finite element method. *Chin. J. Geophys.* **2015**, *58*, 2839–2850.
18. N′ed′elec, J.C. Mixed finite elements in R3. *Numer. Math.* **1980**, *35*, 315–341. [CrossRef]
19. Mukherjee, S.; Everett, M.E. 3D controlled-source electromagnetic 455 edge-based finite element modeling of conductive and permeable heterogeneities. *Geophysics* **2011**, *76*, F215–F226. [CrossRef]
20. Silva, N.V.; Morgan, J.V.; MacGregor, L.; Warner, M. A finite element multifrontal method for 3D CSEM modeling in the frequency domain. *Geophysics* **2012**, *77*, E101–E115. [CrossRef]
21. Castillo-Reyes, O.; Amor-Martin, A.; Botella, A.; Anquez, P.; García-Castillo, L.E. Tailored meshing for parallel 3D electromagnetic modeling using high-order edge elements. *J. Comput. Sci.* **2022**, *63*, 101813. [CrossRef]
22. Bíró, O.; Preis, K.; Richter, K.R. On the use of the magnetic vector potential in the nodal and edge finite element analysis of 3D magnetostatic problems. *IEEE Trans. Magn.* **1996**, *32*, 651–654. [CrossRef]
23. Biró, O.; Preis, K. Partial tree-gauging of second order edge element vector potential formulations. In Proceedings of the Computational Electromagnetics, Montreal, QC, Canada, 28 June–2 July 2015.
24. Zhdanov, M.S. *Geophysical Electromagnetic Theory and Methods*; Elsevier: Amsterdam, The Netherlands, 2009.
25. Jin, J. *Finite Element Method in Electromagnetics*; Wiley-IEEE Press: New York, NY, USA, 2002.
26. Ward, S.H.; Hohmann, G.W. *Electromagnetic Theory for Geophysical Applications*; SEG: Oklahoma City, OK, USA, 1988.
27. Anderson, W.L. A hybrid fast Hankel transform algorithm for electromagnetic modeling. *Geophysics* **1989**, *54*, 263–266. [CrossRef]
28. Zhdanov, M.S.; Keller, G. *The Geoelectrical Methods in Geophysical Exploration*; Elsevier: Amsterdam, The Netherlands, 1994.
29. Guptasarma, D.; Singh, B. New digital linear filters for Hankel J0 and J1 transforms. *Geophys. Prospect.* **1997**, *54*, 263–266.

Article

Numerical Investigation on the Influence of Breakwater and the Sediment Transport in Shantou Offshore Area

Yuxi Wu [1], Kui Zhu [2,*], Hao Qin [1], Yang Wang [3], Zhaolong Sun [2], Runxiang Jiang [2], Wanhu Wang [3], Jiaji Yi [3], Hongbing Wang [3] and Enjin Zhao [1]

[1] College of Marine Science and Technology, China University of Geosciences, Wuhan 430074, China
[2] College of Electrical Engineering, Naval University of Engineering, Wuhan 430033, China
[3] Haikou Marine Geological Survey Center, China Geological Survey, Haikou 570100, China
* Correspondence: zeek@cug.edu.cn

Abstract: The coastline of Shantou is tortuous, while the hydrodynamic environment is complicated. In this paper, the hydrodynamic model is established by the FVCOM (Finite Volume Coastal Ocean Model); the open boundary conditions such as water level, river, and wind field are the input; and the model is verified by tidal harmonic function. According to the previous research, the typhoon wind field with a 10-year return period is selected for storm surge simulation. When there is a bank, the accumulated water on the land cannot enter the ocean due to the block of the bank but accumulates on the inner side of the bank, resulting in higher accumulated water, but less than 0.5 m. In the aspect of sediment deposition, a sediment transport model is established to analyze the sediment deposition in Shantou Port and its surrounding waters. Some reasonable suggestions are put forward for the sediment deposition in Shantou. According to the simulation results, the following conclusions can be drawn: (1) In the case of typhoon storm surge in the return period of 10 years, the bank can effectively protect the inland. Still, accumulated water will collect near the bank. (2) The offshore water level will rise by 0.4 m after adding a bank. (3) The sediment in Shantou Bay mainly comes from the ocean sediment caused by tides, and the largest sedimentation occurs in the main channel.

Keywords: numerical simulation; hydrodynamic force; sediment transport; embankment

Citation: Wu, Y.; Zhu, K.; Qin, H.; Wang, Y.; Sun, Z.; Jiang, R.; Wang, W.; Yi, J.; Wang, H.; Zhao, E. Numerical Investigation on the Influence of Breakwater and the Sediment Transport in Shantou Offshore Area. *Appl. Sci.* **2023**, *13*, 3011. https://doi.org/10.3390/app13053011

Academic Editor: Jianhong Ye

Received: 4 January 2023
Revised: 6 February 2023
Accepted: 15 February 2023
Published: 26 February 2023

1. Introduction

The coastline of China is twisty and turning, and foreign trade in coastal areas prospers. The economy develops rapidly but is also threatened by marine disasters [1], among which the storm surge brings the most severe losses [2–4]. The post-disaster reconstruction after the storm surge requires a large number of materials and personnel investment. Storm surges are generally caused by typhoons [5]. In summer, the southeast coast of China is easily attacked by tropical cyclones, causing massive casualties. As a typical coastal special economic zone in China, Shantou has experienced many typhoons landing [6]. The accompanying storm surges cause heavy financial losses and often cause thousands or even tens of thousands of casualties.

The formation and development mechanism of storm surge was put forward in the early 20th century. Firstly, the theoretical storm surge model was put forward, which was constantly supplemented and improved, and its mechanism was deeply discussed [7]. With the rapid development of geodetic techniques, the research on numerical models has been deepened. Since the 21st century, many numerical models for storm surge models have existed. For example, the POM (Princeton Ocean Model) and the improved ECOM (Estuary Coast and Ocean Model) of Princeton University in the United States are three-dimensional baroclinic models; the Delft-3D model of Delft Water Conservancy Research Institute in the Netherlands; HAMSOM (Hamburg Shelf Ocean Model) developed by the University of Hamburg, Germany; and the FVCOM (An Unstructured Grid Finite Volume Coastal Ocean

Model) of the Massachusetts Institute of Technology are all models used previously [8–10]. The storm surge model in this paper is based on the FVCOM model and is coupled with the typhoon data of a 10-year return period has been verified in the reference.

In order to resist the storm surge, many scholars put forward coastal port facilities protection measures and disaster reduction measures [11–13]. Currently, there are mainly two kinds of protection against storm surges: one is to set up a bank, and the other is to build an ecological vegetation coast [14]. The former is a traditional protective measure that can weaken the waves caused by the storm surge. It has a positive effect in weakening the storm surge, but its influence range is limited, and it has little impact on the sea area far away from the shoreline [15–17]; the latter is that blueprint of green seawall recently put forward, and the previous study have shown that salt marsh vegetation can attenuate tidal current velocity, and a model that can be used to study the interaction mechanism between wetland vegetation and hydrodynamic forces during storm surge transit has been established [18]. In this paper, we consider the effectiveness of setting up the embankment to weaken the storm surge under the typhoon storm surge model with a 10-year return period.

No matter whether the embankment is set up or the vegetation is covered, it belongs to human activities, which will destroy the original coastline in this area, leading to the reduction of the sea area and the tidal volume in the bay, thus affecting the sediment carrying capacity of the sea area, accelerating the sediment deposition rate in the bay [19], and changing the coastline shape again, forming the positive feedback of sediment deposition. Therefore, coastal cities are affected by human activities in bays, estuaries, and ports, and sediment deposition often occurs. It is necessary to analyze and evaluate the sediment in the port area to protect the topography and ecological environment of the basin. Tides usually control the deposition, so the sediment model needs to be coupled with the hydrodynamic model for overall analysis [20–23]. Currently, the widely used sediment transport models include the MIKE model, MOHID model, ECOM, FVCOM, Delft3D model, etc. [24–26]. The sediment models at the port are complicated and are affected by three factors: topography, hydrodynamics, and sediment conditions. Aiming at the problem of sediment deposition in Shantou Port, many scholars have conducted detailed and precise research on sediment deposition in Shantou Port [27,28]. Still, the research time is relatively old, and due to the interference of human activities, the water area of Shantou Port is constantly shrinking, and the hydrodynamic conditions are continually changing. Therefore, it is necessary to build a new model of sediment deposition in Shantou Port in time.

In this paper, the hydrodynamic model of Shantou offshore is established, and the model is verified by the tidal harmonic constant. Based on the hydrodynamic model, according to the study of the Shantou storm surge model by Tang Yue Zhao, a 10-year return period typhoon storm surge model is established. The situation of storm surge increasing water and causing disasters in the 10-year return period is analyzed in the circumstance of deposition. At the same time, the sediment transport model is established and verified. The sediment deposition in Shantou Bay and surrounding waters are analyzed using the model, and the causes of sediment deposition are analyzed.

2. Model Introduction and Verification

2.1. Hydrodynamic Model

This paper uses FVCOM to simulate the ocean hydrodynamic forces, which is a prediction model for the ocean hydrodynamic environment in offshore areas and estuaries [29]. FVCOM adopts the finite volume method, uses unstructured triangular mesh to solve the original equation discretely, and encrypts the research area locally. On the one hand, the finite volume method can discretely solve the variables of the original equation. On the other hand, it can ensure efficient calculation when the equation is integrated, which fully uses the respective advantages of the finite element method and the finite difference method [30]. In the Cartesian coordinate system, the governing equations of FVCOM include the continuity equation, momentum conservation equation in three directions, temperature and salinity conservation equation, and density equation.

$$\frac{\partial u}{\partial x}+\frac{\partial v}{\partial y}+\frac{\partial w}{\partial z}=0 \tag{1}$$

$$\frac{\partial u}{\partial t}+u\frac{\partial u}{\partial x}+v\frac{\partial u}{\partial y}+w\frac{\partial u}{\partial y}-fv=-\frac{1}{\rho_0}\frac{\partial P}{\partial x}+\frac{\partial}{\partial z}(K_m\frac{\partial u}{\partial z})+F_u \tag{2}$$

$$\frac{\partial v}{\partial t}+u\frac{\partial v}{\partial x}+v\frac{\partial v}{\partial y}+w\frac{\partial v}{\partial z}+fu=-\frac{1}{\rho_0}\frac{\partial P}{\partial y}+\frac{\partial}{\partial z}(K_m\frac{\partial u}{\partial z})+F_v \tag{3}$$

$$\frac{\partial P}{\partial z}=-\rho g \tag{4}$$

$$\frac{\partial T}{\partial t}+u\frac{\partial T}{\partial x}+v\frac{\partial T}{\partial y}+w\frac{\partial T}{\partial z}=\frac{\partial}{\partial z}(K_h\frac{\partial T}{\partial z})+F_T \tag{5}$$

$$\frac{\partial S}{\partial t}+u\frac{\partial S}{\partial x}+v\frac{\partial S}{\partial y}+w\frac{\partial S}{\partial z}=\frac{\partial}{\partial z}(K_h\frac{\partial S}{\partial z})+F_S \tag{6}$$

$$\rho=\rho(T,S) \tag{7}$$

Equation (1) is a continuity equation, where x, y, and z represent the positions in the east-west direction, north-south direction, and vertical direction in the Cartesian coordinate system. u, v, and w represent the velocity components in three directions in the Cartesian coordinate system. Equations (2)–(4) are momentum conservation in the east-west direction, north-south direction, and vertical direction, respectively, and F represents the parameter of geostrophic deflection force; ρ is the density of local seawater and ρ_0 is the density of reference position; P is pressure; K_m is the vertical eddy viscosity coefficient; G is the acceleration of gravity; F_u, F_v, F_T, and F_S are diffusion terms of momentum, temperature, and salinity in two horizontal directions, respectively. Equations (5)–(7) are the temperature equation, salinity equation, and density equation, respectively. T is temperature, S is salinity, and K_h is the vertical diffusion coefficient of turbulent kinetic energy.

In the actual simulation process, the natural seabed terrain is often uneven. In order to avoid the false flow and diffusion caused by the rough bottom of the model, it is necessary to transform the σ coordinate vertically to smooth the bottom.

$$\sigma=\frac{z-\zeta}{H+\zeta}=\frac{z-\zeta}{D} \tag{8}$$

Equation (8) is the conversion between the σ coordinate and Cartesian coordinate, where z is the specific position of the Cartesian coordinate system or the height of the water surface relative to the still water, H is the distance from the reference surface to the seabed, and D is the height of the total water column. Under the condition of this coordinate system, the variation range of σ is from 0, indicated by seawater, to -1 at the bottom of the ocean, which avoids the unevenness of the bottom in the original Cartesian coordinate system.

2.2. Typhoon Model

According to the previous research, this paper uses the Holland typhoon model to construct the typhoon wind field with a frequency of 10 years. The model typhoon center pressure is 947 hPa, and the maximum wind speed is 47.26 m/s. The governing equation of the model is as follows:

$$P_r=P_c+(P_n-P_c)\exp(-(R_{\max}/r)^B) \tag{9}$$

$$V_r=[\frac{B}{\rho_a}(\frac{R_{\max}}{r})^B(P_n-P_c)\exp[-(\frac{R_{\max}}{r})^B]+(\frac{rf}{2})^2]^{0.5}-\frac{rf}{2} \tag{10}$$

Equation (9) is the pressure distribution formula of the Holland typhoon model, where P_c is the lowest pressure in the typhoon center and P_n is the background pressure, at 1013 hPa; $R_{\max}$ is the maximum wind speed radius of the typhoon; r is the distance from somewhere to the typhoon center; B is the scale parameter of the typhoon model. Equation (10) is the wind speed distribution formula of the typhoon model, where the air density (ρ_a) is 1.2 kg/m^3; f is the parameter of Coriolis force.

There are many formulas to determine the maximum wind speed radius of the typhoon, such as Graham, Tang Jian, Jiang Zhihui, etc. [31–33], and all put forward corresponding formulas. In this study, the procedure proposed by Willoughby [34] is adopted to determine the maximum wind speed radius of a typhoon:

$$R_{\max} = 51.6 \exp(-0.0223 V_{\max} + 0.0281\varphi) \tag{11}$$

where φ is the latitude of the place where the typhoon center is located.

The maximum wind speed ($V_{\max}$) of the typhoon is constructed according to the relationship between wind speed and air pressure proposed by Atkinson-Hollidy [35], and the empirical formula is given:

$$V_{\max} = 3.7237 \times (P_n - P_c)^{0.6065} \tag{12}$$

2.3. Sediment Model

This paper uses MIKE software's ST (sediment transport) module to calculate sediment transport. The open boundary of the sediment model is driven by the calculation result of the FVCOM hydrodynamic model, and the wind field data are reanalyzed by ECMWF. MIKE FVCOM and Mike FVCOM use unstructured grids and finite volume methods to solve discrete integrals so they can share the same grid system. The governing equation of the sediment model is as follows:

$$\frac{\partial \bar{c}}{\partial t} + u\frac{\partial \bar{c}}{\partial x} + v\frac{\partial \bar{c}}{\partial y} = \frac{1}{h}\frac{\partial}{\partial x}(hD_x\frac{\partial \bar{c}}{\partial x}) + \frac{1}{h}\frac{\partial}{\partial y}(hD_y\frac{\partial \bar{c}}{\partial y}) + \frac{F_S}{h} \tag{13}$$

Equation (13) is the sediment transport formula of the model, where the average sediment concentration is; u and v are the average vertical velocities in the x and y directions, respectively; h is the water depth there; D_x and D_y are turbulent diffusion coefficients of sediment in two directions, respectively; F_S is a function term of sediment source.

Sediment source function F_S is defined by the following formula:

$$F_s = \begin{cases} E(\tau_b/\tau_{ce} - 1)^n & \tau_b > \tau_{ce} \\ 0 & \tau_{ce} \geq \tau_b \geq \tau_{cd} \\ \omega_s c_b p_d & \tau_b < \tau_{cd} \end{cases} \tag{14}$$

where E the sediment scouring coefficient of the bottom of the river is bed; ω_s is the settling velocity of sediment, which is related to the shape, size, and content of sediment particles. p_d is the possible settlement probability; τ_b is the shear stress at the bottom of the bed; τ_{ce} is the critical shear stress initiated by surface sediment; τ_{cd} is the acute shear stress at which suspended sediment begins to settle.

The range of the sediment transport model (Figure 1) selects Shantou Port, including the Rongjiang River and Hanjiang River and its surrounding waters. Based on the hydrodynamic model, rivers are added. The study area is narrowed. Thus, the running time of the model is shortened, and the accuracy of the key research area is improved.

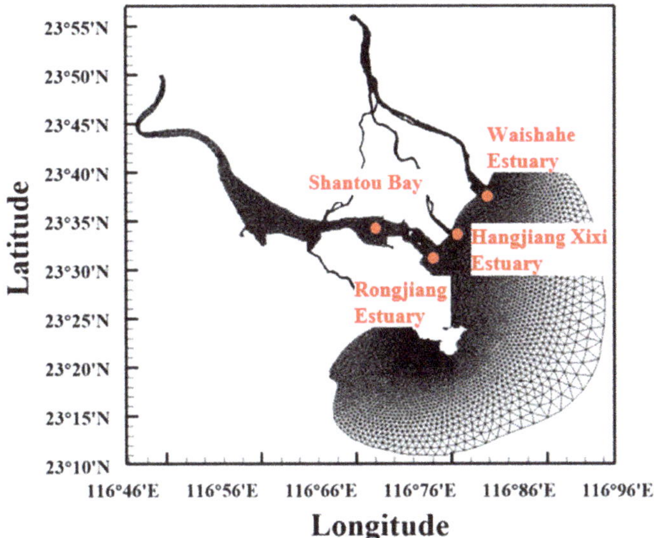

Figure 1. Grid of sediment transport model.

3. Verification of the Model

3.1. Verification of Tidal Harmonic Constant

The tidal harmonic constant of the interpolated hydrodynamic model is verified to prove the validity of the model. In this paper, three long-term tidal stations in the study area are selected, namely Yunaowan Tidal Station (117°06′ E, 23°24′ N), Shantou Tidal Station (116°44′ E, 23°20′ N), and Haimen Tidal Station (116°37′ E, 23°11′ N). In the model of the previous section, the tidal level data of three tide points are derived. The measured tidal level data and the derived tidal level data are calculated by the T-tide program, and the tidal harmonic constants of the corresponding positions are obtained. The simulation results are compared with the measured results' tidal harmonic constants of the four main tidal components K1, O1, M2, and S2. See Table 1 for the verification results of the model. It can be seen from the table that the amplitude error of other tidal components is less than 4 cm, and the late angle error is less than 12 degrees, except for M2, which is slightly larger. The mistake of the Shantou tide gauge station is somewhat more prominent, influenced by its geographical location. Generally speaking, the tidal harmonic constant error between the simulated and the measured results is within an acceptable range, which means that the model has sure accuracy and credibility.

3.2. Verification of Deposition Rate

The sediment transport model is verified by calculating the change of sediment thickness in Shantou Port in the simulation results (Figure 2). The simulation is from 1 January 2015 to 31 December 2019. Shantou Port's natural sediment deposition and surrounding waters are simulated without human intervention. Through calculation, the average deposition thickness of Shantou Port in five years is 0.416 m, and the average annual deposition thickness is 8.3 cm. The siltation speed of Shantou Port has been impacted by human activity throughout history, as evidenced by available data [27]. It can be roughly estimated that, if the initial conditions, such as water area and sediment content of Shantou Port, are constant, the average deposition thickness from 2015 to 2019 should be about 9 cm, slightly higher than the model simulation results. We are assuming that the initial conditions, such as water area and sediment content of Shantou Port, are constant at the beginning of the People's Republic of China. The average deposition thickness from 2015 to 2019 should be about 9 cm, slightly higher than the model simulation results. However, considering the breakwater construction, part of the sediment entering the port will settle at the estuary, resulting in the real sedimentation rate being lower than the estimated value. Therefore, the research model is more accurate.

Table 1. Verification of tidal harmonic constant.

Constituent	Tide Station	Measured Amplitude (°)	Simulated Amplitude (°)	Error (m)	Measured Epoch (°)	Simulated Epoch (°)	Error (m)
K1	Yunaowan	36.04	36.93	0.89	287.02	286.31	0.71
	Shantou	32.13	22.89	9.24	301.74	293.57	8.17
	Haimen	33.42	32.35	1.07	290.62	288.48	2.14
O1	Yunaowan	25.43	25.53	0.10	241.82	241.10	0.72
	Shantou	22.16	15.62	6.54	254.45	250.87	3.58
	Haimen	22.72	22.03	0.69	242.10	242.80	0.70
M2	Yunaowan	57.94	68.79	10.85	7.27	10.70	3.43
	Shantou	32.89	49.40	16.51	26.38	20.30	6.08
	Haimen	26.51	35.57	9.06	11.35	15.26	3.91
S2	Yunaowan	10.79	14.79	4.00	73.42	73.78	0.36
	Shantou	7.66	8.52	0.86	100.68	87.19	13.49
	Haimen	4.13	9.16	5.03	95.86	86.66	9.20

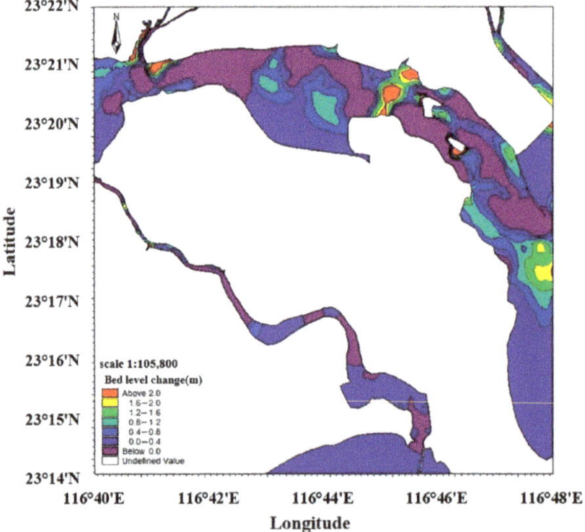

Figure 2. Changes of the bed bottom in the Shantou Port in five years.

4. Model Area and Data

The study area is from 116°22′ E west to 117°48′ E in the east and from 22°68′ N in the south to 23°73′ N in the north, including Shantou's offshore and surrounding waters. We used SMS (surface-water modeling system) software to lay unstructured triangular grids. The density of open boundary nodes decreases gradually from land to sea, which can improve computational efficiency while encrypting the critical research areas. The grid resolution in the coastal part of the open boundary is high, and the grid resolution in the main study area near the coastal area of Shantou City is up to 100 m. The research area grid contains 55,790 computing nodes and 108,354 irregular triangular grids (Figure 3). Considering that the research object belongs to the ocean surface, it is divided into five σ layers vertically.

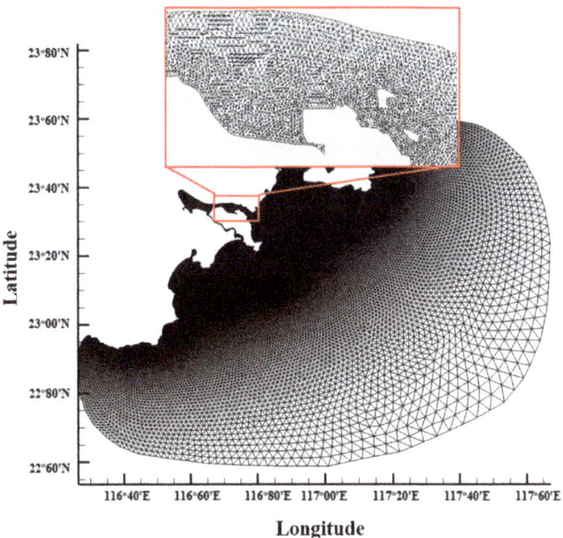

Figure 3. Grid setting in Shantou offshore (the red box in the figure is the coastal area of Shantou City where the grid resolution is up to 100 m).

High-precision data are an essential condition for the model to obtain reliable results. The water depth data in this paper adopt the global seabed topographic data of ETOPO1 provided by the NOAA (National Ocean and Atmospheric Administration) and the chart-sounding data with the number 81001. The ETOPO1 data have the highest accuracy among ETOPO data, with a spatial resolution of $1/60^\circ \times 1/60^\circ$. The bathymetric data of the chart are the measured data, which have a high density in the coastal areas. The datum plane of ETOPO1 data is the global mean sea level, while the datum plane of the chart is the local mean tide height datum plane. After conversion, the tide level starting surface of Shantou City is uniformly used as the datum plane, Figure 4 shows the water level in Shantou offshore and Figure 5 shows the study area after inserting topographic data.

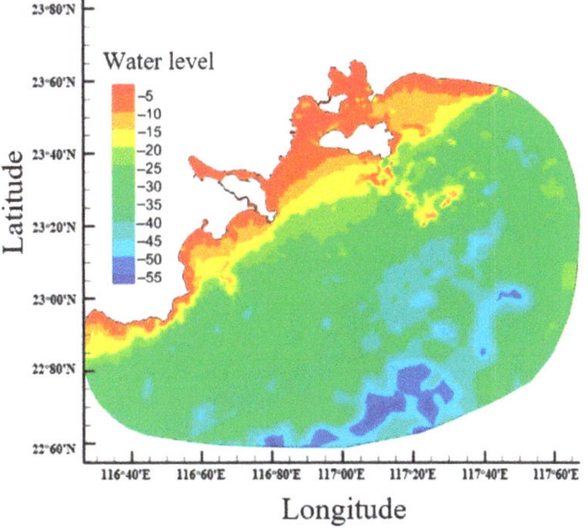

Figure 4. Bathymetric chart (The figure shows the water level in Shantou offshore. The colder color means a deeper water level).

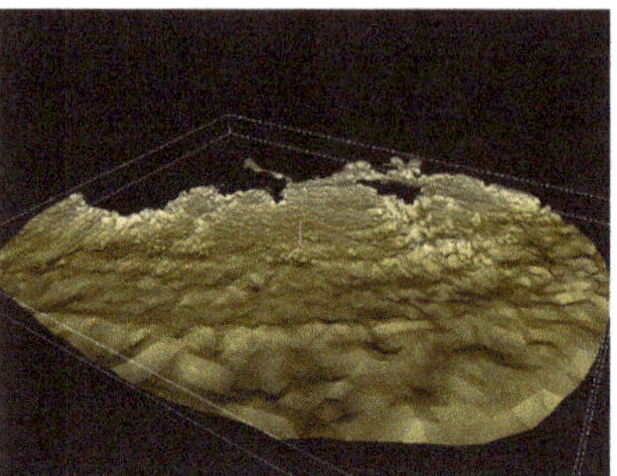

Figure 5. Terrain of the study area.

5. Results of Typhoon Storm Surge Simulation

5.1. Analysis of Typhoon Storm Surge Disasters with Different Intensities

According to the statistics of tropical cyclones in Shantou since the founding of the People's Republic of China, Tang Yue Zhao designed storm surge models in Shantou with different return periods combined with the Holland typhoon model and simulated storm surge with parametric typhoon models. It was found that southeast-west typhoons were the most common, and the typhoon center pressures corresponding to the return periods of 10 years, 50 years, 100 years, and 1000 years were 947 hPa, 937 hPa, 920 hPa, and 905 hPa, respectively [36]. The maximum water increase in different return periods is obtained from the four stations of Nan'ao Station, Chaoyang District, Chenghai District, and Shantou Port as shown in Figure 6.

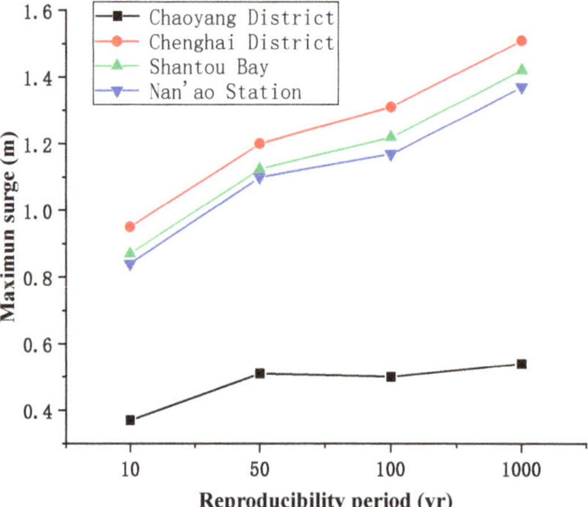

Figure 6. The maximum water increase in different return periods is obtained from the four sta-tions of Nan'ao Station, Chaoyang District, Chenghai District, and Shantou Bay (The maximum water increase is shows an increasing trend with the increase of return period years).

In actual working conditions, the probability of a storm surge of more than 100 years is low, and the port project cannot cope with the storm surge of a 100-year return period [37,38]. According to the research results of Tang Yue, this paper selects the typhoon moving from southeast to northwest in the 10-year return period and analyzes the protection effect of setting the bank dike on Shantou. To make the simulation results of typhoons and storms more intuitive, the north latitude range is 23°33′ N to 23°45′ N and the west longitude from 116°66′ E to 116°84′ E is selected for local simulation. Some grids in the main urban area of Shantou extend inland based on the original coastline. At the same time, Shantou's coast is artificial, so the actual shoreline is changed to a bank embankment with a height of 1.5 m. When the external water level is less than 1.5 m, it is regarded as a solid boundary where no water exchange occurs. The standard simulation calculation is carried out when it is higher than 1.5 m. The extended part is shown in Figure 7. The grid extending to the land is in the red line in the figure. The land elevation data are obtained by combining the extracted remote sensing data with the measured elevation data. The red line in the figure is the original shoreline, the dam is set in the model, and its height is set to be 1.5 m higher than the average sea level.

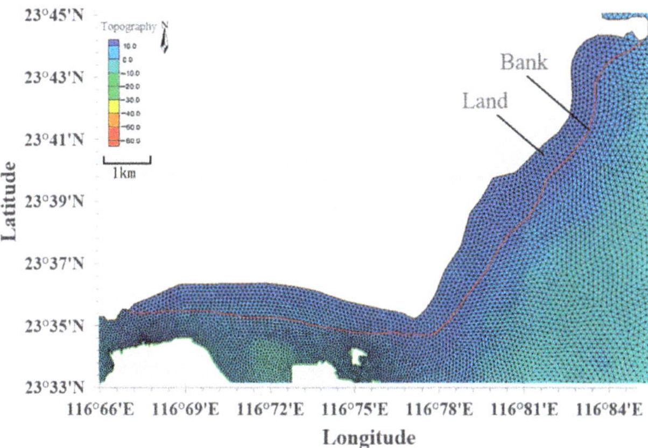

Figure 7. Urban coastal grid expansion (The red line in the figure is the original shoreline, the dam is set in the model, and its height is set to be 1.5 m higher than the average sea level).

Figure 8 shows the distribution of water level and flow field in the Shantou sea area and the inundation of storm surge on the coast and land under the southeast-northwest route during the 10-year return period of typhoon storm surge. The center of the typhoon landed in Chenghai District, Shantou City. Under this path, the hurricane hit Shantou City head-on, which has an excellent representative significance. The simulation time is 72 h, of which 48 to 72 h is the transit time of typhoons.

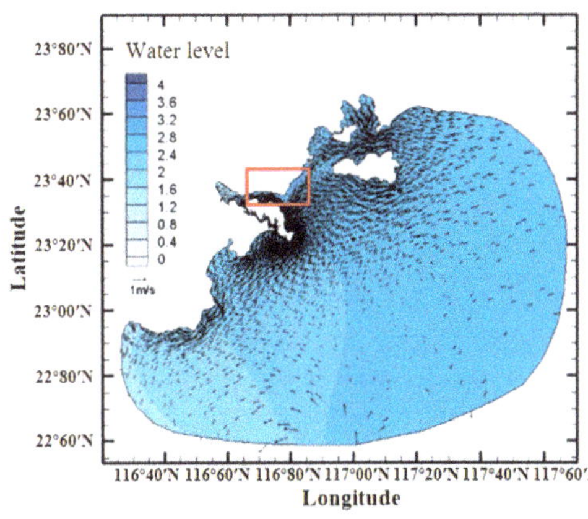

(**a**) Water level distribution with velocity vector in the whole study area at 2 h during storm surge

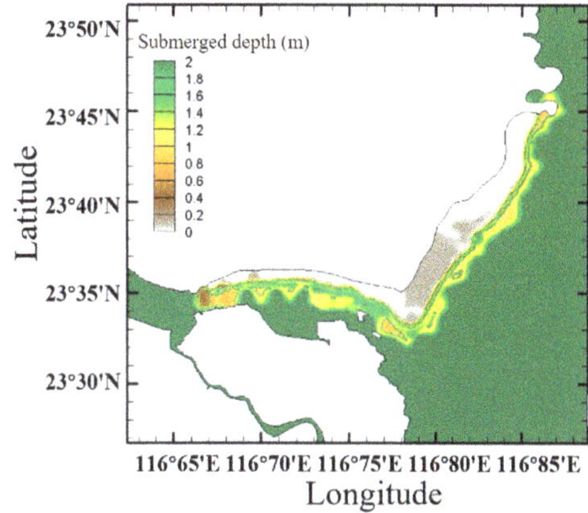

(**b**) The inundation range and depth of Rongjiang River Estuary and main urban area at 2 h during storm surge

Figure 8. *Cont.*

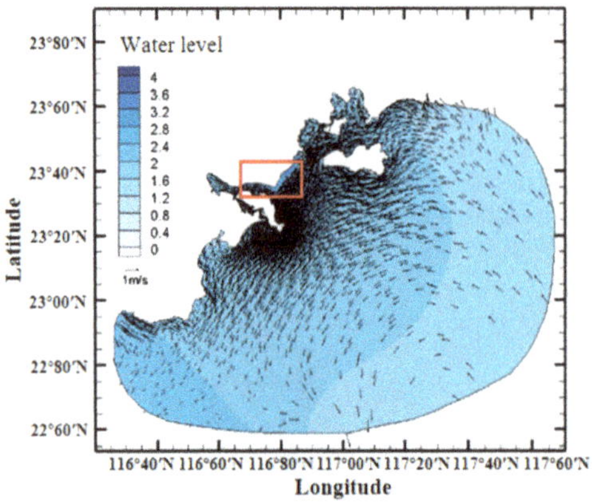

(c) Water level distribution with velocity vector in the whole study area at 8 h during storm surge

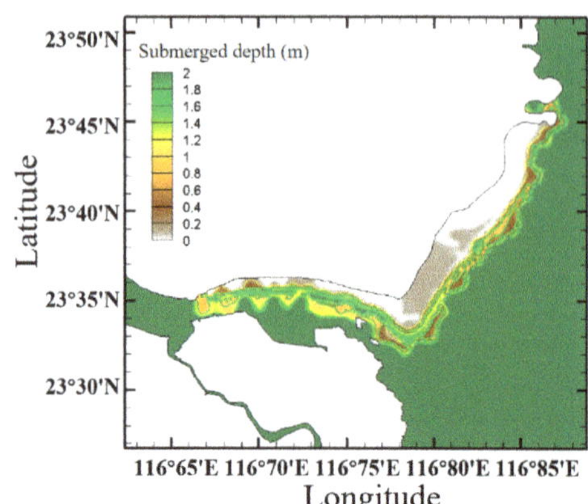

(d) The inundation range and depth of Rongjiang River Estuary and main urban area at 8 h during storm surge

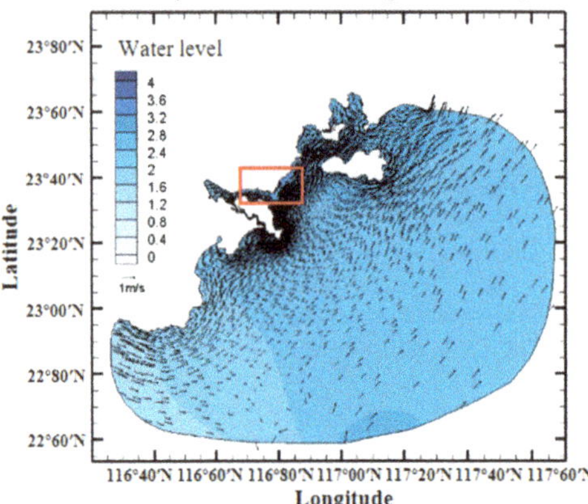

(e) Water level distribution with velocity vector in the whole study area at 14 h during storm surge

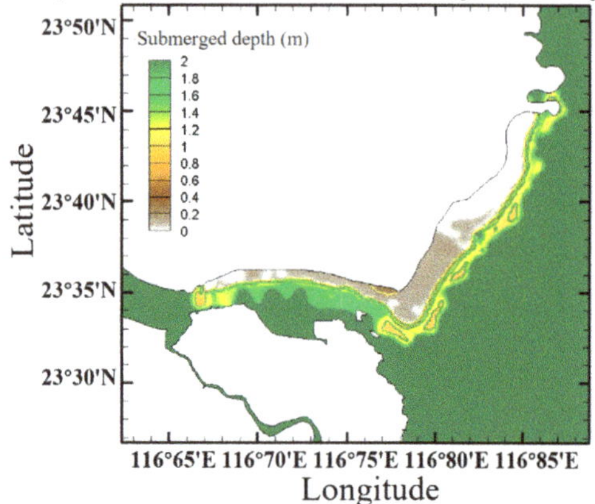

(f) The inundation range and depth of Rongjiang River Estuary and main urban area at 14 h during storm surge

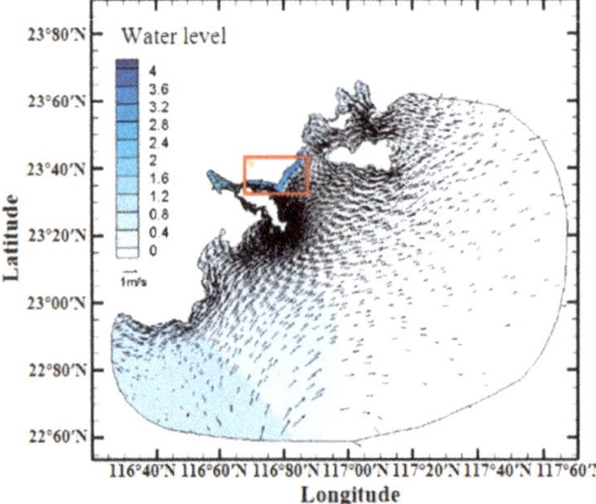

(g) Water level distribution with velocity vector in the whole study area at 20 h during storm surge

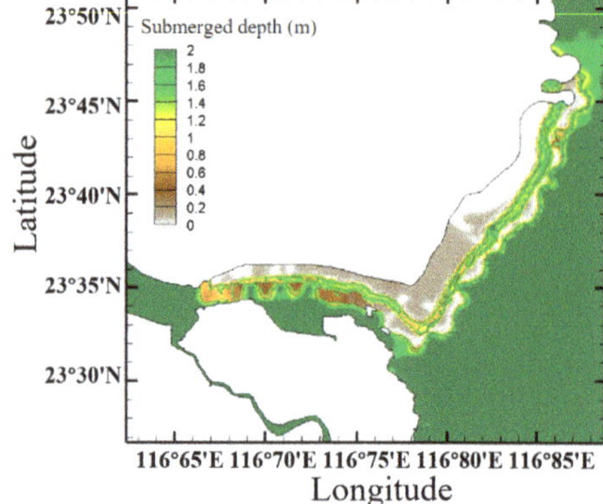

(h) The inundation range and depth of Rongjiang River Estuary and main urban area at 20 h during storm surge

Figure 8. Tidal current field and inundation of typhoon storm in 10-year return period (The red box on the left is the simulation area on the right).

In Figure 8, the left part is the water level distribution (based on the mean sea level of the Yellow Sea) and velocity vector diagram in the study area, and the right part is the inundation range and inundation depth near the Rongjiang Estuary and the main urban areas. The typhoon reached its maximum inundation 14 h ago, and the deepest accumulated water reached 0.38 m. During the 2 h of storm surge, the typhoon center did not match the study area. Still, under the wind, the water level in the study area obviously increased, especially in the northeast of Nan'ao Island, where the water level rose by more than 0.5 m. At 8 h, the typhoon center entered the study area. The water level gradient in the study area was high near the shore and low at the far shore, and the water level near the Rongjiang Estuary increased by nearly 1 m. At 14 h, the typhoon landed, and the hurricane was located northwest of the study area. The water level was high near the shore, low on the far shore, high in the northeast, and low in the southwest. The submerged area inside the dam expanded, but the accumulated water depth was still within 0.5 m. At 20 h, the typhoon center had left Shantou. However, the water level was still changing, and its distribution pattern was consistent with that in typical weather but nearly 0.2 m higher than in average temperature. In the Chenghai area, near the ocean dam, the outside water level dropped significantly, and the submerged area on the inside remained unchanged. Near the Rongjiang side, the water level outside the dam also dropped, and the inundation inside the dam no longer expanded. The inundation reached its maximum 14 h before, and the deepest accumulated water was 0.38 m. Since then, there was no further disaster expansion.

5.2. *Analysis of the Protective Effect of Embankment on Storm Surge*

It can be seen from the storm surge simulation in the 10-year-old situation that the bank can effectively keep out a large amount of seawater. When the typhoon intensity is vigorous, the seawater will cross the bank and enter the urban area. At the same time, after the typhoon transits, the seawater enters the landforms water in the bank instead because of the bank's existence. Figure 9 is a comparison chart of urban inundation degree with or without embankment under the condition of typhoon moving from southeast to northwest during the 10-year return period. The time is the maximum storm surge during the typhoon landing, and the height of the embankment is 1.5 m. It can be seen from the figure that the land in the study area is divided into two parts. One part is adjacent to the South China Sea, and the other is adjacent to the Rongjiang River in Shantou Bay, which can be regarded as the left and right parts of the V-shaped land. When there is no bank, the whole land in the study area will be submerged. The right part of the "V" shape will be submerged seriously, and the accumulated water depth can exceed 0.3 m, while the left part of the "V" shape is higher and far from the sea, and the submerged depth is about 0.1 m. When the bank is set, it can block part of the seawater intrusion, but it will form water near it. The bank can block part of the seawater when the storm surge is most substantial and play a protective role in inland areas.

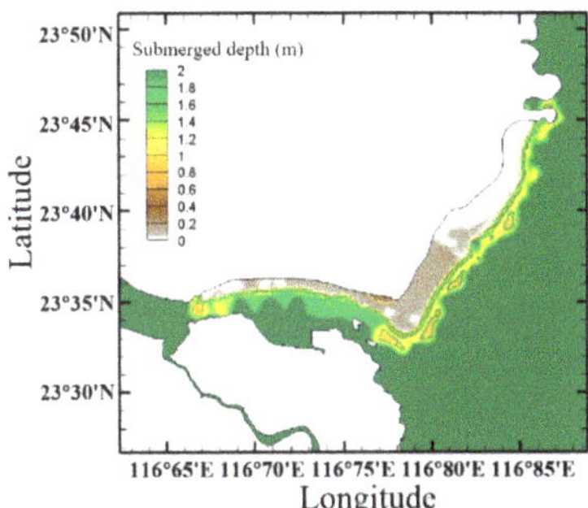

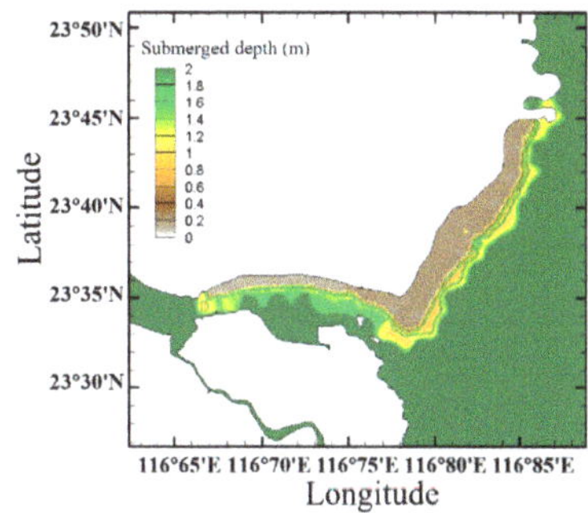

Figure 9. Blocking effect of bank on storm surge (the left picture shows the existence of bank, and the right picture shows the absence of bank).

Because the land topography is high, some seawater can flow back into the sea after the storm surge. At the same time, the water increase caused by the storm surge in the ocean will continue to fall back. However, due to the topographic conditions and embankment, part of the seawater will be left on the land. Figure 10 shows the water accumulation on the land when the typhoon is far away from the storm surge. The simulation results show that, in the case of no embankment, the left part of the V-shape is relatively high, and all the backward seawater flows back to the ocean. In contrast, the right part of the V-shape is relatively flat, and the amount of backward seawater is relatively large, so only some seawater can flow back. When there is a bank, the accumulated water on the land cannot enter the ocean due to the block of the bank but accumulates on the inner side of the bank, resulting in higher accumulated water, but less than 0.5 m.

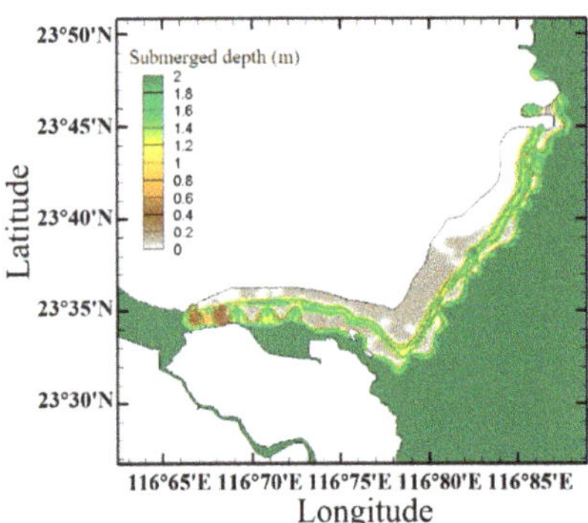

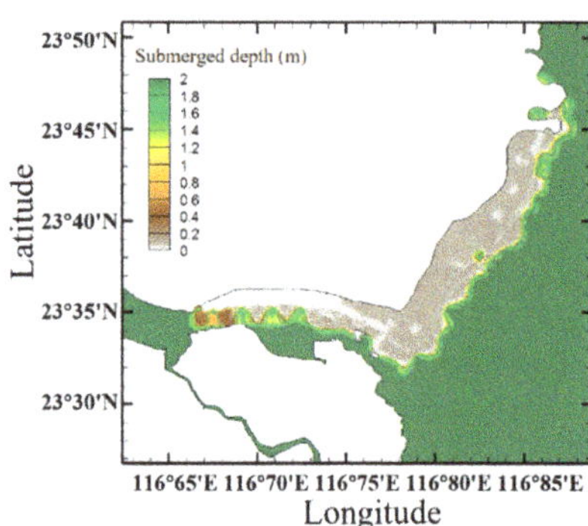

Figure 10. Water accumulation on land after typhoon transit (the left picture shows the existence of bank, and the right picture shows the absence of bank).

In summary, when a storm surge occurs, the bank can effectively block some seawater from entering the inland and play a better protective role. However, after the storm surge, the seawater entering the land will be blocked by the bank and gathered near the bank, which needs timely dredging.

In order to specifically analyze the blocking effect of the embankment on seawater during the storm surge, the right outside section of the V-shaped embankment was selected, and the changed seawater level and seawater velocity with the connecting distance were analyzed in both conditions. Take the landing time of the typhoon moving southeast-northwest in the 10-year return period as an example. The comparison of water levels in the sea area outside the embankment is shown in Figure 11. In the figure, a section with a length of 1000 m is selected at the shore bank on the seaside. The red line segment represents the offshore water level when there is a shore bank, and the blue line represents the offshore water level when there is no shore bank. Here, to visually compare the height of the water level, the location of the coastline is taken as the starting surface of the water level, namely 0 m. When there is a bank, the water level near the coastline can reach 1.4 m in this case. The seawater level stays relatively high with the increased offshore distance. However, when the distance is 400–500 m, the water level drops by nearly 0.4 m, which is consistent with the water level without a bank. Without a bank, the water level at the coastline reached 1.03 m under the action of storm surge, which means that the land was submerged by 1.03 m deep seawater at this time. Therefore, it can be said that the embankment significantly impacts the seawater level within 500 m, which is embodied in allowing for the high seawater to gather near the shore.

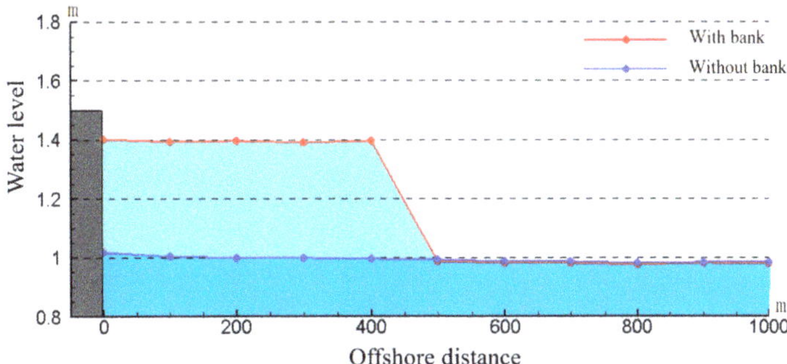

Figure 11. Comparison of water level in the sea area outside the embankment.

The result of comparing the velocity of Shanghai water in the same profile is shown in Figure 12. When there is a bank, because of the barrier of the bank near the coastline, the seawater has no velocity component perpendicular to the shoreline, and the velocity there is small, only 0.21 m/s. With the increased distance from the shore, the velocity of seawater increases in an oscillating way, and the velocity can reach the maximum value of 0.49 m/s at 300 m ashore. After that, the flow rate decreases, reaching the lowest value at 500 m. After 600 m, it becomes stable, and the flow rate is between 0.3 m/s and 0.4 m/s. Under the condition of no bank, the flow rate of seawater in the sea area changes little, and the flow rate on the coastline is faster than that under the condition of no bank. At 600 m, there is a bank gradually approaching the flow rate under the condition of no bank, and at 800 m, the two flow rates are the same, and the changing trend is the same after that.

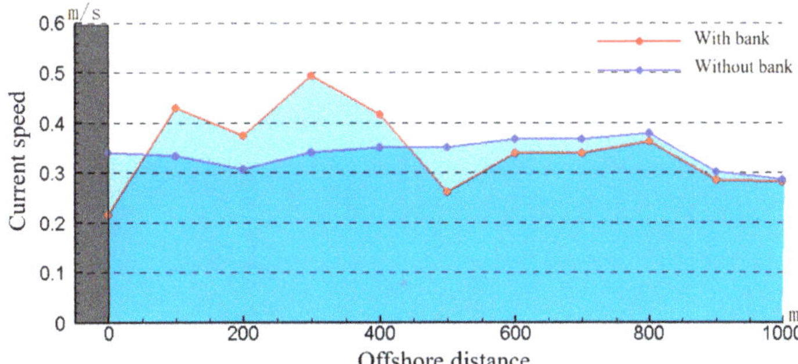

Figure 12. Comparison of velocity in sea area outside embankment.

According to the simulation results, given the typhoon with the intensity of a ten-year return period, the shore dike can effectively block a large amount of seawater when the storm surge occurs. However, due to the function of the shore dike, the offshore water level within 500 m will rise by an additional 0.4 m. When the storm surge is over, the shore dike will cause the seawater entering the land to accumulate water at the shore dike, which needs to be dredged in time. Embankment also has an impact on the velocity of the external sea area within 600 m during a storm surge, resulting in an overall increase of 0.16 m/s in seawater velocity. Generally speaking, the embankment will only work on a specific range of offshore areas.

6. Sediment Deposition in Shantou

6.1. Simulation Results of Main Siltation Areas

Figure 13 shows the sediment deposition in the study area over five years. It can be seen that there are four main siltation areas, which are located at Waishahe Estuary,

Hanjiang Xixi Estuary, Rongjiang Estuary, and Shantou Port waterway. From the shape of the sedimentary area, the sediments in the three estuaries are all fan-shaped because of the exact sediment deposition mechanism. The Hanxixi Estuary is in a regular fan shape; the Waishahe Estuary and Rongjiang Estuary show an irregular sector, with thicker deposits in the southwest of Waishahe Estuary and vaster deposits in the northeast of Rongjiang Estuary. There are several siltation areas in Shantou Bay, but the siltation degree is lighter than that at the estuary. The siltation in the bay can be divided into two parts: The first is the siltation area near the inland river. The main feature of deposition here is that the deposition range is extensive, but the deposition thickness is small. The second part of the siltation is concentrated on the side near the ocean, showing the trend of cutting off the river. This part of siltation is concentrated and small in area but thick in thickness.

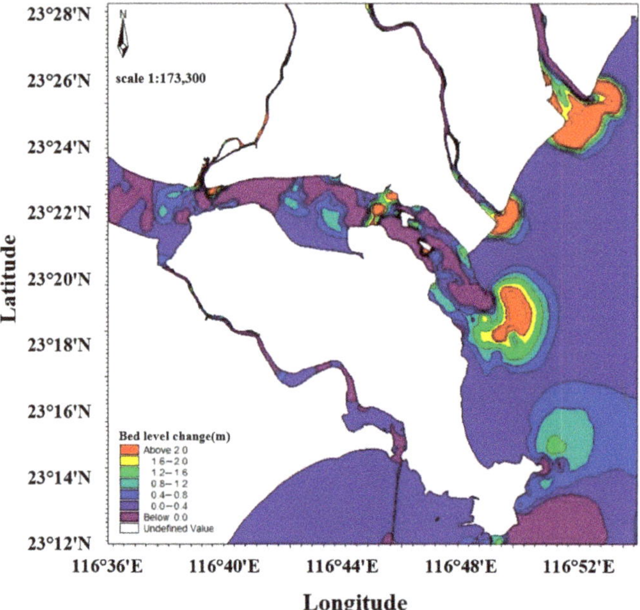

Figure 13. Overall siltation in the study area.

Figures 14–17 shows the enlarged views and 5-year changes of siltation thickness of Waishahe Estuary, Hanjiang Xixi Estuary, Rongjiang Estuary, and Shantou Port Channel, respectively, to facilitate the analysis of siltation in the study area. Hanxi River and Waisha River are both branches of the Hanjiang River, so the law of sediment deposition is the same, and the variation curve of sediment thickness in five years is similar.

However, the runoff of Hanjiang Xixi is smaller than that of Waisha River, so the sediment is easier to deposit, and the final deposition thickness is higher than that of the Waisha River Estuary. The central axis of the area where the sediment of Hanxixi Estuary exceeds 1 m is about 1560 m, and the sediment area is about 0.96 km^2.

Tides control Rongjiang Estuary and Shantou Bay, and the short-term siltation rates of Rongjiang Estuary and Shantou Bay show periodic changes. A period of 15 days is consistent with the astronomical tide period, indicating that it is highly correlated with the astronomical tide, which is in line with reality. Because of the breakwater in the Rongjiang Estuary, the sediment deposition is different under the ebb and flow tide, forming two settlement centers. The reference point selected by Rongjiang Estuary is located at the sedimentation center outside the sand dike. Different from the Waisha River and Hanjiang Xixi River, the sedimentation rate at the sand dike of Rongjiang Estuary is very uniform and generally increases linearly. Two representative locations in Shantou Port are selected for analysis. The red line segment is the silting area near the sea, and the blue line is the main silting center on the inland river. The sedimentation trend of the two places is the same, but the sedimentation rate near the seaside is faster. The siltation rate in Shantou Bay is decreasing yearly because the siltation in Shantou Bay is more controlled by tides, while the siltation at Rongjiang Estuary and the siltation at the channel reduce the tide carrying sediment into the bay. Therefore, unlike Rongjiang Estuary, the siltation rate in Shantou Bay, which is also controlled by tides, is decreasing yearly.

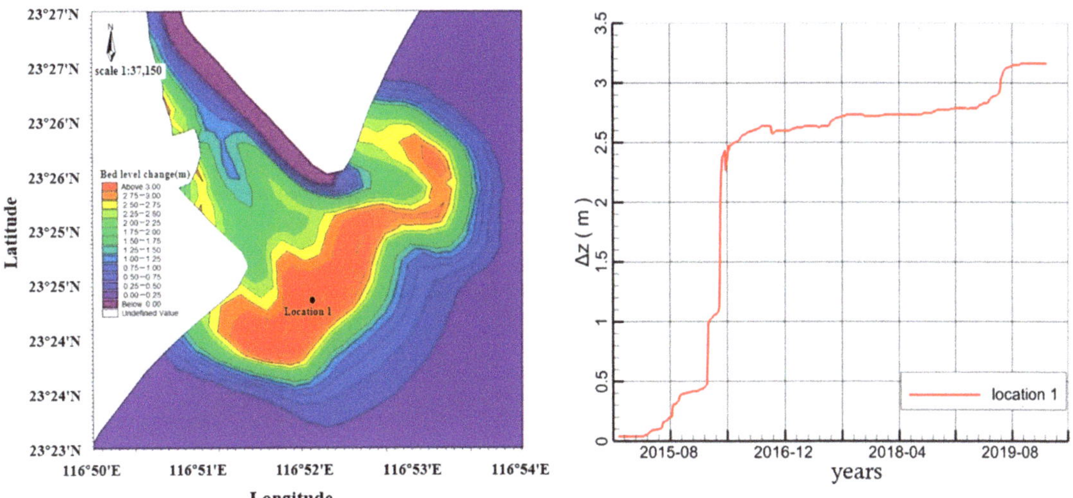

Figure 14. Changes of sedimentation thickness of Waishahe in 5 years.

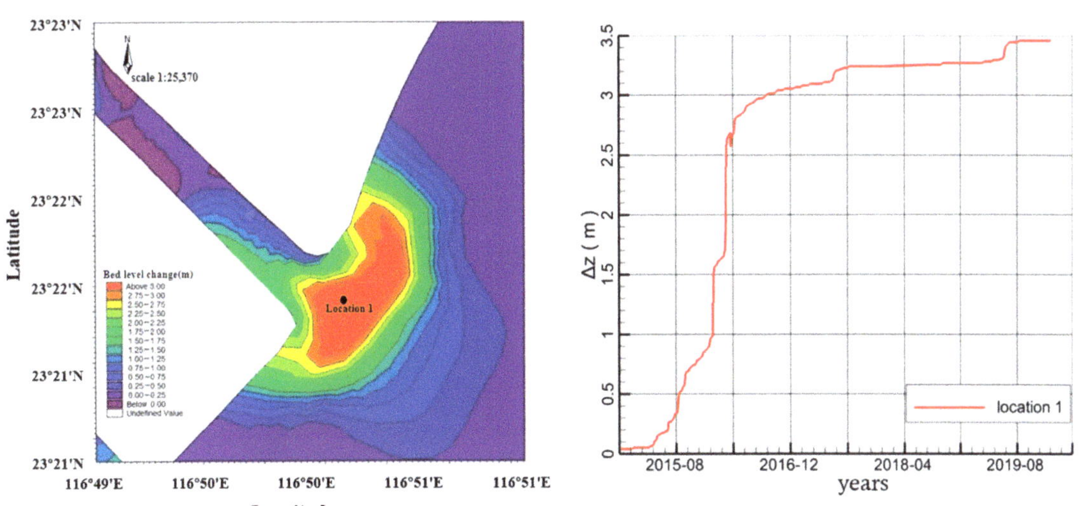

Figure 15. Changes of sedimentation thickness in Hanjiang Xixi Estuary in 5 years.

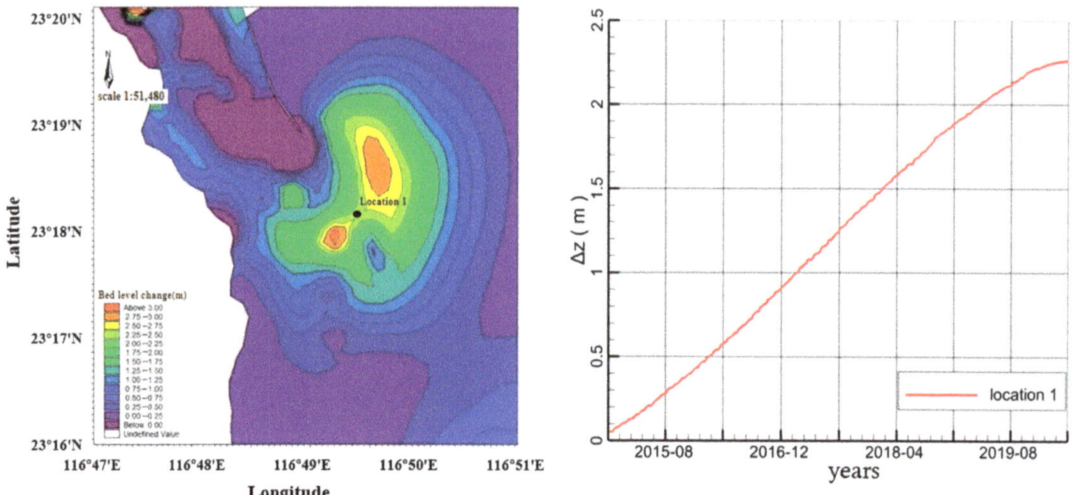

Figure 16. Changes of sedimentation thickness in Rongjiang Estuary in 5 years.

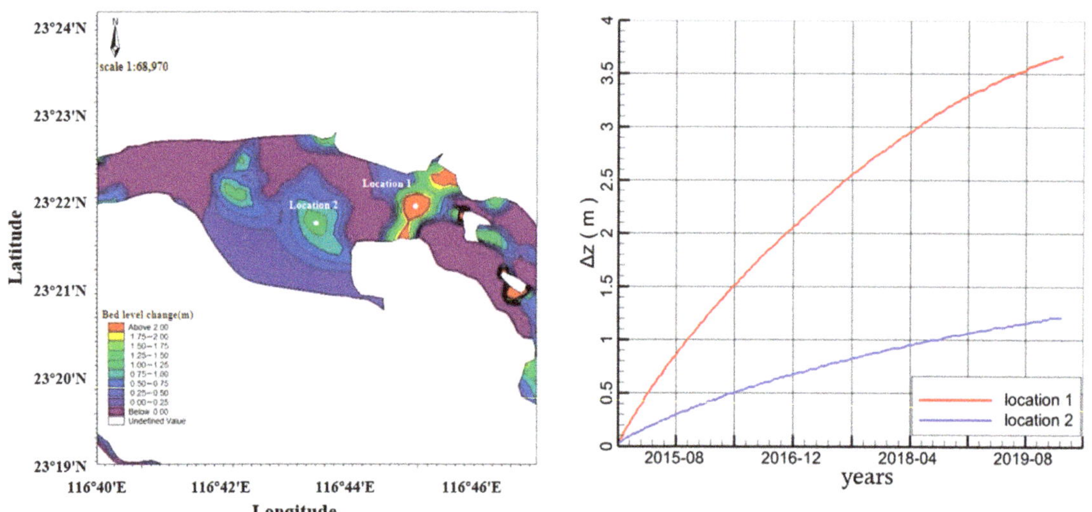

Figure 17. Changes of sedimentation thickness in Shantou Bay in 5 years.

6.2. Discuss the Causes of Siltation in Shantou Bay

Although the siltation range of Shantou Bay is not extensive, the siltation by sea survey will block the river, affect the regular shipping operation, and is not conducive to Shantou's foreign economy and trade. Therefore, it is of far-reaching significance to analyze the causes of siltation in Shantou Bay for the future development of Shantou. Figure 17 shows that the siltation in Shantou Bay is controlled by the tide, which is the decisive factor, but the source of siltation needs further discussion. When rivers flood, the water flow is large, and the sediment content is significant, so the siltation in the study area changes significantly at this time. Therefore, this paper analyzes the influence of rivers on sediment deposition in Shantou Bay from the changes of sediment bodies in Shantou Bay under the historical flood discharge time of each river.

In terms of sediment sources, the sediment sources of Shantou Port mainly include three parts:

- Sediment directly from the Rongjiang River;
- Sediment from the Hanjiang River through Meixi;
- Sediment from the ocean under the action of the tide.

Therefore, the historical flood discharge time of the Rongjiang River and Hanjiang Rivers is selected to analyze the siltation areas of Shantou Bay one by one.

For the Rongjiang River, a typical sluice opening and discharging process of the Rongjiang River from 15 August 2016 to 20 August 2016 was selected for analysis. Figure 18 shows the thickness of the siltation body before and after the flood peak here. It can be seen from Figure 18 that after the large-scale flood discharge in the upper reaches of the Rongjiang River, and the maximum siltation area near the inland riverside has increased to some extent. In contrast, the siltation near the seaside has no noticeable change. This shows that the sediment brought by the flood discharge of the Rongjiang River settled before it entered the estuary. It can be judged that Rongjiang's contribution to Shantou Port's siltation will be more negligible if it is not during flood discharge. The siltation contribution of Rongjiang River to Shantou Port mainly occurred during the sluice opening in the upper reaches of Rongjiang River, which mainly acted on the silts near the inland river but made little contribution to the silts near the sea.

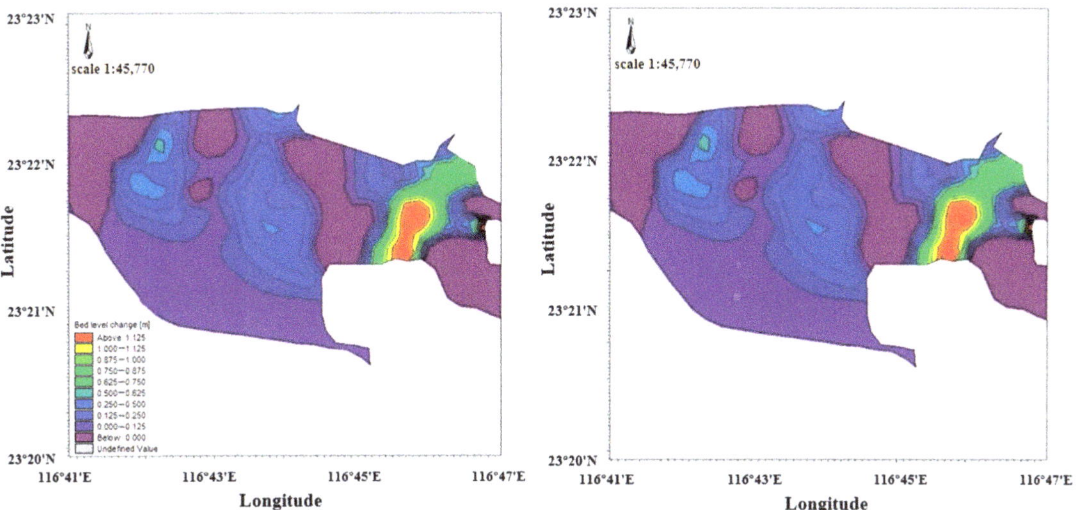

Figure 18. Changes of silts in Shantou Port before and after flood discharge of Rongjiang River.

The flood discharge of the Hanjiang River and Rongjiang River should be around the 20th of that month to avoid the difference in tidal conditions. Figure 19 shows the comparison of sediment deposition in Shantou Port before and after the flood discharge of Hanjiang River, which is 0:00 on 18 October 2016 and 23:00 on 23 October 2016, respectively. During this period, there is no other flood discharge, and it can be seen that the maximum sedimentation area of the whole sediment body has expanded. Meixi connects Han River and Shantou Port, so Meixi, which directly affects sediment deposition in Shantou Port, is selected for analysis. According to the inspection, the flood discharge of Hanjiang River can reach 670 m^3/s, and that of Meixi River is only about 5%, that is, less than 40 m^3/s, which is less than 20% of that of Rongjiang River. The discharge of the Meixi River is much smaller than that of the Rongjiang River, but the degree of siltation after flood discharge is greater. Therefore, it can be judged that the Hanjiang River has contributed to the siltation of Shantou Port.

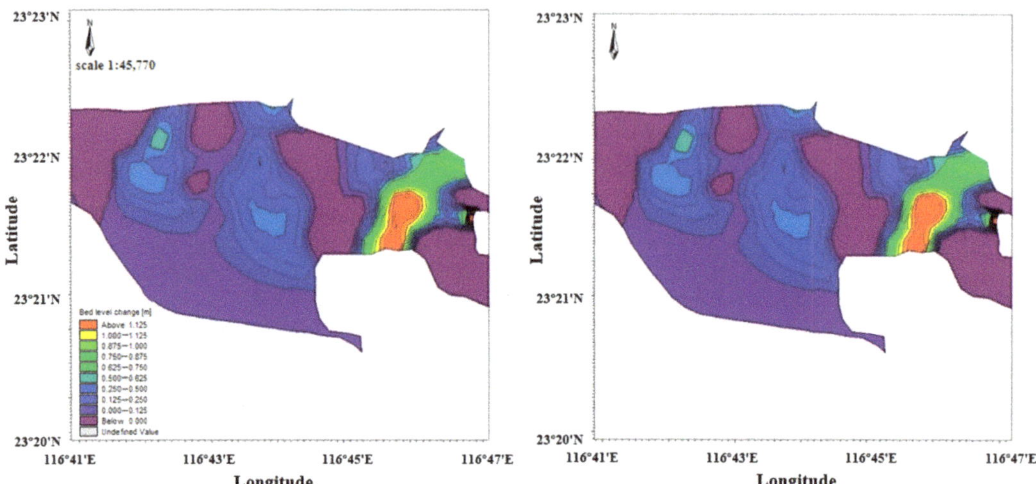

Figure 19. Changes of silts in Shantou Port before and after flood discharge of Hanjiang River.

To further discuss the contribution and deposition mechanism of Hanjiang River to Shantou Port, the siltation of Shantou Port before and after the removal of Hanjiang River sediment in one year, and the siltation at Meixi before and after the flood discharge of Hanjiang River in October 2016 were simulated.

Figure 20 to remove the siltation of Shantou Port before and after one year's sediment load of Hanjiang River, the left figure shows the actual siltation situation. The correct figure shows that the sediment load of Hanjiang River is set to 0, and other conditions are consistent to compare the siltation situation of Shantou Port in one year. The sediment deposition in Shantou Port on the right will be reduced to a certain extent, and the sedimentation in the whole area is relatively slow. However, the overall trend has mostly stayed the same. Therefore, it can be concluded that the Hanjiang River contributes to the siltation of Shantou Port, but it is not the decisive factor.

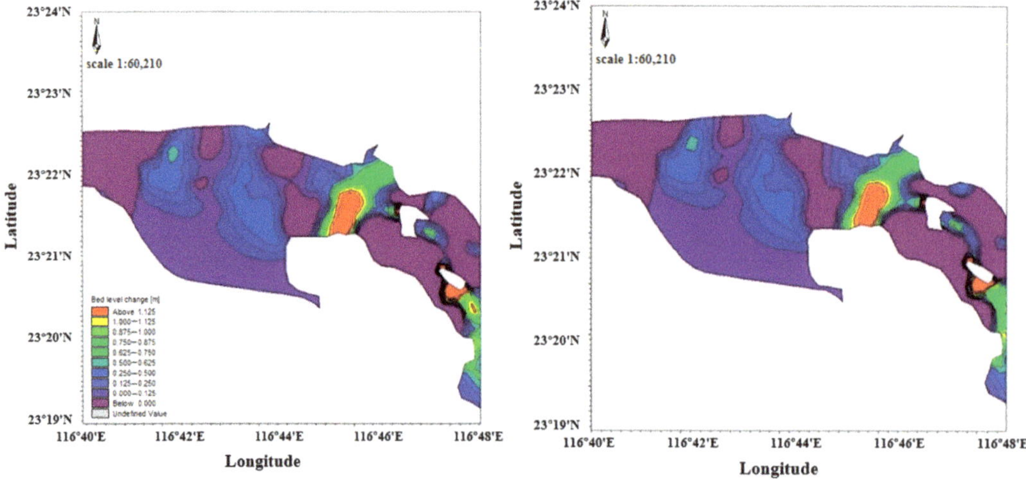

Figure 20. Comparison of Hanjiang River's contribution to Shantou Port's siltation. The left picture shows the real state, and the right picture shows that Hanjiang River's silt is set to 0.

Figure 21 shows the change in sediment deposition in the Meixi Estuary before and after the flood discharge. Before the flood discharge in the left picture, it can be seen that there are two prominent silting areas in the Meixi River when it enters the Rongjiang River. In comparison, the silting thickness in the two areas in the right picture has increased after the flood discharge. According to the hydrodynamic conditions of the Hanjiang River and Rongjiang River, the discharge and velocity of the Hanjiang River are much higher. Combined with Figure 21, it can be concluded that the sediment from the Hanjiang River settles after entering the Rongjiang River through the Meixi River, resulting in a deposition at Meixi Estuary. At the same time, the rest continues to be transported downstream. Most of the sediment brought by the single flood discharge of the Hanjiang River settled at the mouth of the Meixi River, and a small part settled in the Rongjiang River.

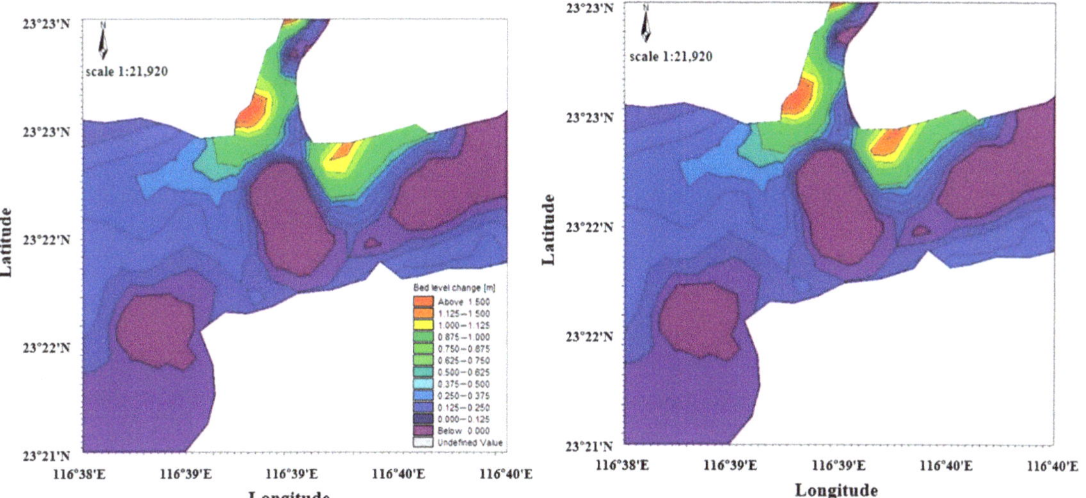

Figure 21. Sediment deposition at Meixi before and after flood discharge.

To sum up, the causes and laws of siltation in Shantou Port are summarized. Sediment sources in Shantou Port include the Rongjiang River, Hanjiang River, and tidal current. Under normal conditions, the Rongjiang River has a small flow rate, slow flow rate, and low sediment load, which does not affect the siltation of Shantou Port. During flood discharge, the sediment of the Rongjiang River mainly settles on the inland river side of Shantou Port. The Hanjiang River has a significant flow rate, a fast flow rate, and a high sediment content, and flows into the Rongjiang River through the Meixi River. However, most of the sediment of the Hanjiang River settles and accumulates at the mouth of the Meixi River after entering the Rongjiang River. Only part of it enters Shantou Port with the Rongjiang River, which contributes to the sedimentation of both parts of Shantou Port, but there are other factors besides this. The decisive factor of sediment deposition in Shantou Port is the sediment entering the port through ocean tides.

7. Conclusions

The main conclusions of this paper are as follows:

(1) For a typhoon with a 10-year return period, the embankment can effectively block a large amount of seawater in the ocean and prevent seawater from entering the land. At the same time, when the storm surge is over, the seawater entering the ground will not flow back to the ocean smoothly, and water will accumulate at the bank, quickly leading to the aggravation of the post-disaster cleaning task.

(2) Due to the effect of the embankment, the offshore water level will rise by an additional 0.4 m, resulting in an overall increase of the seawater velocity by 0.16 m/s.
(3) The main siltation areas of Shantou are located in various estuaries and Shantou Port Channel. The siltation of the Xixi Estuary and Waisha Estuary of Hanjiang River is controlled by runoff, while tides prevent the siltation of Shantou Port Channel and Rongjiang Estuary.
(4) The decisive factor of sediment deposition in Shantou Port is the sediment entering the port through ocean tides. During the flood discharge of the Rongjiang River, most of the residue is deposited on the inland river side of Shantou Port. During the flood discharge of the Hanjiang River, Meixi played a role in the siltation of Shantou Port, contributing to the siltation of two parts of Shantou Port, and Hanjiang River contributed more to the siltation of Shantou Port than Rongjiang River.

Author Contributions: Data curation, E.Z. and W.W.; formal analysis, Y.W. (Yuxi Wu) and E.Z.; funding acquisition, Y.W. (Yang Wang) and R.J.; investigation, J.Y., K.Z., H.W. and H.Q.; project administration, Y.W. (Yang Wang) and Z.S.; writing—original draft, Y.W. (Yuxi Wu); writing—review and editing, K.Z. All authors have read and agreed to the published version of the manuscript.

Funding: This research was funded by National Natural Science Foundation of China (Grant No. 52001286; 52101332; 52202427), National Key Research and Development Program of China (Grant No. 2021YFC3101800), Shenzhen Science and Technology Program (Grant No. KCXFZ20211020164015024) and Shenzhen Fundamental Research Program (Grant No. JCYJ20200109110220482).

Institutional Review Board Statement: Not applicable.

Informed Consent Statement: Not applicable.

Data Availability Statement: Wind field data are from the European Centre for Medium-Range Weather Forecasts. Here is the link: https://cds.climate.copernicus.eu (accessed on 11 July 2021).

Acknowledgments: European Centre for Medium-Range Weather Forecasts; the China Meteorological Administration.

Conflicts of Interest: The authors declare no conflict of interest. The funders had no role in the design of the study; in the collection, analyses, or interpretation of data; in the writing of the manuscript, or in the decision to publish the results.

References

1. Hou, Y.; Yin, B.; Guan, C.; Guo, M.; Liu, G.; Hu, P. Research progress and prospect of marine dynamic disasters in China. *Oceanol. Limnol. Sin.* **2020**, *51*, 759–767.
2. Hu, X.; Yu, H.; Yu, M. Causes and comparative analysis of offshore strengthening of Typhoon Hagupit. *Mod. Agric. Sci. Technol.* **2021**, *9*, 187–190.
3. Shan, C. Response and enlightenment of typhoon "Tian Ge". *Labor Prot.* **2019**, *9*, 39–41.
4. Luo, W.; Song, L.; He, R.; Li, J. Investigation report on urban complex disasters caused by typhoon Mangosteen. *Zhejiang Water Conserv. Sci. Technol.* **2019**, *47*, 16–18.
5. Feng, S. *Introduction to Storm Surge*; Science Press: Beijing, China, 1982.
6. Cai, W. Characteristics of storm surge along Shantou coast. *Sci. Technol. Commun.* **2013**, *5*, 113+111.
7. Bian, J. Numerical Simulation of Coastal Typhoon Storm Surge and Extreme Value Analysis of Water Increase. Master's Thesis, Yangzhou University, Yangzhou, China, 2019.
8. Shen, Y. Numerical Simulation of Storm Surge in Bohai Sea. Master's Thesis, Dalian University of Technology, Dalian, China, 2018.
9. Wang, Y.; Mao, X.; Jiang, W. Long-term hazard analysis of destructive storm surges using the ADCIRC-SWAN model: A case study of Bohai Sea, China. *Int. J. Appl. Earth Obs. Geoinf.* **2018**, *73*, 52–62. [CrossRef]
10. Suh, S.W.; Lee, H.Y.; Kim, H.J.; Fleming, J.G. An efficient early warning system for typhoon storm surge based on time-varying advisories by coupled ADCIRC and SWAN. *Ocean. Dyn.* **2015**, *65*, 617–646. [CrossRef]
11. Zaman, S.; Mondal, M.S. Risk-based determination of polder height against storm surge hazard in the south-west coastal area of Bangladesh. *Prog. Disaster Sci.* **2020**, *8*, 100131. [CrossRef]
12. Jelesnianshi, C.P.; Chen, J.; Wilson, A. *SLOSH: Sea, Lake and Overland Surges from Hurricanes*; NOAA Technical Report; NWS: Miami, FL, USA, 1992; Volume 48, p. 711992.
13. Watson, C.; Johnson, E. Design, implementation and operation of a modular integrated tropical cyclone hazard model: Caribbean disaster mitigation project. In Proceedings of the 23rd Conference on Hurricanes and Tropical Meteorology, Dallas, TX, USA, 10–15 January 1999.

14. Shu, G. Green seawall blueprint for preventing future storm surge disasters. *Science* **2020**, *106*, 14–16.
15. Dai, L.; Zhang, Y. Yancheng Erosion Coastal Protection Design. *Prog. Water Conserv. Hydropower Sci. Technol.* **2009**, *3*, 50–53.
16. Wang, C. *Research on the Construction of Coastal Protection Embankment in Weifang Binhai Development Zone*; Shandong University: Jinan, China, 2013.
17. Shi, S.; Yang, B.; Jiang, W. Numerical simulations of compound flooding caused by storm surge and heavy rain with the presence of urban drainage system, coastal dam and tide gates: A case study of Xiangshan, China. *Coast. Eng.* **2022**, *172*, 104064. [CrossRef]
18. Ma, G.; Han, Y.; Niroomandi, A.; Lou, S.; Liu, S. Numerical study of sediment transport on a tidal flat with a patch of vegetation. *Ocean. Dyn.* **2015**, *65*, 203–222. [CrossRef]
19. Arslan, M.; JingCheng, H.; Wajid, I.M.; Ali, S.A.; Muhammad, A.; Maryam, Y. Impact of Sediment Deposition on Flood Carrying Capacity of an Alluvial Channel: A Case Study of the Lower Indus Basin. *Water* **2022**, *14*, 3321.
20. Zhao, E.J.; Dong, Y.K.; Tang, Y.Z.; Cui, L. Numerical study on hydrodynamic load and vibration of pipeline exerted by submarine debris flow. *Ocean. Eng.* **2021**, *239*, 109754. [CrossRef]
21. Zhao, E.J.; Shi, B.; Qu, K.; Dong, W.B.; Zhang, J. Experimental and Numerical Investigation of Local Scour around Submarine Piggyback Pipeline under Steady Currents. *J. Ocean. Univ. China* **2018**, *17*, 244–256. [CrossRef]
22. Zhao, E.J.; Qu, K.; Mu, L. Numerical study of morphological response of the sandy bed after tsunami-like wave overtopping an impermeable seawall. *Ocean. Eng.* **2019**, *186*, 106076. [CrossRef]
23. Zhao, E.J.; Dong, Y.K.; Tang, Y.Z.; Sun, J.K. Numerical investigation of hydrodynamic characteristics and local scour mechanism around submarine pipelines under joint effect of solitary waves and currents. *Ocean. Eng.* **2021**, *222*, 108553. [CrossRef]
24. Qi, Y. *Numerical Simulation Analysis of Sediment Deposition in Yazidung Reservoir and Study on Sediment Regulation Scheme*; Xi'an University of Technology: Xi'an, China, 2019.
25. Christelle, A.; Jean-Roch, N.; Philip, M.; Irene, P.; Remo, C. Modelling the influence of Tidal Energy Converters on sediment dynamics in Banks Strait, Tasmania. *Renew. Energy* **2022**, *88*, 1105–1119.
26. Yuan, S.; Yuan, P.; Si, X.; Tan, J.; Wang, S.; Liu, X. Numerical simulation of tidal current in Bohai Sea and Yellow Sea based on FVCOM. *Bull. Ocean. Lakes* **2020**, *2*, 10–18.
27. Chu, L.; Ceng, Z. Sedimentation characteristics and development trend of Shantou Port. *Trop. Geogr.* **1983**, *3*, 1–7.
28. Huang, L. Summary of Sediment Source and Phase I Regulation Project of Shantou Port Outer Channel. *Waterw. Eng.* **2001**, *7*, 55–57.
29. Chen, C.; Beardsley, R.; Cowles, G. An unstructured grid, finite-volume coastal ocean model (FVCOM) system. *Oceanography* **2006**, *9*, 78–89. [CrossRef]
30. Li, R. *Numerical Solution of Partial Differential Equations*; Higher Education Press: Beijing, China, 2005.
31. Graham, H.E.; Nunn, D.E. *National Hurricane Research Project, Report No.33: Meteorological Conditions Pertinent to Standard Project Hurricane Atlantic and Gulf Coasts of the United States*; US Department of Commerce and Weather Bureau: Washington, DC, USA, 1959.
32. Tang, J.; Shi, J.; Li, X.; Deng, B.; Jin, M. Numerical simulation of typhoon waves based on typhoon wind field model. *Ocean. Lake Bull.* **2013**, *2*, 24–30.
33. Jiang, Z.; Hua, F.; Qu, P. A new tropical cyclone parameter adjustment scheme. *Adv. Mar. Sci.* **2008**, *1*, 1–7.
34. Willoughby, H.E.; Rahn, M.E. Parametric representation of the primary hurricane vortex. Part I: Observations and evaluation of the Holland (1980) model. *Mon. Weather Rev.* **2004**, *132*, 3033–3048. [CrossRef]
35. Atkinson, G.D.; Holliday, C.R. Tropical cyclone minimum sea level pressure/maximum sustained wind relationship for the western North Pacific. *Mon. Weather Rev.* **1977**, *105*, 421–427. [CrossRef]
36. Tang, Y.; Wang, Y.; Zhao, E.; Yi, J.; Feng, K.; Wang, H.; Wang, W. Study on Hydrodynamic Characteristics and Environmental Response in Shantou Offshore Area. *J. Mar. Sci. Eng.* **2021**, *9*, 912. [CrossRef]
37. Gao, J.; Ma, X.; Dong, G.; Chen, H.; Liu, Q.; Zang, J. Investigation on the effects of Bragg reflection on harbor oscillations. *Coast. Eng.* **2021**, *170*, 103977. [CrossRef]
38. Gao, J.; Ma, X.; Zang, J.; Dong, G.; Ma, X.; Zhu, Y.; Zhou, L. Numerical investigation of harbor oscillations induced by focused transient wave groups. *Coast. Eng.* **2020**, *158*, 103670. [CrossRef]

MDPI

Article

Motion Control of Autonomous Underwater Helicopter Based on Linear Active Disturbance Rejection Control with Tracking Differentiator

Haoda Li [1], Xinyu An [1], Rendong Feng [1,2] and Ying Chen [1,*]

[1] Ocean College, Zhejiang University, Zhoushan 316021, China
[2] Hainan Institute, Zhejiang University, Sanya 572025, China
[*] Correspondence: ychen@zju.edu.cn

Abstract: As a new disk-shaped autonomous underwater vehicle (AUV), the autonomous underwater helicopter (AUH) is devoted to subsea operations, usually diving into the seabed and docking with a subsea docking system. Due to the motion control's performance, the AUH's stability and steady-state accuracy are affected remarkably while docking. Moreover, considering the difficulties of hydrodynamic modeling of AUHs, the classical model-based control method is unsuitable for AUHs. Moreover, there is a large gap between the hydrodynamic simulation results and real situations. Hence, based on the data-driven principle, the linear active disturbance rejection control with a tracking differentiator (LADRC-TD) algorithm is employed for AUH depths and heading control. As the simulation experiments prove, LADRC and LADRC-TD have better anti-interference performance when compared with PID. According to the pool experiments, overshoots of the LADRC-TD are 20 cm and 3° for the depth control and heading control, respectively, which are superior to PID and LADRC. Meanwhile, the steady-state accuracy of the LADRC-TD is ±21 cm and ±2.5° for the depth and heading control, respectively, which is inferior to PID and the same as LADRC.

Keywords: autonomous underwater helicopter; active disturbance rejection control; attitude control; depth control

Citation: Li, H.; An, X.; Feng, R.; Chen, Y. Motion Control of Autonomous Underwater Helicopter Based on Linear Active Disturbance Rejection Control with Tracking Differentiator. *Appl. Sci.* **2023**, *13*, 3836. https://doi.org/10.3390/app13063836

Academic Editors: Enjin Zhao, Hao Qin and Lin Mu

Received: 2 February 2023
Revised: 28 February 2023
Accepted: 13 March 2023
Published: 17 March 2023

1. Introduction

Most of the marine resources, such as deep-sea manganese nodules, cobalt-rich hulls, and hydrothermal sulfides, exist on the seabed, so seabed development and exploration are one application trend of underwater robots. As a new disk-shaped autonomous underwater vehicle (AUV), the autonomous underwater helicopter (AUH) is dedicated to subsea operations, with many advantages, such as seabed residence, fixed-point hovering, high maneuverability, and anti-flow stability [1,2]. As to the work pattern (Figure 1), the AUH generally flies from one subsea docking system (SDS) to another, completing tasks that can include communications, equipment maintenance, and charging.

When the AUH needs to upload data to the sea surface or be charged, it approaches the SDS at a low speed [3], and the docking is then performed through the acoustic-optical guiding system. The controller's performance of the AUH significantly affects its stability and success rate of docking onto an SDS.

As most AUVs have dynamic characteristics, such as tight coupling, strong nonlinear effects, and parameter uncertainty, there remains a difficulty in obtaining an accurate dynamic model. When considering the uncertainty of model parameters and the influence of unknown external perturbation, robust control methods, including the back-stepping and sliding mode controls, are adopted in AUV motion control. With the three-dimensional path following the error model being established based on the virtual guidance method, Wang [4] proposed a path following a robust control system using command-filtered back-stepping control, neural networks, and adaptive control techniques. Moreover, the authors from [5] proposed a back-stepping controller for tracking the desired depth of the AUV's system, which lies on a vertical surface. At present, sliding mode control [6,7] has been widely studied and applied due to its simple design and easy implementation. Zhang [8] designed a terminal sliding mode variable structure control system with fast convergence speeds, high control precision, and strong robustness. Kantapon [9] developed a sliding mode heading control system for over-actuated, hover-capable AUVs that operate over a range of forward speeds. The proposed control system was also proven in field trials to enhance robust vehicle performance, even when subjected to external disturbances.

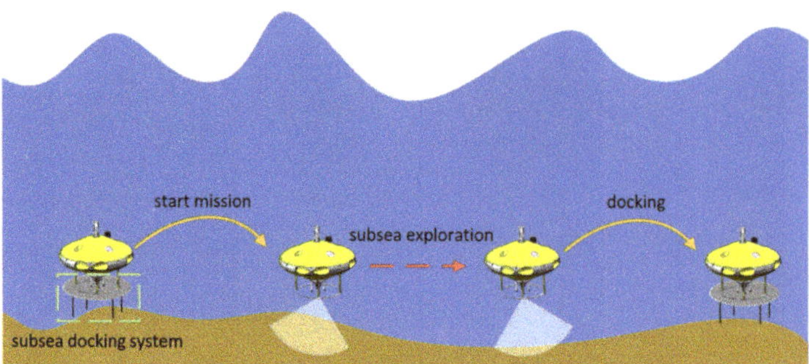

Figure 1. Work pattern of AUH.

The hydrodynamic performance of an autonomous underwater vehicle (AUH) can be greatly affected by various factors, including the flow channel, external communication mechanism, external sensors, and other components. Although many hydrodynamic simulations of AUHs use simplified models [10–12], these models may not accurately capture the complex interactions between the vehicle and its environment. There is a large gap between the hydrodynamic simulation based on computational fluid dynamics (CFD) technology and the real situation; thus, it is unsuitable to use classical model-based control methods for the accurate control of AUHs [13,14]. Currently, researchers have gradually applied data-driven control methods to underwater vehicle control, such as model-free adaptive control (MFAC) [15], active disturbance rejection control (ADRC) [16], and fuzzy control [17] due to their advantages of having simple parameter adjustments and no model parameter limitations. Strictly speaking, the motion control of the AUH is a nonlinear system, and the system parameters will vary with the working depth and cruising speed. The authors currently use a neuroadaptive learning algorithm [18] and two new integral robust algorithms to deal with the nonlinear system with disturbance [19]. Furthermore, the authors introduced a global differentiator based on higher-order sliding modes (HOSM) and dynamic gains to solve the trajectory tracking problem [20] and proposed an improved active disturbance rejection control method for the output tracking of uncertain plants with unknown control directions [21].

For the docking of AUHs, its accurate low-speed motion control is the basis. Considering the requirements of this task, more attention should be paid to the steady-state accuracy, overshoot, and anti-interference of AUHs. In view of the excellent anti-interference performance of ADRC, the authors plan to use the linear active disturbance rejection control with tracking differentiator (LADRC-TD) algorithm in this paper, aiming at realizing AUH motion control at a low speed. According to practical experience, the pitch angle does not show a dramatic change at low speeds. As exhibited in Figure 2, the pitch angle varies within $\pm 5^\circ$ when the AUH is moving at a speed of 0.2 m/s. Moreover, the simulation and pool experiments were conducted to verify the practicability and anti-interference performance of the LADRC-TD algorithm. This work's main contribution is the application of the LADRC-TD to the heading and depth control of the AUH—a new disk-type AUV—and demonstrating its superiority over LADRC and PID through simulations and pool experiments.

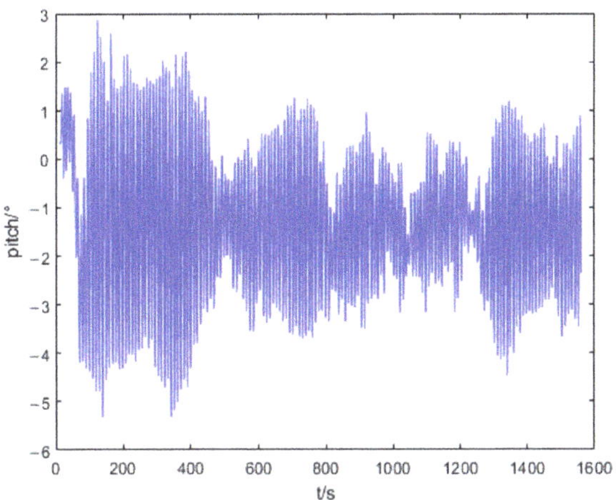

Figure 2. Pitch variation of the AUH at 0.2 m/s.

Section 2 introduces the basic physical parameters and components of the Z-AUH, the software and hardware models of the control system, and the AUH's dynamics and kinematics models. In Section 3, the algorithm principle and the controller design of the LADRC-TD are described. In Section 4, the effects of the PID, LADRC, and LADRC-TD algorithms are compared by using simulation experiments, and disturbing force and torque are added to the depth control and heading control, respectively. In Section 5, the excellent performance of the LADRC-TD is highlighted by the fixed-point pool experiment and dynamic experiment and is mirrored by the simulation results. Furthermore, due to the fixed-point experiment, the steady-state accuracy and overshoot of the PID, LADRC, and LADRC-TD in the depth control and heading control are also obtained. Finally, the conclusions are presented in Section 6.

2. Overview and Modeling of the AUH

2.1. Overview of the AUH

Inspired by the undersea stingray, our team developed a disk-shaped underwater vehicle, namely the AUH, which can take off, land on the seafloor, and hover at any altitude [22]. The following Table 1 exhibits the physical parameters and components of the Z-AUH.

Table 1. The physical parameters and components of Z-AUH.

Mass	About 800 kg
Diameter	2 m
Net buoyancy	70 N
Propeller layout	Four propellers on the horizon, and two propellers on the vertical, as shown in Figure 3
Maximum Thrust	200 N per propeller
Component	Inertial measurement unit (IMU), depth altimeter, sonar, radio, Iridium, Beidou navigation system, buoyancy adjustment, optical camera, ultra-short baseline (USBL)

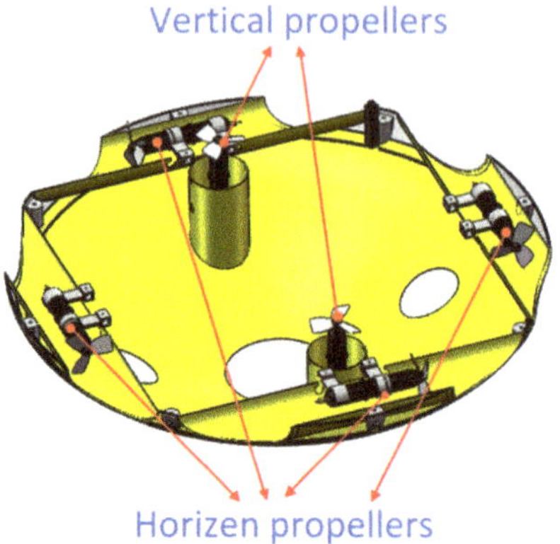

Figure 3. Propellers layout.

The control software and hardware structures of the AUH are presented in Figure 4. The hardware of the control system comprises four parts: propellers, depth altimeter, IMU, and the main control panel, ARM I. MX6Q. Among them, the propellers are the executive components of the control system. As exhibited in Figure 3, there are six propellers, including two vertical propellers and four horizontal propellers. The vertical propellers are mainly responsible for the vertical movement and the pitch angle of the AUH, while the horizontal propellers are responsible for the movement and heading on a horizontal plane. The depth altimeter and IMU take charge of the detection and transmission of the current state information of the depth, height, and attitude angle, respectively. Moreover, the depth altimeter and IMU follow the RS232 and RS485 protocols, respectively. The main control panel accepts the state information sent back by the IMU and depth altimeter, making decisions, calculating the thrust of each propeller timely by the LADRC algorithm, and giving instructions to the propellers through the CAN protocol. In terms of the software, the main control panel is equipped with MOOS, a commonly used underwater robot control system. As presented in this paper, four program modules comprising Depth Height, IMUServer, Marine LADRC, and CANDevice, are employed to manage the depth altimeter, IMU, LADRC-TD controller, and propellers, respectively, and the subscribe–publish mechanism of MOOS is used. By combining the aforementioned four modules with the MOOSDB module, the AUH motion control software architecture is built.

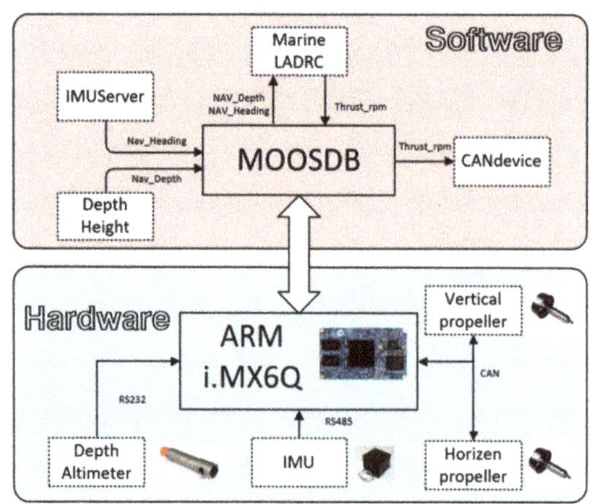

Figure 4. Control software and hardware structures of the AUH.

2.2. Modeling of AUH

In order to establish the simple dynamic and kinematic models of the AUH, the body-fixed coordinate system and the inertial coordinate system are defined. Figure 5 exhibits the coordinate system of the AUH mentioned in this paper.

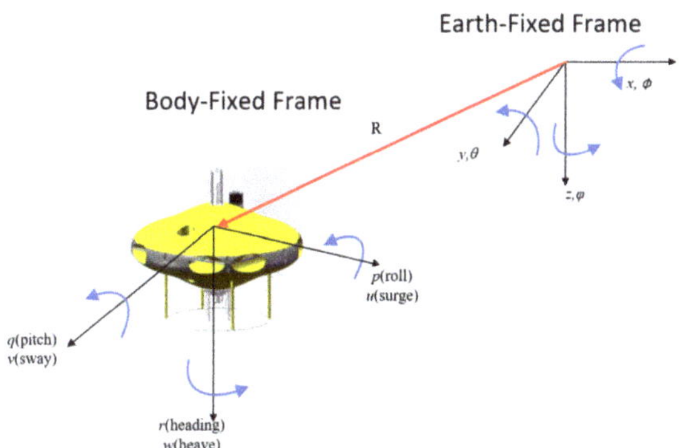

Figure 5. Coordinate system of AUH.

In Figure 5, the absolute linear position in the inertial coordinate system is $\xi = [x, y, z]^T$, and the angular position is $\boldsymbol{\eta} = [\phi, \theta, \varphi]^T$. In the body-fixed frame, the origin is deemed as the center of mass of the AUH. Moreover, in the ontological coordinate system, the linear velocity is defined as $\boldsymbol{V} = [u, v, w]^T$, and the angular velocity is $\boldsymbol{v} = [p, q, r]^T$.

As the depth and heading control of the AUH at low speeds are mainly considered in this paper, the simplified kinematic and dynamic models are as follows [7].

$$\begin{cases} \dot{X} = u\cos\varphi - v\sin\varphi \\ \dot{Y} = u\sin\varphi + v\cos\varphi \\ \dot{Z} = w \\ \dot{\Psi} = r \end{cases} \tag{1}$$

$$\begin{cases} \dot{w} = -\frac{Z_w}{m+Z_{\dot{w}}}w - \frac{Z_{|w|w}}{m+Z_{\dot{w}}}|w|w + \frac{1}{m+Z_{\dot{w}}}(\tau_Z + \tau_{env}) + \frac{W-B}{m+Z_{\dot{w}}} \\ \dot{\varphi} = r \\ \dot{r} = -\frac{N_r}{I_z+N_{\dot{r}}}r - \frac{N_{|r|r}}{I_z+N_{\dot{r}}}|r|r + \frac{1}{I_z+N_{\dot{r}}}(\tau_N + \tau_{env}) + \frac{Y_{\dot{v}}-X_{\dot{u}}}{I_z+N_{\dot{r}}}uv \end{cases} \tag{2}$$

where m refers to the mass of AUH; $Z_{\dot{w}}$ and $N_{\dot{r}}$ represent the hydrodynamic-added mass and hydrodynamic-added mass moment of inertia, respectively. Z_w and N_r are the linear damping coefficients; $Z_{|w|w}$ and $N_{|r|r}$ are the second-order damping coefficients. I_z denotes the moment of inertia about the z-axis; τ_Z and τ_N are the thrust of the vertical propellers and the horizontal propellers, respectively; τ_{env} refers to the environmental disturbing force. W represents gravity and B denotes buoyancy.

3. Design of AUH Controller

Han proposed ADRC [23], which involves disturbance estimation and compensation. The model uncertainty of the system can be considered as the internal disturbance, while the external disturbance is the total disturbance of the system. The nonlinear error feedback law includes a compensation component that estimates and compensates for the total disturbance in real time without distinguishing between the internal and external disturbances. This component simplifies the controlled object to a series of integrators, making it easy to construct an ideal controller. Essentially, this compensation component has an anti-interference effect. However, ADRC is initially in a nonlinear form and has numerous parameters, making parameter setting and implementation challenging in engineering.

LADRC [24] is an extension of ADRC, which uses a linear extended state observer (LESO) to estimate the state and disturbance of the system. It has a simpler structure than ADRC and is easier to analyze and set parameters.

3.1. LESO

The LESO is a linear observer that extends the state-space of the system to include the disturbance variables. It estimates the state and disturbance variables simultaneously, which improves the accuracy of the control system. LADRC has been applied in various fields, such as motor control, robotics, and power electronics. It is a promising control method for systems with uncertainties and disturbances.

In this paper, the controllers designed an all-adopt second-order LESO, and the same is true for the heading controller and depth controller. Hence, the design of the depth controller is taken as an example here.

First, ζ_1, u_1, b_1 are designed as below.

$$\begin{cases} \zeta_1 = \frac{Z_w w - Z_{|w|w}|w|w + \tau_{env} + W - B}{m + Z_{\dot{w}}} \\ u_1 = \tau_Z \\ b_1 = \frac{1}{m + Z_{\dot{w}}} \end{cases} \tag{3}$$

Afterwards, Equation (2) can be written as follows.

$$\ddot{w} = b_1 u_1 + \xi_1 \tag{4}$$

where, ζ_1 refers to the generalized total disturbance of the system, including internal uncertainty and unknown external disturbance. Later, a second-order state-space expression can be obtained by extending ζ_1 to a state variable x_3. In stable systems, it is generally assumed that the differential of x_3 is bounded and defined that $\dot{x}_3 = k$.

$$\begin{cases} y_1 = x_1 \\ \dot{x}_1 = x_2 \\ \dot{x}_2 = x_3 + b_1 u \\ \dot{x}_3 = k \end{cases} \tag{5}$$

In Equation (5), y_1 represents the AUH's depth in the current state, which can be detected by a depth altimeter. The observer [25] is established based on the above state-space equation.

$$\begin{cases} e = \hat{x}_1 - y_1 \\ \dot{\hat{x}}_1 = \hat{x}_2 - l_1 e \\ \dot{\hat{x}}_2 = \hat{x}_3 - l_2 e + b_1 u \\ \dot{\hat{x}}_3 = -l_3 e \end{cases} \tag{6}$$

In Equation (6), $\hat{x}_1$, $\hat{x}_2$, and $\hat{x}_3$ refer to the estimations of x_1, x_2, and x_3, respectively. l_1, l_2, and l_3 represent the gains of the observer. To sum up, Equation (6) is the second-order extended state observer used by the depth controller. Moreover, to simplify the tuning process, the observer gains are parameterized as [24]

$$L = [l_1, l_2, l_3]^T = [3w_0, 3{w_0}^2, {w_0}^3]^T \tag{7}$$

where w_0 refers to the bandwidth of the system.

The linear state error feedback (LSEF) control law is mainly used to generate actual control signals and improve control accuracy by combining with the compensation quantity of the disturbance estimation above. In this paper, the PD control law is adopted, and the output of each propeller is calculated [24].

$$\begin{cases} e_1 = \overline{v} - \hat{x}_1 \\ e_2 = \hat{x}_2 \\ u_0 = K_p e_1 - K_d e_2 \\ u = \frac{u_0 - \hat{x}_3}{b_1} \end{cases} \tag{8}$$

where $\overline{v}$ refers to the desired signal and u denotes the output of the controller.

Generally, LADRC is composed of LESO and LSEF. However, classical LADRC methods often exhibit overshoots when working with large spans, which can be addressed by incorporating a tracking differentiator (TD). In order to mitigate this issue, this paper applies signal processing techniques and a TD in the control system's input signal to enhance the transition process and improve control performance.

3.2. Tracking Differentiator

A tracking differentiator is a type of control algorithm used in control systems to estimate the derivative of the input signal of the system. It is a high-frequency filter that can accurately estimate the derivative of the input signal, even in the presence of noise and disturbances. The tracking differentiator can be used in various control applications, such as motion control, robotics, and aerospace systems. It is particularly useful when the input signal is noisy or when the system is subject to disturbances that can affect the accuracy of the derivative estimation.

In classical control theory, the differential signal for a given signal v is generally represented, as seen below [23].

$$y = w(s)v = \frac{s}{Ts+1}v = \frac{1}{T}(v - \frac{1}{Ts+1}) \tag{9}$$

Equation (8) can be written as follows.

$$w(s) = \frac{s}{\tau^2 s + 2\tau s + 1} = \frac{{r_a}^2}{(s + r_a)^2} s \tag{10}$$

Running it on a computer, we have to discretize Equation (9) [23].

$$\begin{cases} m_1(n) = m_1(n-1) + m_2(n-1)h \\ m_2(n) = m_2(n-1) + \left[{r_a}^2(m_1(n) - \overline{v}(n)) - 2r_a m_2(n-1)\right]h \end{cases} \tag{11}$$

where h refers to the sampling step size and r_a denotes the speed factor. The larger the r_a, the faster the signal tracking speed. By combining LADRC with TD, the output of the controller LADRC-TD is u'.

$$\begin{cases} e_1{}' = m_1 - \hat{x}_1 \\ e_2{}' = \hat{x}_2 - m_2 \\ u_0{}' = K_p \cdot e_1{}' - K_d \cdot e_2{}' \\ u' = \frac{u_0{}' - \hat{x}_3}{b_1} \end{cases} \tag{12}$$

In summary, the block diagram of the LADRC-TD algorithm can be obtained (Figure 6).

In the case of the AUH, the dynamics can be decoupled into vertical and horizontal movements, where the vertical propellers control the depth of the vehicle, and the horizontal propellers control the heading. By decoupling the dynamics, it becomes easier to design control systems that can effectively control the vehicle's motion in each direction. This approach helps to simplify the control system design and ensure that each subsystem is optimized for its specific task, resulting in improved overall control performance. In the control system described in the paper, the LADRC-TD is used for both the depth controller and the heading controller of the AUH. The control flow for the AUH can be summarized as follows (Figure 7).

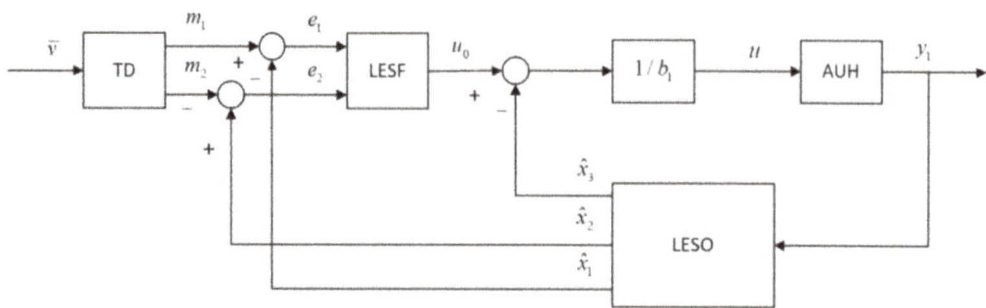

Figure 6. Block diagram of LADRC-TD.

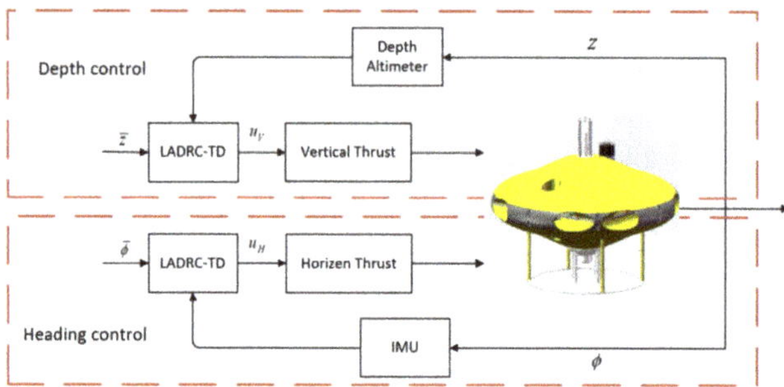

Figure 7. Control flow of the AUH.

4. Numerical Simulations

In the simulation experiments, the main objective was to evaluate the performance of the LADRC-TD algorithm for the heading and depth control of the AUH and to compare it with the LADRC and PID algorithms. The simulation conditions for all three algorithms were kept consistent to ensure the fairness of the experiments.

4.1. Heading Control Simulations

To reflect the anti-interference performance of the LADRC and LADRC-TD, 5 NM torque was added to the AUH as a disturbance, with a holding time of 4 s and a cycle of 20 s. In order to simulate the impact of the underwater turbulence on the AUH, a torque was applied that considered the asymmetry of the components outside the AUH shell, such as USBL and Iradia. However, it should be noted that the turbulence was not expected to be significant. Therefore, a torque with a reasonable amplitude of 5NM was selected for this simulation.

As presented in Figure 8a, a sinusoidal signal is used as the system input, aiming at testing the tracking performance of each algorithm. Moreover, PID exhibits large fluctuations in the heading tracking when under disturbance, while LADRC and LADRC-TD have small fluctuations, proving their certain anti-interference abilities. According to Figure 8b,c, both PID and LADRC show an overshoot under the step response. Although LADRC-TD has a slightly lower response speed, it shows no overshoot. Simultaneously, PID also exhibits an obvious fluctuation when under disturbance, and LADRC and LADRC-TD have similar performances, as shown in Figure 8a.

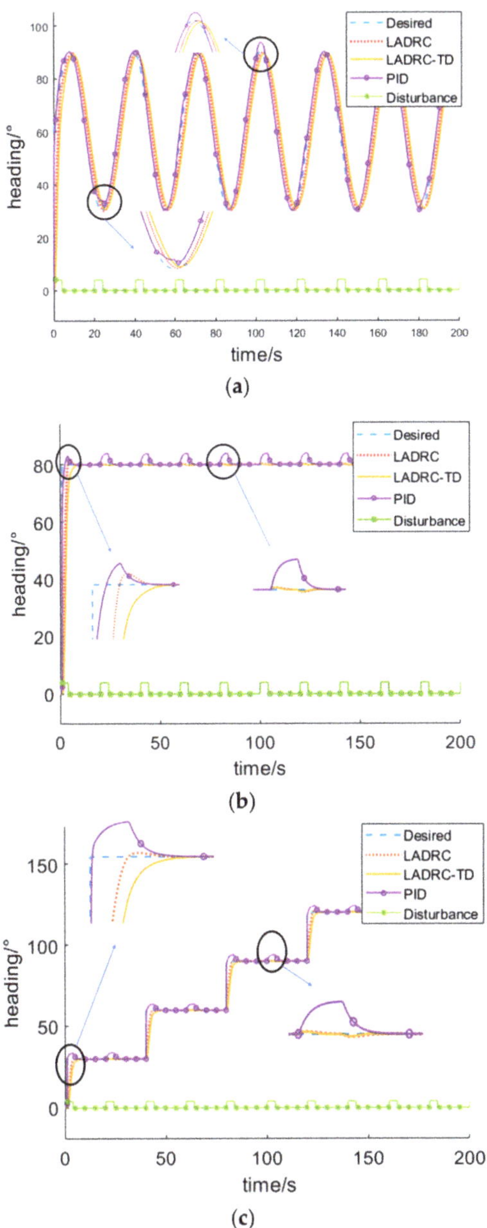

Figure 8. Simulations of heading control with PID, LADRC, and LADRC-TD; (**a**) heading tracking with the sinusoidal signal; (**b**) long-span-step response; (**c**) multi-step response.

4.2. Depth Control Simulations

Similar to the previous simulation, the AUH was interfered with by a 10 N co-directional force, with a holding time of 4 s and a cycle of 20 s. In practice, the depth control of the AUH is mostly a step response and therefore, the long-span-step and multi-step simulations are performed in this section (Figure 9).

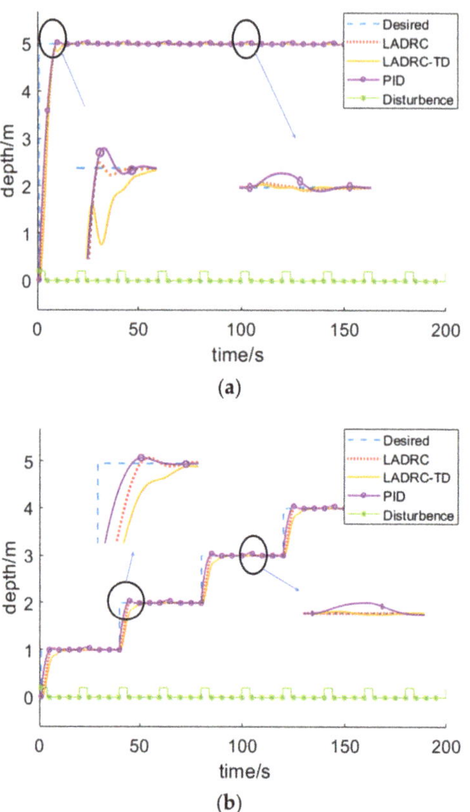

Figure 9. Simulations of depth control with PID, LADRC, and LADRC-TD. (**a**) Long-span-step response; (**b**) multi-step response.

Similar to the heading control simulation, PID and LADRC demonstrated an overshoot in the multi-step responses. Moreover, LADRC and LADRC-TD exhibit small fluctuations, while PID fluctuates wildly under disturbance. Although the response speed of LADRC-TD is slightly lower, it shows no overshoot.

Because of the limitation of the simulations, the accuracy advantage of the LADRC-TD algorithm in the heading control and depth control cannot be reflected when compared with other algorithms. However, the simulations demonstrate that the anti-interference performance of LADRC-TD and LADRC is superior to PID in heading control and depth control. Moreover, LADRC-TD performs better than LADRC under the step response, as LADRC-TD shows no overshoot phenomenon. Therefore, the LADRC-TD algorithm is employed by the AUH for depth control and heading control in practical applications. In the next section, the corresponding steady-state accuracy of each algorithm can be obtained by virtue of the pool experiments.

5. Experiments and Results

5.1. Pool Experiments

The experiments were conducted in a round pool with a diameter of 45 m and a depth of 6 m at the Ocean College of Zhejiang University (Figure 10). There were two experiments carried out: a fixed-point experiment and a dynamic experiment. During the fixed-point experiment, the AUH received signals for depth and heading adjustments and maintained its position in the horizontal plane while translating vertically. In the dynamic experiment, the AUH moved back and forth between two points at a low speed, and the outcomes of PID and LADRC-TD were compared.

Figure 10. The pool experiment of the AUH.

In the tuning process, K_p should be first tuned by increasing it gradually until an oscillation is observed. Then, K_d will be increased until a small overshoot is achieved. Following this, K_i is tuned to reduce steady-state accuracy. Finally, subtle adjustments are made to the parameters to optimize the dynamic response and steady-state performance. The tuning process for LADRC and LADRC-TD is similar. The bandwidth w_0 should be tuned first. In theory, a larger value of w_0 would result in better performance. However, the appropriate value of w_0 is dependent on factors such as the sampling frequency and the level of noise in the sensors. As the sampling frequency of the IMU and a depth altimeter are all set to 1 Hz, the optimal value of w_0 may be smaller than anticipated. Then, K_p and K_d should be tuned until the system is stable, which is similar to PID and finally, b_1 and r_a should be fine-tuned to optimize the performance.

As exhibited in Figure 11a, for the heading control experiment, overshoot exits in all PID, LADRC, and LADRC-TD controllers, with overshoot amounts of 12°, 6°, and 3°, respectively. The response time of LADRC is slightly shorter than that of LADRC-TD, conforming to the simulation results. However, the response time of PID is significantly longer than that of LADRC and LADRC-TD, which differs from what is observed in the simulation. In practice, to reduce the oscillations, the parameter K_p must be reduced, thus increasing the response time compared to the simulations. The steady-state accuracy of LADRC and LADRC-TD is about ±2.5°, and that of PID is ±3°.

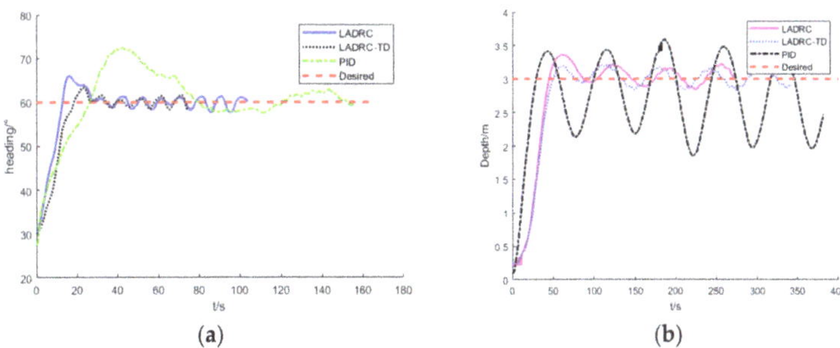

Figure 11. Fixed-point experiment; (**a**) step response in heading control; (**b**) step response in the depth control.

As for the depth control shown in Figure 11b, the overshoot of PID is 43 cm, and the steady-state accuracy is ±1.1 m. The steady-state accuracy of LADRC is similar to that of LADRC-TD, namely approximately ±21 cm. However, the overshoot of LADRC is 41 cm, and that of LADRC-TD is merely 20 cm around.

In the dynamic experiment, the AUH needs to go back and forth between two points by three times at 0.2 m/s, thereby simulating the docking process when the AUH is close to the SDS.

As observed from the comparison of the heading tracking of LADRC-TD and PID (Figure 12a,b), LADRC-TD has a better tracking performance, shorter response time, and lower overshoot. It is worth noting that the heading angle range is from 0 to 360 degrees, and when the angle turns from 300 to 380 degrees, it appears to be discontinuous (Figure 12b) because it first turns from 300 to 0 degrees and then from 0 to 20 degrees Due to the excellent heading tracking performance of the LADRC-TD, the three trajectories of the AUH show high similarity, as presented in Figure 12c. On the contrary, the three trajectories under PID control (Figure 12d) are chaotic because PID cannot accurately track the heading change in a timely manner. Although the trajectory of LADRC-TD is further away from the target trajectory than PID, this paper focuses on the heading tracking performance of different algorithms, which is different from the trajectory tracking.

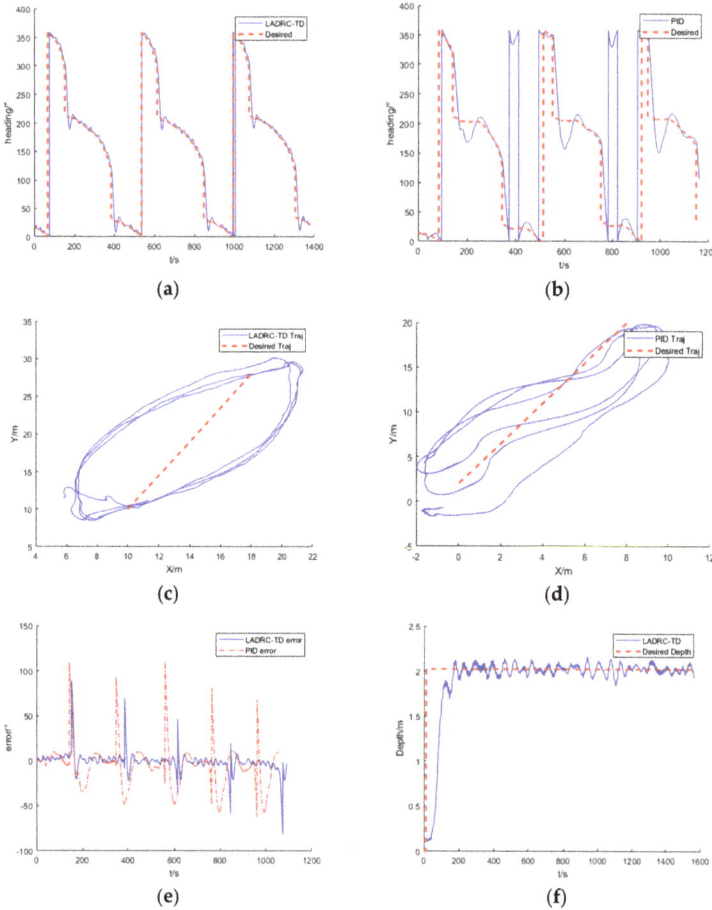

Figure 12. Dynamic experiment: (**a**) heading tracking with LADRC-TD; (**b**) heading tracking with PID; (**c**) trajectory when using LADRC-TD in heading control; (**d**) trajectory when using PID in heading control; (**e**) heading tracking error of PID and LADRC-TD in dynamic experiment; (**f**) depth-keeping with LADRC-TD during AUH proceeding.

As observed from the tracking error (Figure 12e), the PID heading tracking error is greater than LADRC-TD, the PID adjustment time is about 23 s longer than LADRC-TD, and the fluctuation of the adjustment process is more obvious when compared to LADRC-TD. During the movement of the AUH, the steady-state accuracy of depth control remains within ±21 cm (Figure 12f), demonstrating that the excellent anti-interference performance of LADRC-TD can meet the requirements of practical tasks.

5.2. Results

Based on the simulation and pool experiments, a comprehensive comparison is given to the control performances of PID, LADRC, and LADRC-TD in AUH depth control and heading control in this paper. Among them, PID has poor performance, a sizeable steady-state error in the depth and heading controls, and a worse anti-interference performance when compared to LADRC and LADRC-TD. LADRC demonstrates a performance between the LADRC-TD and PID. The steady-state accuracy and anti-interference of LADRC and LADRC-TD are almost the same. As LADRC-TD is added with tracking differentiation, the overshoot decreases, and the response time is slightly increased by approximately 2 s. However, more attention was paid to the accuracy, overshoot, and anti-interference at low speeds of the algorithms. As discovered, the response time of the LADRC-TD algorithm fully meets the task requirements. Table 2 summarizes the results of each algorithm in the simulations and the fixed-point experiment. In addition, the AUH performed PID and LADRC-TD in the dynamic experiment. As demonstrated, LADRC-TD has excellent dynamic depth-keeping performance and superior heading tracking performance compared to PID.

Table 2. Result summary of each algorithm.

Depth Control	**PID**	**LADRC**	**LADRC-TD**
Overshoot	43 cm	41 cm	20 cm
Anti-interference	worse	baseline	nearly
Steady-state error	±1.1 m	±21 cm	±21 cm
Heading Control	**PID**	**LADRC**	**LADRC-TD**
Overshoot	12°	6°	3°
Anti-interference	worse	baseline	nearly
Steady-state error	±3°	±2.5°	±2.5°

6. Conclusions

Aiming at the accurate motion control of an AUH at low speed, this paper focuses on overshoot, steady-state accuracy, and the anti-interference of an AUH's heading and depth control, which is the basis for docking the AUH onto an SDS. Moreover, based on the data-driven principle, the LADRC-TD algorithm was adopted to resolve the problem in this paper's hydrodynamic modeling of the AUH. By adding differential tracking to the classical LADRC algorithm, the excellent anti-interference performance of classical LADRC can be obtained, and the overshoot of the system can be reduced. According to the simulations and pool experiments, the following conclusions are drawn:

1. The LADRC-TD algorithm has the least overshoot, namely 20 cm and 3° in the depth control and heading control, respectively, which is less than PID and LADRC.
2. According to the simulations, the anti-interference of LADRC-TD is better than PID and nearly the same as LADRC.
3. The steady-state error of LADRC-TD is ±21 cm and ±2.5° in the depth control and heading control, respectively, which is lower than PID and the same as LADRC.

As demonstrated, LADRC-TD has a better control effect than PID and LADRC. The next step is to solve the motion control problem of an AUH at 1 m/s. In practice, the pitch angle of the AUH undergoes significant changes at such speeds, which can lead to instability and negatively impact its cruising capabilities. Therefore, it is crucial to develop effective motion control strategies to mitigate the adverse effects of pitch angle variations and ensure stable and efficient operation of the AUH at 1 m/s.

Author Contributions: Conceptualization, H.L., X.A. and Y.C.; methodology, H.L., X.A. and Y.C.; software, H.L., X.A. and R.F.; validation, H.L., X.A. and R.F.; formal analysis, H.L. and X.A.; investigation, R.F.; resources, Y.C.; data curation, H.L.; writing—original draft preparation, H.L.; writing—review and editing, X.A. and Y.C.; visualization, H.L.; supervision, R.F.; project administration, X.A. and Y.C. All authors have read and agreed to the published version of the manuscript.

Funding: The study was supported financially by the National Natural Science Foundation of China (Grant No. 52001279), the National Key R & D Program of China (NO. 2017YFC0306100).

Acknowledgments: The authors would like to thank Zhikun Wang and other Ocean College people at Zhejiang University for their inspiration and helping experiments.

Conflicts of Interest: The authors declare no conflict of interest.

References

1. Wang, Z.; Liu, X.; Huang, H.; Chen, Y. Development of an Autonomous Underwater Helicopter with High Maneuverability. *Appl. Sci.* **2019**, *9*, 4072. [CrossRef]
2. Peng, S.; Yang, C.; Fan, S.; Zhang, S.; Wang, P.; Xie, Y.; Chen, Y. A Hybrid Underwater Glider for Underwater Docking. In Proceedings of the MTS/IEEE OCEANS 2013, San Diego, CA, USA, 10–13 June 2013; Volume 7.
3. Cai, C.; Xu, B. Development and Test of a Subsea Docking System Applied to an Autonomous Underwater Helicopter. In Proceedings of the MTS/IEEE OCEANS 2022, Hampton Roads, VA, USA, 17–20 October 2022; Volume 7.
4. Wang, J.; Wang, C.; Wei, Y.; Zhang, C. Three-Dimensional Path Following of an Underactuated AUV Based on Neuro-Adaptive Command Filtered Backstepping Control. *IEEE Access* **2018**, *6*, 74355–74365. [CrossRef]
5. Gharesi, N.; Ebrahimi, Z.; Forouzandeh, A.; Arefi, M.M. Extended State Observer-Based Backstepping Control for Depth Tracking of the Underactuated AUV. In Proceedings of the 2017 5th International Conference on Control, Instrumentation, and Automation (ICCIA), Shiraz, Iran, 21–23 November 2017; pp. 354–358.
6. Mat-Noh, M.; Mohd-Mokhtar, R.; Arshad, M.R.; Zain, Z.M.; Khan, Q. Review of Sliding Mode Control Application in Autonomous Underwater Vehicles. *Indian J. Geo-Marine Sci.* **2019**, *48*, 973–984.
7. Du, P.; Yang, W.; Wang, Y.; Hu, R.; Chen, Y.; Huang, S.H. A Novel Adaptive Backstepping Sliding Mode Control for a Lightweight Autonomous Underwater Vehicle with Input Saturation. *Ocean Eng.* **2022**, *263*, 112362. [CrossRef]
8. Zhang, Y.; Gao, L.; Liu, W.; Li, L. Research on Control Method of AUV Terminal Sliding Mode Variable Structure. In Proceedings of the 2017 International Conference on Robotics and Automation Sciences, Hong Kong, China, 26–29 August 2017; pp. 88–93.
9. Tanakitkorn, K.; Phillips, A.B.; Wilson, P.A.; Turnock, S.R. Sliding Mode Heading Control of an Overactuated, Hover-Capable Autonomous Underwater Vehicle with Experimental Verification. *J. F. Robot.* **2018**, *35*, 396–415. [CrossRef]
10. Chen, C.W.; Jiang, Y.; Huang, H.C.; Ji, D.X.; Sun, G.Q.; Yu, Z.; Chen, Y. Computational Fluid Dynamics Study of the Motion Stability of an Autonomous Underwater Helicopter. *Ocean Eng.* **2017**, *143*, 227–239. [CrossRef]
11. An, X.; Chen, Y.; Huang, H. Parametric Design and Optimization of the Profile of Autonomous Underwater Helicopter Based on Nurbs. *J. Mar. Sci. Eng.* **2021**, *9*, 668. [CrossRef]
12. Lin, Y.; Lin, Y.; Guo, J.; Li, H.; Wang, Z.; Chen, Y.; Huang, H. Improvement of Hydrodynamic Performance of the Disk-Shaped Autonomous Underwater Helicopter by Local Shape Modification. *Ocean Eng.* **2022**, *260*, 112056. [CrossRef]
13. Du, P.; Huang, S.H.; Yang, W.; Wang, Y.; Wang, Z.; Hu, R.; Chen, Y. Design of a Disc-Shaped Autonomous Underwater Helicopter with Stable Fins. *J. Mar. Sci. Eng.* **2022**, *10*, 67. [CrossRef]
14. Liu, Z.; Zhou, J.; Wang, Z.; Zhou, H.; Chen, J.; Hu, X.; Chen, Y. Research on an Autonomous Underwater Helicopter with Less Thrusters. *J. Mar. Sci. Eng.* **2022**, *10*, 1444. [CrossRef]
15. Li, X.; Ren, C.; Ma, S.; Zhu, X. Compensated Model-Free Adaptive Tracking Control Scheme for Autonomous Underwater Vehicles via Extended State Observer. *Ocean Eng.* **2020**, *217*, 107976. [CrossRef]
16. Li, H.; He, B.; Yin, Q. Fuzzy Optimized MFAC Based on ADRC in AUV Heading Control. *Electronics* **2019**, *8*, 608. [CrossRef]
17. Wan, L.; Zhang, Y.; Sun, Y.; Li, Y. Thruster-AUV's Motion Adjustment Method Based on Fuzzy Control. In Proceedings of the 2015 International Industrial Informatics and Conputer Engineering Conference, Xi'an, China, 10–11 January 2015; pp. 266–269.
18. Yang, G.; Yao, J.; Dong, Z. Neuroadaptive Learning Algorithm for Constrained Nonlinear Systems with Disturbance Rejection. *Int. J. Robust Nonlinear Control.* **2022**, *32*, 6127–6147. [CrossRef]
19. Yang, G. Asymptotic Tracking with Novel Integral Robust Schemes for Mismatched Uncertain Nonlinear Systems. *Int. J. Robust Nonlinear Control.* **2023**, *33*, 1988–2002. [CrossRef]
20. Oliveira, T.R.; Estrada, A.; Fridman, L.M. Global and Exact HOSM Differentiator with Dynamic Gains for Output-Feedback Sliding Mode Control. *Automatica* **2017**, *81*, 156–163. [CrossRef]
21. Teixeira, A.; Gouvea, J.A.; Zachi, A.R.L.; Rodrigues, V.H.P.; Oliveira, T.R. Monitoring Function-based Active Disturbance Rejection Control for Uncertain Systems with Unknown Control Directions. *Adv. Control Appl.* **2021**, *3*, e66. [CrossRef]

22. Zhou, J.; Huang, H.; Huang, S.H.; Si, Y.; Shi, K.; Quan, X.; Guo, C.; Chen, C.; Wang, Z.; Wang, Y.; et al. AUH, a New Technology for Ocean Exploration. *Engineering* **2022**. [CrossRef]
23. Han, J. From PID to Active Disturbance Rejection Control. *IEEE Trans. Ind. Electron.* **2009**, *56*, 900–906. [CrossRef]
24. Gao, Z. Active Disturbance Rejection Control: A Paradigm Shift in Feedback Control System Design. In Proceedings of the American Control Conference, Minneapolis, MN, USA, 14–16 June 2006; pp. 2399–2405.
25. Gao, Z. Scaling and Bandwidth-Parameterization Based Controller Tuning. In Proceedings of the American Control Conference, Denver, CO, USA, 4–6 June 2003; pp. 4989–4996.

Article

Study on Sedimentary Evolution of the Hanjiang River Delta during the Late Quaternary

Yang Wang [1], Liang Zhou [2], Xiaoming Wan [1], Xiujuan Liu [2,*], Wanhu Wang [1,*] and Jiaji Yi [1]

1 Haikou Marine Geological Survey Center, China Geological Survey, Haikou 570100, China
2 Hubei Key Laboratory of Marine Geological Resources, China University of Geosciences, Wuhan 430074, China; zhouliang2812@stu.ouc.edu.cn
* Correspondence: xjliu@cug.edu.cn (X.L.); wangwanhu@mail.cgs.gov.cn (W.W.)

Abstract: In recent years, coastal areas have been threatened by many potential hazards due to global warming, glacier melting and sea level rise. Understanding their evolutionary history and development trends can help predict disasters and further reduce the corresponding losses. The Hanjiang River delta in the southeastern part of China is the second largest delta in Guangdong Province and has such challenges. Studying the sedimentary evolution and delta initiation of the Hanjiang Delta is beneficial for understanding the response of the Hanjiang Delta to present and future sea level and climate changes. In this research, we drilled a series of cores from the Hanjiang subaqueous delta, which contains information on the sedimentary environment, climate change and sea level change during the late Quaternary. Combined with previous research results and under the constraint of high-precision and high-resolution AMS^{14}C and OSL, we carried out a multi-proxy analysis that included micropaleontology and grain size to obtain information on the sedimentary environment, sea level change and climate change. We then further discussed the initiation of the Hanjiang delta and its primary factors. The Quaternary sediments began depositing in the early Late Pleistocene (MIS5), and three sedimentary cycles can be recognized from bottom to top. The dating results also indicate that the first two cycles were formed during the late Pleistocene, while the last cycle was formed during the Holocene. The initiation of the Hanjiang Delta was indicated by a progradation in the process of a transition from estuary to a typical delta. At this time, the rate of delta progradation seaward was fast, and increasing amounts of sediments moved through the third line of islands into the sea. The barrier–lagoon system began to develop in the estuary of Hanjiang during this period. With the sequential construction of the delta, the lagoon was filled and covered by delta deposition, and the barrier bar moved to the sea; thus, the barrier-coast delta depositional model was established in the study area. Since the last glacial period (LGM), the Hanjiang River Delta and other river deltas in the region seem to have experienced similar evolutionary histories, including the filling of incised paleo-valleys and estuaries in the Early Holocene and deltaic progradation in the Middle to Late Holocene, controlled by sea level change.

Keywords: Hanjiang River Delta; late Quaternary; estuary; deposition

Citation: Wang, Y.; Zhou, L.; Wan, X.; Liu, X.; Wang, W.; Yi, J. Study on Sedimentary Evolution of the Hanjiang River Delta during the Late Quaternary. *Appl. Sci.* **2023**, *13*, 4579. https://doi.org/10.3390/app13074579

Academic Editor: Oleg S. Pokrovsky

Received: 1 February 2023
Revised: 30 March 2023
Accepted: 31 March 2023
Published: 4 April 2023

1. Introduction

Estuarine deltas connect rivers and oceans and collect abundant geological information on land–sea interactions [1,2]; these areas promote the progress and development of human society [3]. However, in recent years, coastal erosion, ground deposition and inundation of coastal lowlands caused by global warming and sea level rise may have affected the sustainable economic development of the estuary delta. The increased potential for natural disasters in estuary delta areas means that we must understand their sedimentary evolution history to develop and utilize them more reasonably.

In the past few decades, the late Quaternary sedimentary evolution and initial delta construction of large river deltas in East Asia and Southeast Asia, including the Yellow

River, Yangtze River, Pearl River, Red River and Mekong River, have been comprehensively studied. However, the small- and medium-sized river deltas in this area have received less attention. The Hanjiang River Delta is a typical representative of the medium-sized river delta in East Asia. Therefore, an analysis of the late Quaternary sedimentary evolution process of the Hanjiang River Delta can provide representative geological information as a reference for the study of late Quaternary sedimentary records, sea level and climate change in small- and medium-sized estuary deltas and their adjacent continental shelf areas.

The study of deltas can be traced back to the end of the 19th century. Galloway (1975) divided the world's major river deltas into three categories: the Mississippi River Delta, which represents river-dominated deltas; the San Francisco River delta in Brazil and the Senegal River Delta in Africa, which represent wave-dominated deltas; and the Ganges Delta and the Basheng-Langa River Delta in Malaysia, which represent tidal-dominated deltas [4]. Since the middle and late 20th century, the concepts of chronology and sequence stratigraphy have been widely used and developed in geology. With the deepening of human research on Earth system science, the stratigraphic structure and sedimentary evolution of postglacial river deltas have become the focus attention of scholars [1]. The latest research suggests that global climate and sea level changes have a certain unity, though regional characteristics can be found [5].

The main factors affecting the evolutionary history of river deltas are sea level changes, monsoon climate and human activities in the late Quaternary [6], which makes the river delta region an excellent place to judge human activities, climate change and sea level changes. Although current research exhibits substantial differences and uncertainties, it has still received extensive attention from scholars inside and outside of China [7–9]. Since the late Quaternary river delta system has certain similarities with the evolution process of the present and future delta regions, to reduce these uncertainties, we need to better understand the evolution process of modern delta regions.

There are different views on the vertical variation in Quaternary sedimentary facies in the Hanjiang Delta. Zhang [10] believed that the Quaternary deposits in the western Shantou-Raoping fault zone can be divided into three parts: the lower part is continental sand and gravel with clay, the middle part is marine clay and sand layer, and the upper part is continental and artificial fine and medium sand accumulation. The Quaternary to the east of the fault zone is also divided into three parts, but the lower and upper parts are marine strata and the middle is continental strata; some sections under the lower marine strata have more than ten meters of continental deposits [10]. For the Hanjiang River Delta, Chen (1984) established five major sedimentary cycles according to the vertical variation in the grain size of the Quaternary sediments. Each cycle shows a sedimentary rhythm of coarse (gravel, sand) and fine (silt, clay) sediments [11].

There are currently four viewpoints on the age of the Hanjiang River Delta: Early Pleistocene, Middle Pleistocene, Early Late Pleistocene and Holocene. There is considerable controversy among these viewpoints [9–11]. We believe that the accuracy of dating needs to be further discussed due to the limitations of current dating techniques. In addition, the drilling data based on previous studies are mostly distributed in the plain area of the Hanjiang River Delta, and there is a relative lack of research on the sea area.

In this research, we used geological drilling information to establish the stratigraphic structure of the sea area outside the Hanjiang Estuary. To define the stratigraphic age before the Holocene and the postglacial delta construction time, the characteristics of late Quaternary sedimentary evolution in the sea area outside the Hanjiang Estuary were analyzed by means of: sediment lithology division; AMS^{14}C and optically stimulated luminescence dating; foraminiferal and ostracod identification; particle size testing; and clay mineral content determination. According to the analysis of strata and sedimentary environment, the period and age of late Quaternary transgression were analyzed, and the paleoenvironment and paleoclimate changes were speculated. These results were combined with the results of previous calculations of regional relative sea level height for a discussion on the characteristics of regional sea level change. Based on accurate dating data and sedimentary facies analysis results, the sedimentary evolution characteristics of the Hanjiang Delta since the late Quaternary are discussed. Importantly, the OSL age of the underlying marine strata of ZK03 borehole also confirms the high sea level phenomenon in the study area during MIS5. In addition, the rich information of sedimentary environment changes in the Hanjiang River Delta revealed in this study, such as sea level changes, can provide the possibility for extracting historical environmental change information. It also provides support for delta land use and future development planning of coastal zone.

2. Materials and Methods

2.1. Geological Context of the Study Area

The Hanjiang River Delta is located below Chaozhou City in northeastern Guangdong Province, China. It is the second largest delta in the coastal area of South China after the Pearl River Delta. The magmatic rocks in the Hanjiang delta area are mainly Yanshanian intrusive rocks and Himalayan magmatic rocks. Yanshanian intrusive rocks are the most widely distributed rocks in the Hanjiang Delta. These rocks intruded into the upper Jurassic volcanic rock series, placing their formation age in the late Jurassic or its later period; hence, they are the products of late Yanshanian magmatic activity. It is generally believed that the Hanjiang Delta accepted Quaternary sediments in the middle and late Pleistocene. According to the absolute age, sedimentary facies and sedimentary cycle analysis, sporopollen climate analysis and sea level change analysis of the Quaternary sediments in the Hanjiang River Delta, Li (1987) divided the Quaternary strata of the Hanjiang River Delta into groups. These groups are the Nanshe Formation in the lower part of the middle section of the upper Pleistocene, Jiali Formation in the upper part of the middle section of the upper Pleistocene, Tuopu Formation in the upper section of the upper Pleistocene, Lianxia Formation in the lower Holocene, Chaozhou Formation in the lower section of the middle Holocene, Chenghai Formation in the upper section of the middle Holocene and Dongli Formation in the upper Holocene [12]. In addition, the Hanjiang delta is located on the southeast side of the second compound uplift belt of the Neocathaysian structure [10]. The Hanjiang River Basin is characterized by hilly areas, and the Hanjiang River is a typical mountainous medium and small river.

2.2. Research Material

From June to August 2020, based on previous papers, monographs and geological survey reports, the Haikou Marine Geological Survey Center of China Geological Survey constructed 9 geological boreholes (ZK01-03, HK01-06) in the sea area outside the Hanjiang River estuary, Guang'ao Bay, Haimen Bay and other areas (Figure 1). The average core rate of core sediment was >95%. After the geological logging of the cores of 9 boreholes was completed, the different lithology strata were divided. A total of more than 600 sediment samples were collected from different lithology strata. Each sediment sample was tested for grain size, and some horizon samples in ZK03, ZK01, HK01 and HK03 were selected to carry out age tests and micropaleontological analyses.

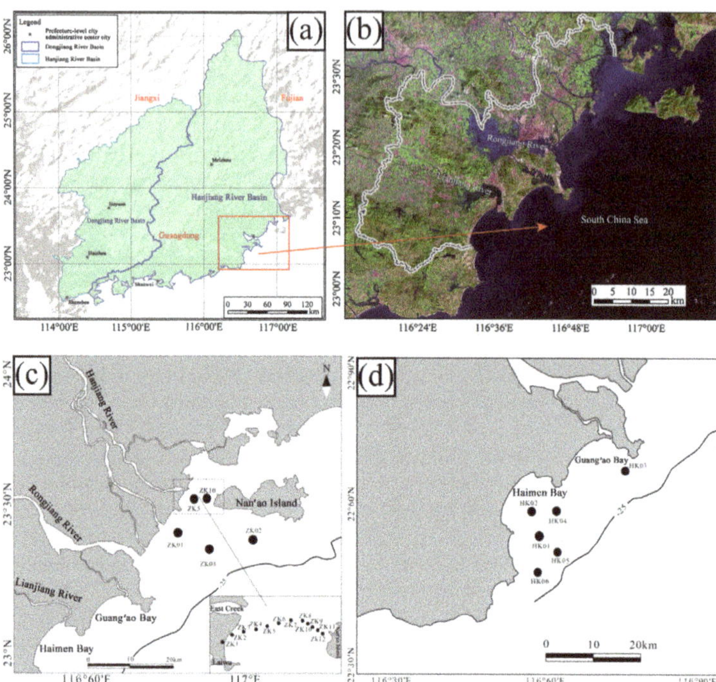

Figure 1. Borehole distribution map of the study area: (**a**) Location map of the study area; (**b**) Remote sensing image of the study area; (**c**) The sea area outside Hanjiang Estuary; (**d**) Haimen Bay and Guang 'ao Bay.

2.3. Experimental Method

2.3.1. Age Determination

1. AMS ^{14}C

The AMS ^{14}C radioactive age is very important to the study of the Quaternary sedimentary evolution of the Hanjiang River Delta. The effective selection of test materials determines the accuracy of the test results. Gastropod shells, plant debris and peat in sediments are effective dating materials. We sent the samples of 8 layers of the ZK03, HK01 and HK03 holes to the State Key Laboratory of Marine Environmental Science of Xiamen University for AMS ^{14}C. The experimental results show that there are 3 foraminifera samples and 5 carbonaceous mud sediment samples. We used the CALIB 5.0 program to correct the AMS ^{14}C radioactive age data of the test results to the calendar age. The marine13 correction curve was selected for marine inorganic carbon samples such as shells. Since the samples were from the coastal zone where the sea and land interact, the organic matter in the silt samples was selected from the mixing curve of the ocean and the Northern Hemisphere atmosphere. According to the research of Southon et al. about the effect of regional marine carbon reservoirs in the South China Sea, we used $\Delta R = -25 \pm 20$ an in the correction process.

2. OSL age

The main dating materials are quartz and feldspar in silt and sandy samples. Under shading conditions, we vertically inserted a black cylindrical sample container (approximately 2 cm in diameter) into the core and immediately covered the sample after sampling to avoid direct exposure to the sun. We sampled eight layers in the ZK03, ZK01, HK01, ZK02 and HK05 boreholes and completed the test work in the State Key Laboratory of Offshore Marine Environmental Science of Xiamen University. The apparatus was a Daybreak 2200, the excitation light source was blue light (488 ± 15 nm) and infrared light (880 ± 60 nm), the signal was detected with a PMT QA9235 photomultiplier tube, and the intensity of the irradiation source was 0.055 Gy/Sec.

2.3.2. Microfossil Analysis

A total of 50 samples were collected from ZK03, HK01, HK03, HK06 and other boreholes. The fossils of foraminifera and ostracods were processed and analyzed, and the biostratigraphic age was divided according to the fossil assemblage and paleoenvironment. The micropaleontological assemblages of these samples were analyzed, and they were sent to the Micropaleontological Laboratory of the School of Earth Sciences, China University of Geosciences (Wuhan) for the preparation and analysis of microfossils. Sample handling complied with the National Standard of the People's Republic of China (GB/T 12763.8–200)—Microfossil analysis procedures in marine survey specifications.

The micropaleontological analysis included six steps: sample drying, 10% H_2O_2 soaking dispersion, standard mesh sieve washing, identification statistics, sample under the microscope and sample slag drying (Figure 2).

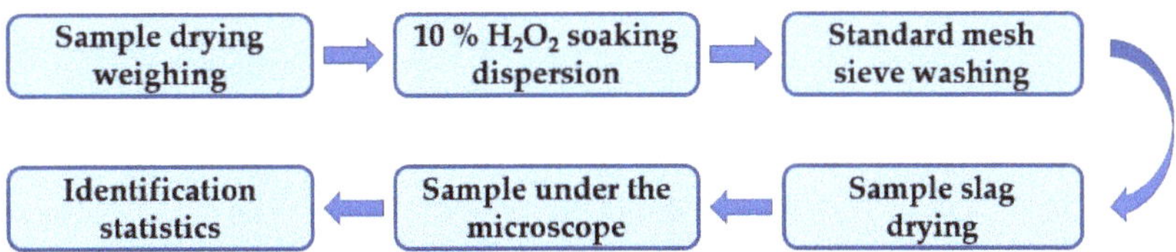

Figure 2. Analysis and processing flow of micropaleontological fossil samples.

The foraminiferal fossil samples were selected under a Nikon SMZ1000 stereomicroscope, and the relative abundance of common species and exinite types (percentage in the total number of samples) and the simple differentiation of species in the samples (number of fossil species contained in each sample) were counted and calculated. If there were too many samples, the reduction method was employed; usually, the number of samples was greater than 300.

2.3.3. Grain Size Test

Grain size is an important structural feature of sediments and the basis for their classification and naming. It is usually used for sedimentary research, sedimentary environment research, material movement, hydrodynamic conditions and particle size trend analysis. The particle size of the sediment should be analyzed by a particle size analyzer. There are three kinds of particle size analyzers: nanometer particle size analyzers, laser particle size analyzers and single particle light resistance particle size analyzers. The laser particle size analyzer used in this study is from the Geological Survey Experimental Center, Institute of Geological Survey, China University of Geosciences, Wuhan. The instrument model is a Mastersizer-3000 laser particle size analyzer produced by the British Malvern Company.

The test process includes two steps: sample pretreatment and on-machine testing. Sample pretreatment removes calcareous components and organic matter and fully utilizes technology to increase the accuracy and reliability of the results. During the last pretreatment, each sample was tested three times. After completion, the test data were checked, and those with poor repeatability were measured again.

The surface sediments are named by the Folk classification of marine sediments. The Udden–Wentworth grade scale, the limit of Wentworth grade and its classification were used to determine the grain size of the borehole sediments. The average grain size (Mz), sorting coefficient (σ), skewness (SK) and kurtosis (KG) were calculated by the Fokker and Ward diagram.

3. Results

3.1. Strata Age

The AMS ^{14}C and OSL dating results show that the deeper the strata are, the older the strata (Tables 1 and 2). The OSL dating data of two sediments at the bottom of the ZK03 borehole showed a reversal phenomenon, which may be due to the pollution of the samples during the sampling process. The age of the Holocene sediments in the upper part of the borehole is approximately 10 ka BP, the age of the lower part of the last glacial river environment or exposed weathered sediments is roughly between 17–30 ka BP, and the oldest strata at the bottom of the Quaternary sediments in the study area is ~90 ka BP.

Table 1. AMS^{14}C results obtained for borehole samples.

Borehole Number	Depth/m	Dating Materials	Average Calendar Age/ka BP	Error/ka
ZK03	34.5	mud	24.1	±0.18
ZK03	37.5	mud	27.9	±0.26
ZK03	39.7	mud	37.4	±0.70
HK01	2	foraminiferan	2.4	±0.03
HK01	8.5	mud	8.1	±0.05
HK01	24.3	mud	30.2	±0.31
HK03	3	foraminiferan	3.2	±0.04
HK03	5.2	foraminiferan	5.0	±0.04

Table 2. OSL results obtained for borehole samples.

Borehole Number	Depth/m	Dating Materials	Age/ka BP	Error/ka
ZK03	40.9	Grey medium fine sand	55.8	±5.8
ZK03	64.7	Dark gray argillaceous fine sand	96.7	±13
ZK03	74.1	Bluish gray muddy silty sand	74.2	±8.2
ZK01	38.0	Grayish yellow medium sand	52.6	±7.5
HK01	26.1	Dark gray medium sand	17.3	±2.3
HK01	33.1	Dark gray sandy clay	55.9	±6.1
ZK02	21.8	Dark gray medium fine sand	67.2	±5.3
HK05	13.5	Grayish yellow sandy clay	16.2	±1.9

3.2. Abundance of Foraminifera and Ostracods

Benthic foraminifers and ostracods are mainly divided into nearshore species (living in estuaries and nearby tidal flats with salinitie of 1–31), inland shelf species (living in continental shelf waters below 50 m with salinitie of 20–31) and euryhaline species (living in open waters above 50 m with salinitie >31). Microscopically, foraminifera fossils were found in a small number of layers in each borehole. Some layers had more foraminifera than others. The abundance of ostracods was low, and they were better preserved. The abundance of fossil species is shown in Tables 3 and 4.

A total of 22 samples were collected from borehole ZK03. We found 18 species of benthic foraminifera, of which 13 were calcareous transparent shells, 5 were porcelain shell types, and no cemented shell type was found. These species were mainly located in the middle of the formation (17 m, 25–29 m, 34.5 m). The dominant species of foraminiferal fauna in the borehole were Asterorotalia subtrispinosa and Ammonia, which had the highest abundance in the 25–27 m layer, with an average of 100/50 g dry sample. This fauna is a typical coastal shallow water assemblage dominated by warm water and contains a certain salty sea–land transitional facies.

Table 3. ZK03, HK03, HK01 and HK06 Statistics of Foraminiferal Abundance in Part of Borehole.

Sample Number	Depth/(m)	Number of Shell Statistics	Bonded Shell		Porcelain Shell									Calcium Transparent Shell																								
			Textularia sp.	*Textularia foliacea*	*Spiroloculina* sp.	*Quinqueloculina laevigata*	*Quinqueloculina seminula*	*Quinqueloculina lamarckiana*	*Quinqueloculina akneriana*	*Quniqueloculina* sp.	*Massilina laevigata*	*Triloculina tricarinata*	*Flintina bradyana*	*Guttulina* sp.	*Glandulina* sp.	*Cibicides* sp.	*Hanzawaia nipponica*	*Hanzawaia convex*	*Rotalidium annectens*	*Rotalinoides compressiusculus*	*Ammonia beccarii* var.	*Ammnoia* spp.	*Ammonia tepida*	*Asterorotalia substrispinosa*	*Asterorotalia diplocava*	*Asterorotalia binhaiensis*	*Eponides* sp.	*Rosalina bradyi*	*Pararotalia inermis*	*Pseudorotalia schroeteriana*	*Cribrononion subincertum*	*Elphidium advenum*	*Elphidium hispidulum*	*Elphidium* sp.	*Elphidium crispum*	*Nonion commune*	*Nonion* sp.
ZK03	17	25	0	0	0	2	0	0	0	2	0	0	0	0	0	0	0	0	0	0	4	5	0	4	0	0	0	0	1	0	0	2	0	4	0	1	0
ZK03	25	385	0	0	0	2	0	2	2	5	1	0	0	0	0	0	0	0	2	0	6	11	2	321	3	5	0	0	2	0	1	7	0	9	0	2	2
ZK03	27	115	0	0	0	0	0	0	0	1	0	0	0	0	0	0	0	0	0	0	5	18	0	78	2	2	0	0	2	0	0	2	0	5	0	0	0
ZK03	29	33	0	0	0	0	0	0	0	0	0	0	0	0	0	0	0	0	0	0	11	14	0	1	0	0	0	0	0	0	0	3	0	4	0	0	0
ZK03	34.5	48	0	0	0	0	0	0	0	0	0	0	0	0	0	0	0	0	0	0	0	3	0	36	0	2	0	0	0	1	0	3	0	2	1	0	0
HK03	3	101	0	0	1	2	0	11	1	24	5	0	0	0	0	0	0	0	1	0	0	0	0	0	0	0	4	0	16	18	0	7	2	6	3	0	0
HK03	5.2	355	1	1	6	13	2	62	11	47	32	3	2	1	0	2	0	2	64	4	7	6	0	0	0	0	8	0	12	0	0	22	5	12	5	16	11
HK01	2	411	0	0	2	12	2	33	5	14	4	2	0	0	2	3	3	0	188	1	2	4	0	0	0	0	0	1	0	0	0	14	5	9	1	92	12
HK01	4	280	0	0	2	13	1	31	9	17	1	4	2	0	0	4	2	0	63	2	3	5	1	0	0	0	0	1	0	0	0	25	26	17	4	36	11
HK06	10.3	9	0	0	0	0	0	0	0	0	0	0	0	0	0	0	0	0	0	1	1	1	0	1	0	0	0	0	0	0	0	3	0	2	0	0	0

Table 4. Statistics of ostracod abundance in some layers of ZK03, HK03, HK01 and HK06 holes.

Drilling Number	Depth/m	*Sinocytheridea longa*	*Sinocytheridea impressa*	*Albileberis sinensis*	*Albileberis* sp.	*Sinocytheridea* sp.	*Pontocythere minuta*	*Stigmatocythere costa*	*Stigmatocythere roesmanisis*	*Bicornucythere bisanensis*	*Neonomoceratina delicata*	*Neomonoceratina chenae*	*Keijella kloempritensis*	*Keijella bisanensis*	*Neocytheretta faceta*	*Pistocythereis* sp.	*Pistocythereis subovata*
ZK03	17	0	0	1	0	0	0	0	0	0	0	0	0	0	0	0	0
ZK03	25	3	1	0	0	1	1	0	0	3	2	2	1	0	0	0	0
ZK03	27	0	0	0	0	0	0	0	0	2	0	0	0	0	0	0	0
ZK03	29	0	0	0	0	0	0	0	0	1	0	0	0	0	0	0	0
ZK03	34.5	17	22	8	2	1	0	0	2	4	3	4	3	2	2	4	2
HK03	3	3	1	0	0	0	1	0	0	2	0	1	0	0	0	0	0
HK03	5.2	47	8	1	3	4	0	1	1	6	4	7	2	3	2	1	2
HK01	2	17	15	8	16	14	3	3	5	21	13	14	13	18	15	9	17
HK01	4	14	6	2	4	4	0	0	0	5	7	4	5	2	4	6	16
HK06	10.3	1	2	2	0	0	0	0	0	0	1	1	0	0	0	0	0

Nine samples were detected in core HK03, and 23 kinds of benthic foraminifera were found in the upper strata (3 m, 5.2 m), 16 of which were calcareous transparent shells and 6 of which were porcelain shells; cemented shells accounted for only 1 genus and 2 species, with an overall abundance of general, an average of 115/50 g dry sample. The dominant species of the foraminiferal fauna in the borehole was Rotalidium annectens, which accounted for 40–50% of the whole group, followed by the porcelain shell type, mainly represented by Quinqueloculina lamarckiana, which accounted for more than 10%. In addition, there were many porcelain shells that were severely worn, and the first-level classification of species could not be seen. High levels of *Pseudorotalia schroeteriana*, a typical macrobenthic foraminifera of the warm-water shallow-water type, were found in calcareous transparent shells. The species composition of the fauna was a typical nearshore shallow water combination dominated by warm water molecules. The degree of shell preservation was poor, with a broken shell rate of approximately 40%.

Seventeen samples were detected in core HK01, and 22 benthic foraminiferal species appeared in the uppermost layer (2 m to 4 m), including 14 calcareous transparent shell types and 8 porcelain shell types; the overall abundance was medium, with an average of 456/50 g dry sample, and the shells were preserved in medium to good conditions. The dominant species of the foraminiferal fauna found in this borehole was *Rotalidium annectens*, accounting for approximately 50–60% of all faunas, and Nonion commune, accounting for approximately 12% of all faunas. Quinqueloculina lamarckiana represented approximately 8% of the total fauna. The species composition of the fauna was dominated by typical warm-water nearshore shallow water assemblages, namely, Rotalidium annectens-Nonion commune.

Two samples were detected in core HK06, and foraminifera appeared only at the bottom of the horizon (10.6 m). Only 6 species of calcareous transparent shells of the benthic type were found, for a total of 9 shells, which were *Rotalinoides Compressiuscula*, *Ammonia beccarii* var., *Ammonia* spp., *Asterorotalia subtrispinosa*, *Elphidium advenum* and *Elphidium* sp. The genera and species of foraminifera found in this layer were relatively simple, and the individuals were small, consisting primarily of common molecules in the nearshore transitional facies combination.

Ostracod fossils were found in only 10 layers of the tested samples, and the differentiation degree was low. The surface ornamentation of most of the selected ostracod shells was clear, and the boreholes and layers where ostracod individuals appeared intensively were ZK03-17 m, ZK03-25 m, ZK03-27 m, ZK03-29 m and ZK03-34.5 m, for a total of 94 individuals; HK03-3 m, HK03-5.2 m, for a total of 100; HK01-2 m, HK01-4 m, for a total of 280; and hK06-10.3 m, for a total of 7. A total of 21 species of 9 genera were identified by the stereomicroscopy observation and photography. *Sinocytheridea longa*, *Albileberis sinensis*, *Albileberis* sp., *Sinocytheridea* sp., *Bicornucythere bisanensis*, *Neomonoceratina chenae*, *Keijella kloempritensis* and *Keijella bisanensis* were the most abundant genera in the whole sampling borehole. The majority of the fauna are common in coastal areas of China, with some genera and species, such as *Stigmatocythere*, seen not only in modern coastal areas of China and the Quaternary but also in the Pliocene–Modern deposits of Indonesia. Additionally, *Bicornucythere bisanensis* is the most common eurythermal species in the marginal seas of the western Pacific and has been reported in Japan, Indonesia and Malaysia [13].

3.3. Particle Size Analysis

The average particle size reflects the average kinetic energy condition of medium transport. The average particle size of the ZK03 sediment is 1.77–173.5 μm (Figure 3), and the particle size changes greatly, indicating that the hydrodynamic conditions in the study area since the late Quaternary have experienced a shock, causing an alternating sedimentary environment. The grain size of the borehole sediments is generally coarse, and the content of the particles above the silt grain size is more than 60%. In the layers of more than 8 m, 18–40 m and 60–70 m, the mud content increases, exceeding 20%, which reflects strong sedimentary hydrodynamic conditions. In the 0–40 m and below-60 m layers, the contents of clay and silt are relatively high. In addition, the sorting coefficient reflects the dispersion and concentration of sediment particles, and a smaller value indicates better sorting. The overall sorting coefficient of the ZK03 sediment is 0.584–3.544, indicating poor sorting. Vertically, the standard deviation of the 0–20 m horizon is 0.6–2.3, indicating poor sorting. The remaining horizon standard deviations are generally 2.0–3.5, indicating poor sorting. Each layer of the whole borehole shows a large fluctuation, reflecting the poor sediment sorting caused by unstable hydrodynamic conditions. Skewness is used to represent the symmetry of the frequency curve, reflecting the proportion of coarse and fine particles in the sediment. The skewness of most samples in the ZK03 borehole is positive or nearly symmetrical, and the range of variation is large. The 0–30 m, 60–66 m and 70–75.8 m layers have positive skewness, and the particle size distribution is biased toward fine-grained components, while only the 30–38 m, 40–45 m, 54–60 m and 66–70 m parts of the sample have a slightly negative deviation, biased toward the coarse-grain end. The peak state is used to illustrate the sharpness or bluntness of the curve compared with the normal frequency curve and reflects the concentration of the particle size distribution. The smaller the value is, the higher the sharpness. The peak state of the ZK03 borehole is mostly above 0.9, which is nearly normal or sharp, indicating that the later sedimentary environment causes less modification to the sediment; thus, the original sedimentary environment of the sediment is represented.

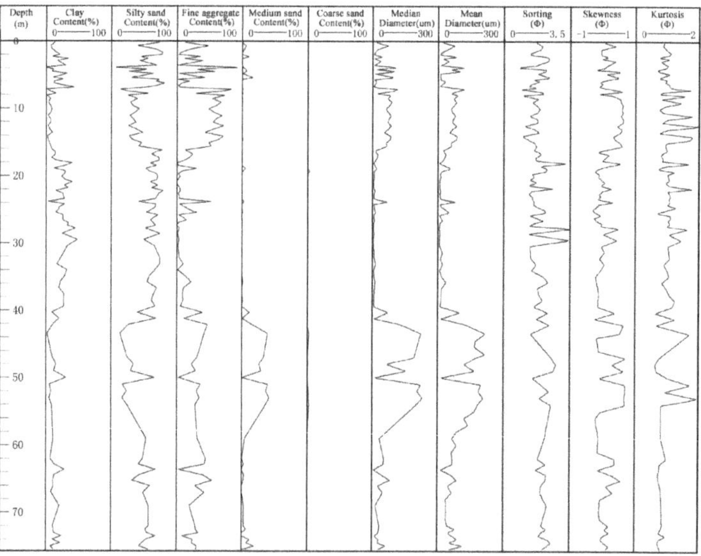

Figure 3. Grain size parameter variation of ZK03 borehole sediment.

Similarly, the content of each component of the ZK01 borehole 0–2 m changes little, and the average particle size is small, indicating that the hydrodynamic conditions are weak. The contents of fine sand and clay at 2–4 m changes greatly, indicating that the hydrodynamic force experienced a large change at that time. The sorting is 2.0–3.0, indicating very poor sorting. The sample areas of 4–34 m from bottom to top are mainly silt and clay, and the average particle size gradually decreases and finally stabilizes, indicating that the hydrodynamic force gradually weakens, and sorting and skewness also reflect this change. The clay content at 34–38 m gradually decreases from top to bottom, mostly less than 20%, while the clay content below 35 m is less than 10%. The content of medium sand and fine sand changes greatly, the sorting is mostly 2–3, and the fluctuation is large, indicating that the hydrodynamic conditions in this period are strong, and the fluctuation is violent and unstable. The average particle size of 38–42 m is small, with positive skewness, and the particle size distribution is biased toward fine-grained components. The average content of 42–50 m clay is less than 10%, the average content of sand is more than 60%, and the sorting coefficient changes greatly, indicating that the hydrodynamic fluctuation is strong, the skewness is negative, and the coarse grain is biased. The 50–52.4 m component is mostly composed of clay and silt. The sand content gradually decreases, and the average particle size decreases to below 50 μm, indicating that the hydrodynamic force is medium (Figure 4).

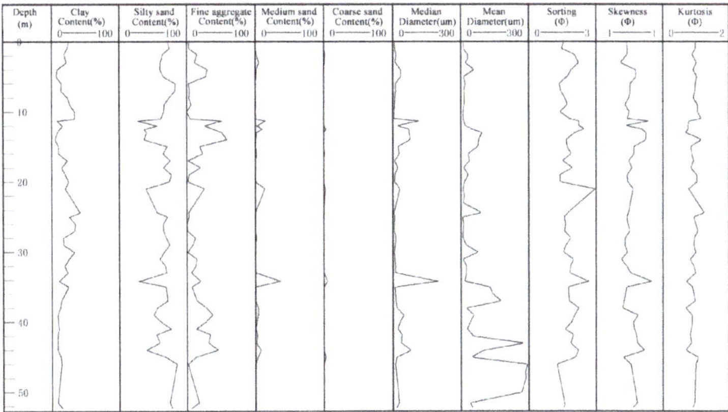

Figure 4. Grain size parameter variation diagram of ZK01 borehole sediment.

On the Pasuge C-M images, Zone 1 generally reflects strongly hydrodynamic riverbed deposition, with mass movement dominated by displacement. Area 2 reflects tidal channel deposition with strong hydrodynamics, which is dominated by saltation and displacement. Zone 3 reflects the sediments of the estuary sandbar, beach and sandbank with medium hydrodynamic strength, which is dominated by saltation. Zone 4 reflects the weak natural levees, crevasse splay, tidal flats and other deposits, which are mainly suspension and displacement ones. Areas 5 and 6 reflect the weak hydrodynamic delta front and front slope, delta plain, flood plain, stagnant water depression and other deposits, which are mainly suspension ones. In the borehole core, the hydrodynamic conditions reflected by the sediment transport mode indicate the sedimentary environment and sedimentary cycle.

The sediment sample points of ZK030–3.9 m mainly fall in the C-M image 6 area, are mainly suspended sediment, and the hydrodynamic force is weak. The 9–4.5 m sediment sample points mainly fall in C-M image area 3, are mainly for saltation deposition, and the hydrodynamic force is strong, which may represent a short regression after river sedimentation; the 4.5–30.4 m sediment sample points mainly fall in C-M image areas 5 and 6, are mainly for sediment transport, and the hydrodynamics force is weak. The 30.4–32.2 m sediment sample points mainly fall in C-M image area 3, are mainly for saltation deposition, and the hydrodynamic force is strong. The 32.2–42.1 m sediment sample points mainly fall in C-M image area 6, being mainly suspended sediment, and the hydrodynamic force is weak; the 42.1–54 m sediment sample points mainly fall in C-M image area 3, being mainly for saltation deposition, and the hydrodynamic force is strong; the 54–75.5 m sediment sample points mainly fall in C-M images 5 and 6, being mainly suspended sediment, and the hydrodynamic force is weak (Figure 5).

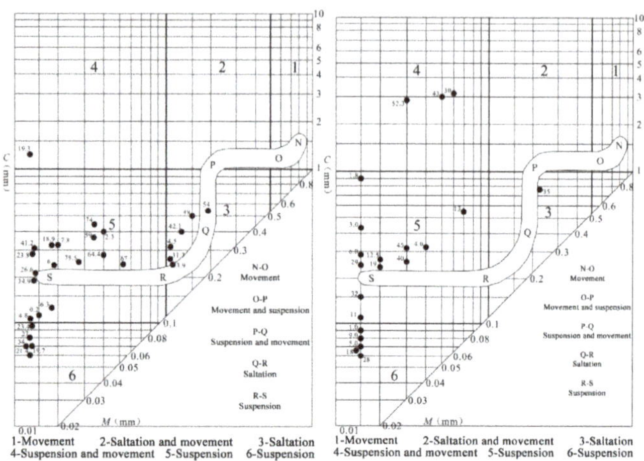

Figure 5. Grain size falling point map of Passeig C-M image of ZK03/ZK01 borehole sediments.

Similarly, the 0–30 m sediment sample points of the ZK01 borehole mainly fall in the 5th and 6th areas of the C-M image, being mainly suspended sediment, and the hydrodynamic force is weak. The 30–37 m sediment sample points mainly fall in C-M image area 3, are mainly for saltation deposition, and the hydrodynamic force is strong; the 37–43 m sediment sample points mainly fall in the C-M image 6 area, are mainly suspended sediment, and the hydrodynamic force is weak. The 43–52.3 m sediment sample points mainly fall in the C-M image 4 area, are mainly suspended load and bed load deposition, and the hydrodynamic force is medium (Figure 5).

4. Discussion

4.1. Age of Earliest Acceptance of Quaternary Sediments: MIS 3 or MIS 5

The most controversial issues in coastal sediment dating concern the MIS3 and MIS5 stages. In principle, the maximum limit for analytical measurements with the AMS radiocarbon dating method is approximately 60 ka. In this study, AMS ^{14}C and OSL dating methods were combined to constrain the earliest sedimentary age of the Quaternary in the Hanjiang Delta in the MIS5 period; the study included lithology, grain size analysis, dating and foraminifer abundance statistics of ZK03 sediments outside the Hanjiang River estuary. Grain size analysis reveals the transition process from marine fine-grained clay ~ silt to coarse sandy sediments, which can better reflect the characteristics of the transition from MIS5 high sea level to MIS4 low sea level. Predecessors believe that the Quaternary sediments in the Hanjiang River Delta began in the middle of the late Pleistocene. On the one hand, because only AMS 14C is used, it is impossible to accurately measure the age of strata greater than 60 ka; on the other hand, the drillings at that time were mainly distributed in the Han River Delta land area. Due to the tectonic background of the overall subsidence of the Han River Delta in the late Quaternary and the obstruction of the third island mounds, the terrain of the Han River Delta area during MIS5 was much higher than the current elevation of the sedimentary basement of the marine strata, and the transgression may not have reached the Han River Delta plain area.

4.2. Sedimentary Evolution Process

The results of OSL dating at the bottom of the ZK03 drilling hole show that the oldest Quaternary stratum in the Hanjiang Delta was in the MIS5 period. The grain size, clastic minerals and paleontological characteristics of sediments show that this section is a flood plain–floodplain deposit. During 90–70 ka BP, the global sea level was in a relatively high sea level period [13]. The strata in the study area during this period were alternately formed by massive sand and mud, composed of mildly layered hard mud and silty sediments, and contained peat fragments, but no shells or marine indicators were found. It is speculated that this section is an estuarine–beach sedimentary environment. During this period, the study area experienced a change in the sedimentary environment from continental facies to marine facies. Due to the barrier of the third island mound, the influence range of this transgression was small, and no record of this transgression was found inside the delta plain.

At approximately 70 ka BP, the Earth entered the last glacial period [13], the seawater withdrew from the delta plain, and a set of medium coarse sand-weathered clay layers was deposited. The global sea level was relatively low during the last glacial period, but there was still a period of relatively high sea level. The second transgression in the study area occurred after the late Pleistocene, and sedimentary strata with silty clay and muddy fine sand as the main sedimentary types were developed, representing the shallow marine sedimentary environment defined by the interactions between river and seawater tidal dynamics. At approximately 30 ka BP, this transgression reached its high sea level period [14–16]. Then, until the arrival of the last glacial maximum (26–17 ka BP), the sea level dropped to its lowest level [12–14], and a set of continental bottom gravel layers developed in the study area. Some boreholes also found porphyritic weathered clay and yellow–white clay sandy gravel layers.

The Hanjiang River Delta began to accept postglacial deposition at approximately 10.4 ka BP, and the sea level began to rise rapidly, reaching the maximum transgression range at approximately 6.3 ka BP, during which time the sedimentary characteristics changed many times. The sedimentary evolution process of the study area since the Holocene can be divided into three stages from the old to new in chronological order. At approximately 10.4–8.5 ka BP, the end of the last glacial period, the global climate began to gradually become warm and humid, and at approximately 9.2 ka BP, after gradually transforming into marine deposits, the estuary environment developed. At approximately 8.5–6.3 ka BP, the composition of foraminiferal species represented a typical coastal shallow water combination, mainly warm water molecules, and contained a certain proportion of euryhaline transitional phase molecules, representing the coastal–shallow sea estuary sedimentary environment. At approximately 6.3–0 ka BP, large-scale regression led to the continuous diffusion of sediments carried by rivers to the open sea, and the strong wave action in the open sea led to the redistribution of sediments from the basin into the open sea. The sediments were distributed along both sides of the estuary and parallel to the coastline, forming a series of barrier sand dams. After the maximum transgression, the study area began to construct the Holocene delta and gradually formed the current land–sea pattern. In addition, dredging has had a certain impact on the deposition of coastal waters. Since we visited and investigated, no dredging has occurred in the coastal waters of the study area. Only the channel in the Rongjiang Estuary has been dredged on a small scale, which has little effect on the sedimentary evolution of the whole study area.

4.3. Holocene Delta Formation and Barrier–Lagoon System

After the Hanjiang Delta region reached the Holocene maximum transgression at approximately 6.3 ka BP [17], the sea began to retreat. Under the dynamic marine background of weak tidal action and strong wave action in the bay, coupled with the blocking effect of the third row of island mounds in the bay, the river action in the outer sea area is relatively weak. Strong offshore waves distribute and redistribute the sediments along both sides of the estuary, forming a series of barrier bars. As the delta advances to the sea, sediment gradually accumulates outside the third island mound to form a coastal sand bank, which is a coastal sandy sedimentary body composed of a large number of intertidal organisms, shells and coarse-grained sediments. Due to the barrier between the coastal sandbanks and the third island mound, the sea water in the inner edge is prevented from entering the open sea. Tidal action causes this sea water to enter the inner edge of the third island mound along the intermittent sandbanks. This situation, coupled with the low-lying terrain in the western region, leads to tidal influx, forming the lagoon inside the third island mound. The Holocene barrier–lagoon system developed in this area, and then the sedimentary evolution of the barrier coastal delta began. In the late Holocene, the change in the sea level began to stabilize. As the delta continued to advance to the sea, the barrier–lagoon system also continued to move to the sea.

From the last glacial period to the early Holocene, the sea level began to rise rapidly, and the early Holocene transgression led it to reach the maximum range [18–20]. At this time, the main sedimentary environment of the Hanjiang Delta was mainly shore–shallow sea and shallow bay. The low-density flow of the Hanjiang River into the sea was a plane jet, which led to a relatively stable water environment after the sediments of the Hanjiang River entered the estuary. A large number of river sediments began to be deposited there, and with hydrodynamic action, they spread to the open sea, forming the main body of the delta plain dominated by river action [21]. At this time, the barrier coastal delta was in the formation period, the main sediment lithology was the continental margin alluvial plain channel sand body and the underwater front sand body, and the fine-grained sediments diffused to the open sea via marine dynamic action (Figure 6).

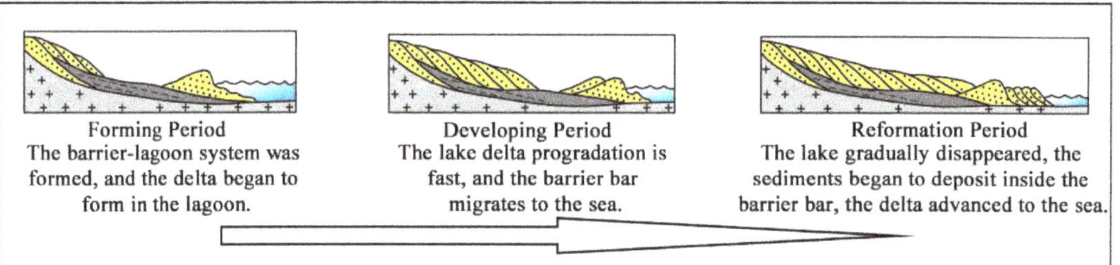

Figure 6. Evolution stage of barrier coastal delta.

After the maximum transgression, the delta began to advance to the sea, and a large amount of sediment was transported out to the sea through the third island mound. The barrier–lagoon system began to develop, inhibiting the outward diffusion of river sediments and accelerating delta construction. In the eastern part of the lagoon, due to the shallow water body, a large number of river sediments were deposited, and delta construction advanced rapidly. Delta deposits dominated by river action were developed. On the west side of the lagoon, due to the relatively low-lying terrain, the tidal water influx deepened the water body in the area, and the lagoon swamp sedimentary environment became completely different from that in the eastern region. Inside the lagoon, a certain scale of fan-shaped delta plain deposits dominated by river sand developed. Outside the barrier, a large amount of sediment accumulates under the action of oceanic dynamics, forming a large number of sandbars that were distributed parallel to the coastline and advanced toward the sea (Figure 6).

In the late Holocene, the sea level tended to be stable and equal to the present sea level. The Hanjiang River Delta advanced to the open sea beyond the barrier, and the strong wave action hindered the delta from advancing to the sea; this delta still exists inside the lagoon delta plain. At this time, the delta had been pushed out of the third island hills, and the sediments in the basin continued to accumulate due to wave action, forming a new barrier sand bar parallel to the coastline (Figure 6) and gradually evolving into the current barrier coast of the Hanjiang Delta.

5. Conclusions

Through the lithology description of the ZK03, ZK01 and HK01 borehole sediments in the Hanjiang River Delta and the identification of micropaleontological fossils and particle size tests, combined with previous research results and under the constraints of high-resolution and high-precision AMS^{14} C and OSL age framework, the late Quaternary sedimentary evolution process of the Hanjiang River Delta was discussed. The main research results are as follows:

(1) The study area began to accept Quaternary sediments in the early Late Pleistocene (MIS5). A total of 10 sedimentary units developed from the bottom to the top of the Late Quaternary sediments, which mainly experienced a floodplain environment and estuary–beach environment in the early Late Pleistocene; a floodplain environment, coastal shallow sea environment, bar–lagoon environment, coastal shallow sea environment, and floodplain environment in the middle and late Pleistocene; and an estuary environment, bar–lagoon environment and delta sedimentary environment in the Holocene. From bottom to top, the study area is mainly divided into three sedimentary cycles. Each sedimentary cycle shows a sedimentary rhythm from coarse to fine from bottom to top, starting with medium and coarse sand or gravel deposition and ending with muddy silt or clay deposition. The first and second cycles belong to the late Pleistocene, and the third cycle belongs to the Holocene.

(2) After the maximum transgression in the Holocene, the progradation rate of the Hanjiang River Delta was fast, and increasing amounts of sediments converged to the sea through the third row of island mounds. Under the dynamic marine background of weak tidal action and strong wave action in the bay, coupled with the blocking effect of the third island mound in the bay, the sediments imported into the open sea were redistributed along both sides of the estuary and parallel to the coastline, forming a series of barrier–lagoon systems. With the continuous construction of the delta, the lagoon water body was filled with delta deposits, and the barrier bar moved to the sea. In this cycle, the study area developed a unique barrier–lagoon coastal delta deposit.

Author Contributions: Data curation, Y.W.; formal analysis, Y.W. and L.Z.; funding acquisition, Y.W. and X.W.; investigation, J.Y. and W.W.; methodology, L.Z.; project administration, X.L. and X.W.; writing—original draft, Y.W.; writing—review and editing, Y.W., X.L. and X.W.; visualization, W.W. and J.Y.; supervision, X.L. All authors have read and agreed to the published version of the manuscript.

Funding: Comprehensive Geological Survey of Chaoshan Coastal Zone, grant number DD20208013; Comprehensive Survey of Natural Resources in Huizhou-Shanwei Coastal Zone, grant number DD20230415.

Acknowledgments: Chaoshan Coastal Zone, grant number DD20208013; Huizhou-Shanwei Coastal Zone, grant number DD20230415.

Conflicts of Interest: The authors declare no conflict of interest. The funders had no role in the design of the study; in the collection, analyses, or interpretation of data; in the writing of the manuscript; or in the decision to publish the results.

References

1. Stanley, D.J.; Warne, A.G. Worldwide initiation of Holocene marinedeltas by Decelera-tion of sea-level rise. *Science* **1994**, *265*, 228–231. [CrossRef] [PubMed]
2. Maloney, J.M.; Bentley, S.J.; Xu, K.; Obelcz, J.; Georgiou, I.Y.; Miner, M.D. Mississippi River subaqueous delta is entering a stage of retrogradation. *Mar. Geol.* **2018**, *400*, 12–23. [CrossRef]
3. Sun, Z.; Li, G.; Yin, Y. The Yangtze River deposition in southern Yellow Sea during Marine Oxygen Isotope Stage 3 and its implications for sea-level changes. *Quat. Res.* **2015**, *83*, 204–215. [CrossRef]
4. Galloway, W.E. Process framework for describing the morphologic and stratigraphic evolution of deltaic depositional systems. *Houst. Geol. Soc.* **1975**, 87–98.
5. Bard, E.; Hamelin, B.; Arnold, M.; Montaggioni, L.; Cabioch, G.; Faure, G.; Rougerie, F. Deglacial sea-level record from Tahiti corals and the timing of global meltwater discharge. *Nature* **1996**, *382*, 241–244. [CrossRef]
6. Fan, D.; Wu, Y.; Zhang, Y.; Burr, G.; Huo, M.; Li, J. South Flank of the Yangtze Delta: Past, present, and future. *Mar. Geol.* **2017**, *392*, 78–93. [CrossRef]
7. Zhang, Q.; Liu, C.; Zhu, C. Impact of environmental changes on human activities during the holocene in the Changjiang (Yangtze) river delta region. *Mar. Geol. Quat. Geol.* **2004**, *24*, 9–15.

8. Wang, Z.; Saito, Y.; Zhan, Q.; Nian, X.; Pan, D.; Wang, L.; Chen, T.; Xie, J.; Li, X.; Jiang, X. Three-dimensional evolution of the Yangtze River mouth, China during the Holocene: Impacts of sea level, climate and human activity. *Earth-Sci. Rev.* **2018**, *185*, 938–955. [CrossRef]
9. Han, M.K.; Wu, L.; Liu, Y.F.; Mimura, N. Impacts of sea-level rise and human activities on the evolution of the pearl River Delta, South China. *Proc. Mar. Sci.* **2000**, *2*, 237–246.
10. Zhang, H. Faulting and the Formation and Development of Hanjiang Delta. *Acta Oceanol. Sin.* **1983**, *5*, 202–211.
11. Chen, G. Preliminary understanding of the Holocene transgression and regression in the Hanjiang and Rongjiang rever deltas. *Mar. Sci. Bull.* **1984**, *3*, 39–44.
12. Li, P.R.; Huang, Z.G.; Zong, Y.Q. *Hanjiang Delta*; China Ocean Press: Beijing, China, 1987.
13. Sun, J.-l.; Xu, H.L.; Wu, P.; Wu, Y.-B.; Qiu, X.-L.; Zhan, W.-H. Late Quaternary sedimentological characteristics and sedimentary environment evolution in sea area between Na'ao and Chenghai, eastern Guangdong. *J. Trop. Oceanogr.* **2007**, *26*, 30–36.
14. Cao, Q. *Characteristics of Palaeoceanographical and Palaeoenvironmental Evolution since the Pleistocene in the Okinawa Trough*; The Institute of Oceanology, Chinese Academy of Sciences: Beijing, China, 2002.
15. Lea, D.W.; Martin, P.A.; Pak, D.K.; Spero, H.J. Reconstructing a 350 ky History of Sea Level Using Planktonic Mg/Ca and Oxygen Isotope Records from a Cocos Ridge Core. *Quat. Sci. Rev.* **2002**, *21*, 283–293. [CrossRef]
16. Fairbridge, R.W. Eustatic changes in sea level. *Phys. Chem. Earth* **1961**, *4*, 99–185. [CrossRef]
17. Siddall, M.; Rohling, E.J.; Almogi-Labin, A.; Hemleben, C.; Meischner, D.; Schmelzer, I.; Smeed, D.A. Sea-level fluctuations during the last glacial cycle. *Nature* **2003**, *423*, 853–858. [CrossRef] [PubMed]
18. Zhao, B.; Wang, Z.; Chen, J.; Chen, Z. Marine sediment records and relative sea level change during Late Pleistocene in the Changjiang delta area and adjacent continental shelf. *Quat. Int.* **2008**, *186*, 164–172. [CrossRef]
19. Zong, Y.Q. Postglacial Stratigraphy and Sea-Level Changes in the Han River Delta, China. *J. Coast. Res.* **1992**, *8*, 1–28.
20. Nguyen, V.L.; Ta, T.; Saito, Y. Early Holocene initiation of the Mekong River delta, Vietnam, and the response to Holocene sea-level changes detected from DT1 core analyses. *Sediment. Geol.* **2010**, *230*, 146–155. [CrossRef]
21. Hori, K.; Saito, Y. An early Holocene sea-level jump and delta initiation. *Geophys. Res. Lett.* **2007**, *34*, L18401. [CrossRef]

Article

A Pilot Study of Stacked Autoencoders for Ship Mode Classification

Ji-Yoon Kim [1] and Jin-Seok Oh [2,*]

[1] Romantique, Contents AI Research Center, 27 Daeyeong-ro, Busan 49227, Republic of Korea
[2] Division of Marine System Engineering, Korea Maritime and Ocean University, 727 Taejong-ro, Yeongdo-gu, Busan 49112, Republic of Korea
* Correspondence: ojs@kmou.ac.kr; Tel.: +82-410-4283

Abstract: With the evolution of the shipping market, artificial intelligence research using ship data is being actively conducted. Smart ships and reducing ship greenhouse gas emissions are among the most actively researched topics in the maritime transport industry. Owing to the massive advances in information and communications technology, the internet of things, and big data technologies, smart ships have emerged as a very promising proposition. Numerous methodologies and network architectures can smoothly collect data from ships that are currently in operation, as is currently done in research on reducing ship fuel consumption by deep learning or conventional methods. Many extensive studies of stacked autoencoders have been carried out in the past few years. However, prior studies have not addressed the development of algorithms or deep learning-based models to classify the operating states of ships. In this paper, we propose for the first time a deep learning-based stacked autoencoder model that can classify the operating state of a ship broadly into the categories of At Sea, Stand By, and In Port, using actual ship power load data. In order to maximize the model's performance, the stacked autoencoder architecture, number of hidden layers, and number of neurons contained in each layer were measured by performance metrics such as true positive rate, false positive rate, Matthews correlation coefficient, and accuracy. It was found that the model's performance was not always improved by increasing its complexity, so the feasibility of developing and utilizing an efficient model was verified by comparing it to real data. The best-performing model had a (5–128) structure with latent layer size 9. It achieved a true positive rate of 0.9035, a false positive rate of 0.0541, a Matthews correlation coefficient of 0.9054, and an accuracy of 0.9612, clearly demonstrating that deep learning can be used to analyze ship operating modes.

Keywords: ship mode; autoencoder; ship mode classification; deep-learning model

Citation: Kim, J.-Y.; Oh, J.-S. A Pilot Study of Stacked Autoencoders for Ship Mode Classification. *Appl. Sci.* **2023**, *13*, 5491. https://doi.org/10.3390/app13095491

Academic Editors: Enjin Zhao, Hao Qin and Lin Mu

Received: 20 February 2023
Revised: 17 April 2023
Accepted: 26 April 2023
Published: 28 April 2023

1. Introduction

In the maritime transport industry, smart ships and reducing ship greenhouse gas emissions have been actively researched [1]. Smart ships have emerged as information and communications technology, the internet of things, and big data technologies have advanced. Different from a conventional ship, a smart ship is characterized by its ability to use data collected by sensors installed within the ship to self-navigate or to provide appropriate information to assist in the decisions of crew members operating the ship [2]. Studies on reducing ship greenhouse gas emissions have mainly focused on reducing ship fuel consumption and eco-friendly ships that do not use oil.

Research on smart ships has been actively pursued, and various methodologies and network architectures that can smoothly collect data from ships in operation have been proposed. This research has included areas such as big data collection systems [3], cyber security considerations [4], data management to reduce learning costs [5], framework structures for index systems [6], surveys of architectures and applications [7], and priority items for smart shipping [8]. Furthermore, various data on actual ships is being collected, which is used in research on reducing ship fuel consumption by deep learning or conventional methods to perform knowledge-free path planning with reinforcement learning [9], energy-saving auto path planning algorithms [10], energy-saving management systems for smart ships [11], power scheduling for saving energy with reinforcement learning [12], and forecasting ship fuel consumption [13]. However, the application of ship operation mode classification remains unresearched.

Due to the characteristics of ship operations, the operating state of a ship can be broadly classified as At Sea, Stand By, or In Port. In the At Sea state, all of the devices on the ship are connected and powered, and the load changes in each device are small. Stand By refers to the state in which the ship is entering or exiting a port, and it is characterized by large fluctuations in the total power consumption due to changes in the ship speed and the use of auxiliary devices. Lastly, In Port refers to the operating state in which a ship has entered a port and cargo is being loaded or unloaded from the ship. In this state, fewer auxiliary devices are being powered on the ship, and total power consumption and power fluctuations are low. The power fluctuation characteristics of each ship type are as follows:

- Container Ship: During the In Port state, the total power consumption and power fluctuations are large if many reefer containers are being carried.
- LNG Ship: Cargo pumps are used when loading or unloading crude oil. Therefore, the ship is under the highest power load during the In Port state.
- Bulk Ship: Bulk ships with cargo cranes installed are under a very large power load during the In Port state.

The operating states of ships, which are presently being acquired in large quantities, have not, however, been adequately researched. As such, researchers are faced with the problem of needing to label data manually based on ship voyage information to use the collected ship data. Research on ship state classification models that can classify the operating states of ships is required to overcome this issue.

Prior studies have not addressed the development of algorithms or deep learning-based models to classify the operating states of ships. Autoencoders are used in handwriting recognition [14], anomaly detection [15], fault diagnosis [16], and fraud diagnosis [17] and produce better results in comparison with existing algorithms. Thus, we selected the autoencoder model for classifying ship data.

However, the performance of the stacked autoencoder model depends on proper control of the components. When utilizing an autoencoder-based classification model, model design considerations include:

- Which structure has better performance?
- What are the appropriate values for the number of hidden layers and the number of neurons in each layer to achieve better performance?
- What size of latent layer is suitable? The size of the latent layer has a significant effect on the performance of the classification model.

In order to address these issues, we conducted comparative experiments on the structure of the classification model, the appropriate values for the number of hidden layers and the number of neurons in each layer, and changes in the size of latent layers. We find the best model to classify the ship's operating state as either At Sea, Stand By, or In Port using actual ship power load data.

This paper makes the following contributions: First, the structure of the first stacked autoencoder model using actual ship data is presented. Second, the performance change of the model according to the components of the stacked autoencoder was investigated. In particular, since there is no previous study that uses actual ship data, we design and perform performance comparison experiments of classification models according to structural changes of the stacked autoencoder. Third, the experimental results are analyzed.

2. Related Works

In the past several years, many studies have applied autoencoder models to solve practical problems. An autoencoder [18] is a learning neural network model that approximates input and output values to reconstruct them the same, and the main purpose of the autoencoder is to learn informative representations of data using an unsupervised method. Types of autoencoders include the stacked autoencoder, sparse autoencoder, denoising autoencoder, contractive autoencoder, and variational autoencoder [19].

The stacked autoencoder can learn efficiently to create robust features from training data. Research has been conducted on the benefits of stacked autoencoders to solve their problems. Ghosh et al. [20] used a stacked autoencoder model to classify human emotional data and achieved good results in categorizing human emotions with a spectrogram dataset. Ghosh's approach was able to produce better results compared to traditional methods, which could not distinguish between happy and angry people.

Ambaw et al. [21] compared different conventional neural networks, support vector machines, and stacked autoencoders on the recognition of continuous phase frequency-shift keying under carrier frequency offset, noisy, and fast-fading channel conditions with the based model. In this study, the three features selected for recognition were the approximate entropy of the received signal, the approximate entropy of the received signal's phase, and the approximate entropy of the instantaneous frequency of the received signal. It was found that the stacked autoencoder performed better than support vector machines and traditional neural networks; a stacked autoencoder model can give a better accuracy result for most signal to noise ratios.

Singh et al. [22] proposed a stacked autoencoder model to reduce complexity and processing time for detecting epilepsy. The model classified epileptic data into normal, ictal, and preictal. He selected machine learning algorithms such as Bayes Net, Naïve Bayes, multilayer perceptron, radial basis function neural networks, and the C4.5 decision tree classifier as comparison models. He proved that the stacked autoencoder model had the best performance score with the least processing time.

Law et al. [23] suggested using a cascade of two types of networks, stacked autoencoders and extreme learning machines, for multi-label classification to enhance a stacked autoencoder's performance. The proposed model was compared with eleven other algorithms with seven datasets. However, she claimed that the model had promising performance.

Aouedi et al. [24] introduced a stacked sparse autoencoder model to integrate feature extraction and classification processes. The model uses denoising and a dropout technique to enhance feature extraction performance and prevent overfitting. It was proven that the model produced a better output than conventional models.

Deperlioglu [25] built a stacked autoencoder classification model for heart sound analysis. Traditional methods use data transforms to get the S1 and S2 segments of heart sounds. Deperlioglu's novel approach was to utilize only a stacked autoencoder model to get segments of heart sounds for direct classification. This model was compared with conventional algorithms. The proposed model's performance was similar to prior models. According to this study, a stacked autoencoder can be used in the medical field with efficient and effective classification results.

Gokhale et al. [26] compared a proposed stacked autoencoder model with seven previously established algorithms using ten datasets to find the key genes for cancer. Traditional gene selection approaches using statistical or feature selection methods have accuracy problems. However, the proposed stacked autoencoder-based framework outperformed conventional methods in this study.

Arafa et al. [27] introduced a reduced noise-autoencoder for solving the problem of imbalanced data in genomic datasets. Arafa's approach was able to solve the dimensionality problem with the stacked autoencoder with feature reduction and create new low-dimensional data. In addition, the accuracy score was improved by more than eight percent.

3. Theoretical Background

3.1. Autoencoder

An autoencoder consists of an encoder, a latent layer, and a decoder, as shown in Figure 1. The encoder is called a recognition network and extracts the features of the original data entered as input. The layer that stores the extracted features during this task is called the latent layer. The decoder is known as a generative network, and it converts the features into output. Through this process, the autoencoder can reorganize the core information in the input data.

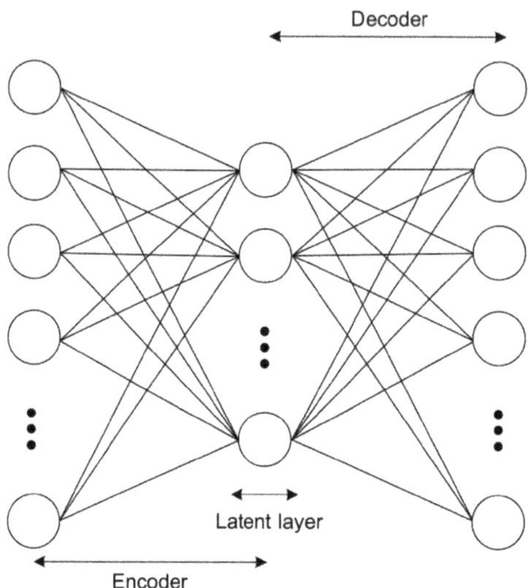

Figure 1. Autoencoder.

The encoder of the autoencoder can be defined as

$$h = \sigma(W_e x + b_e)$$

Here, x is the input data, and W_e and b_e are the eth weight value matrix and bias vectors, respectively. σ is the activation function. h is the encoder output. The decoder of the autoencoder can be expressed as

$$\hat{x} = \sigma(W_d h + b_d)$$

In the decoder, the encoder output h is used as an input variable. W_d and b_d are the dth weight value matrix and bias vectors, respectively. σ is the activation function. $\hat{x}$ is the decoder output.

The autoencoder learns in order to make the decoder's output value as similar to the input value as possible. Therefore, minimizing the difference between the input value and the decoder output value by adjusting the parameters during the training of the autoencoder model is important. As such, selecting a loss function that is suitable for the goals of the appropriate model is also critical. If the mean square error is used as the loss function, it can be expressed as follows:

$$L(x, \hat{x}) = \frac{1}{N} \sum_{i=1}^{N} (\hat{x} - x)^2$$

Here, x is the input value, and $\hat{x}$ is the decoder output value. N is the Nth term.

3.2. SAE

An SAE [28], also known as a deep autoencoder, is organized as shown in Figure 2. It has a structure in which multiple hidden layers are contained in the encoder and decoder, and its structure is symmetrical in relation to the latent layer. The latent layer is located between the encoder and the decoder, as in an autoencoder, and it stores the feature data that are acquired from the encoder.

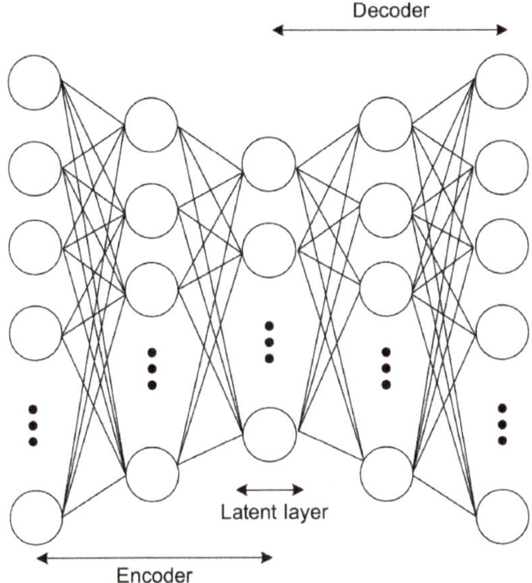

Figure 2. SAE.

The SAE model is trained using the greedy layer-wise training methodology [29]. This methodology was proposed to determine the optimal parameters of an SAE, and it has been proven to be effective in learning an SAE with multiple hidden layers. This methodology can also reduce the network size of the SAE model and increase the training speed. Furthermore, it has the advantage of potentially reducing the risk of overfitting.

3.3. Dataset

Real ship data are not open-access data. Thus, we collected data from a real ship. The data in this study were those of a 13,000 TEU container ship used in actual operations. It is propelled by one MAN-Burmeister and Wain diesel engine and has four 3480 kW generators. Table 1 lists the specifications of the target ship.

Table 1. Specifications of a container ship.

Ship Type	Container Ship
Length	365 m
Width	48 m
Draft	10.8 m
Engine Output	79,106 BHP
Generator Output	3480 kW × 4
Maximum Speed	23.0 knots
TEU	13,154 TEU

The power load of the ship continuously fluctuates according to the operating state of the ship [30]. Furthermore, the steering equipment installed on the ship is powered by an electric motor, and the power load consumed by the electric motor is affected by the external resistance of the ship's hull [31]. Therefore, this study collected data on the electric load, which is the total electric load of the ship, as well as data that indicate the external resistance of the ship, such as its heading, rudder angle, water depth, water speed, wind angle, wind speed, and ship speed. Table 2 shows the types of data measured in this study.

Table 2. Various types of data measured.

Kind of Data	Unit	Range
Electric Load	kW	585–4377
Heading	°	0–360
Rudder Angle	°	−35–35
Water Depth	m	0–839
Water Speed	m/s	−4–6
Wind Angle	°	0–360
Wind Speed	m/s	0–47
Ship Speed	knots	0–19

The data were measured in 10 min intervals, and a total of 30,340 data values were collected. Figure 3 shows the collected ship power load data. Changes in the power load occurred as the ship was operated.

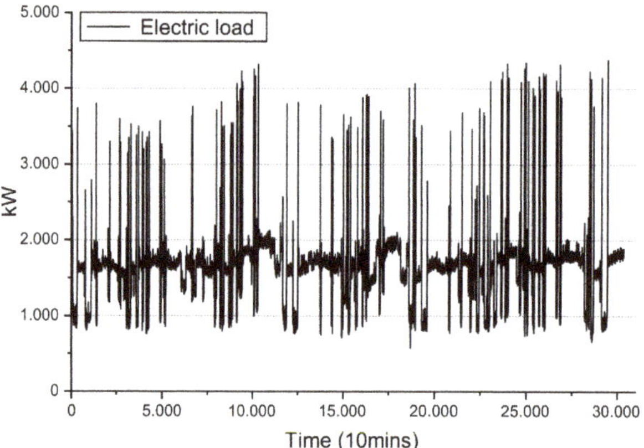

Figure 3. Ship electric load.

Table 3 shows the number of data points according to the ship's state. The ship is most often operated At Sea, and the least amount of time is spent in Stand By. In addition, by the number of data points measured, there are twice as many instances of the In Port state compared to the Stand By state.

Table 3. Number of data collected for each state of the ship.

State of the Ship	Number of Data
In port	5324
Stand by	2582
At sea	22,434

4. Approach

This section describes the design approach for the SAE that classifies ship operating modes. In particular, we hoped to determine whether the structure of the SAE model and the size of its latent layer affected the operating mode classification performance.

4.1. Overview

An overview of the process of this research is given below.

- Training: The stacked autoencoder is sufficient to extract features and reconstruct. However, there is a need to find the structure of the model that is the most efficient. The stacked autoencoder models are trained to classify the features of input data, while the Adam optimizer is utilized as the loss function. In the training process, five structures, which are composed of depth and size, were trained by latent layer values (3, 6, 9, and 12).
- Selection of models for comparison: After completing the training process, the true positive rate (TPR), false positive rate (FPR), accuracy, and Matthews correlation coefficient (MCC) were used to select the best performance model for each structure. Thus, only five models were kept for a final comparison. Trained models classified ships as being in one of three states: In Port, Stand By, and At Sea. In this process, the MCC was considered to be the most important of all the evaluation metrics.
- Performance comparison: Five models from the previous step of the process were compared. For comparison, we used the same evaluation metrics and a confusion matrix. The confusion matrix is a tabular summary for users to check precisely the correct and incorrect results made by classification models. By using this technique, the performance of the classification models was analyzed in detail.

4.2. Model Design

The SAE described in Section 2 was used in the operating mode classification model. To improve the performance of the SAE model, this study considered two aspects: the structure of the model and the size of the latent layer. Here, the model structure refers to the number of hidden layers within the encoder and decoder and the number of neurons used in each hidden layer. The size of the latent layer refers to the number of neurons in the latent layer. Figure 4 shows the basic structure of the model. The encoder and decoder were arranged in a symmetric form, and a softmax layer [32] was added to the output part of the decoder for classification.

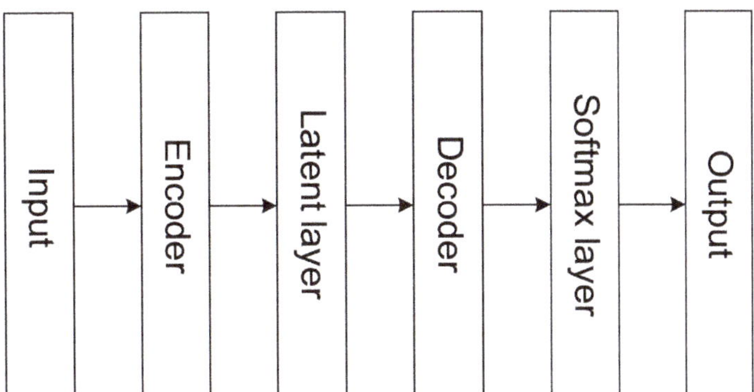

Figure 4. SAE model structure.

The softmax layer uses a softmax activation function, which is employed in a deep learning-based model, to produce a classification model that classifies data into three or more classes [33]. If the input vector is z, the softmax activation function can be expressed as follows:

$$softmax(z) = \frac{\exp(a_i)}{\sum_{j=i}^{k} \exp(a_j)}$$

Here, k is the number of classes that should be output by the multi-class classifier, and $\exp(a_i)$ is the standard exponential function of the ith input vector. Lastly, $\exp(a_j)$ is the standard exponential function of the jth output vector.

4.3. Evaluation Metrics

This study selected accuracy, TPR, FPR, and MCC as the evaluation metrics to compare the performance of the models. The accuracy is the ratio of the number of data points that the multi-class classification model correctly classifies to the overall number of data points. TPR is the level at which the actual correct answer is clearly predicted. FPR is used to evaluate multi-class classification performance. Lastly, MCC has the advantage of being able to express the confusion matrix of a multi-class classification model in a balanced way; it is also a balanced evaluation metric that is good for representing the performance of such models [34,35]. MCC was considered to be the most important of all the evaluation metrics because this study presents research on a model that performs multi-class classification. The evaluation metrics can be expressed as shown in the following equations:

$$\text{True Positive Rate (TPR)} = \frac{TP}{TP + FN}$$

$$\text{False Positive Rate (FPR)} = \frac{FP}{FP + TN}$$

$$Accuracy = \frac{TP + TN}{TP + FP + TN + FN}$$

$$\text{MCC} = \frac{c \times s - \sum_k^K p_k \times t_k}{\sqrt{\left(s^2 - \sum_k^K {p_k}^2\right) \times \left(s^2 - \sum_k^K {t_k}^2\right)}}$$

Here, TP is the number of true positives, TN is the number of true negatives, FP is the number of false positives, and FN is the number of false negatives. c is the number of samples correctly predicted out of the number of samples, and s is the number of samples. K is the total number of classes, and k is an individual class (a class from 1 to K). p_k is the number of times class k was predicted, and t_k is the number of times class k truly occurred.

4.4. Composition of Models for Comparison Experiments

We performed experiments on SAE models that had various structures and latent layer sizes to find the best SAE model. The hidden layers of the model were all fully connected, and a rectified linear unit (ReLU) activation function was used. A total of five model structures were employed in the comparison experiments. We used (depth, size) to express the model structures and depict the composition of the models. Here, depth refers to the number of hidden layers in the encoder and decoder, including the latent layer. Size refers to the number of neurons in the first hidden layer of the encoder. The encoder had a structure in which the number of hidden layer neurons decreased by half compared to that in the previous layer. Additionally, the decoder structure was symmetrical with the encoder structure. Table 4 shows the structures of the models used in the comparison experiments.

Table 4. Structures of the models for comparison.

Structure (Depth, Size)	Neurons in Each Layer
(5, 32)	32–16-size of the latent layer-16–32
(5, 64)	64–32-size of the latent layer-32–64
(5, 128)	128–64-size of the latent layer-64–128
(7, 64)	64–32–16-size of the latent layer-16–32–64
(7, 128)	128–64–32-size of the latent layer-32–64–128

Here, to find the exact size of the latent layer, we performed experiments that changed the latent layer values to 3, 6, 9, and 12. In addition, the collected dataset was divided into training, testing, and validation datasets to prevent the overfitting problem when training the classification model. Further, 60% of the entire dataset was used as the training dataset, and 20% was used as the testing dataset. The last 20% of the dataset was utilized as the validation dataset. Furthermore, an Adam optimizer [36] was employed as the optimization function in the training of all models, and a value of 1×10^{-4} was used as the learning rate. The evaluation experiments for comparison were performed with the Python, Scikit-Learn, and TensorFlow libraries.

5. Experimental Results and Discussion

5.1. Experimental Results

This section assesses the performance of the models by comparing the model structures used in the comparison experiments. For this analysis, 200 epochs of training were performed on all models, and the batch size was 64. Additionally, the evaluation metrics were used to evaluate the performance of each model structure, and the latent layer size suitable for each model structure was found. Comparative evaluations were also performed on the models that showed the best performance for each model structure.

Table 5 lists the evaluation results for the (5–32) structure. The accuracy performance is stable regardless of the latent layer size. However, the FPR and MCC values are affected by the latent layer size. A latent layer size of 9 produces the best performance in the (5–32) structure.

Table 5. Evaluation results of the (5–32) structure.

Latent Layer Size	TPR	FPR	MCC	Accuracy
3	0.8656	0.0742	0.8798	0.9510
6	0.7848	0.0499	0.8831	0.9522
9	0.9541	0.0974	0.8877	0.9541
12	0.8392	0.0597	0.8836	0.9525

Table 6 shows the evaluation results for the (5–64) structure. Different from the evaluation results for the (5–32) structure, there is a significant difference in accuracy. The performance rapidly worsens when the latent layer size is 6. However, the FPR's performance is exceptional. The performance comparison results obtained using MCC indicate that the best performance occurs when the latent layer size is 12.

Table 6. Evaluation results of the (5–64) structure.

Latent Layer Size	TPR	FPR	MCC	Accuracy
3	0.9141	0.0867	0.8692	0.9469
6	0.4924	0.0355	0.8470	0.9372
9	0.9556	0.1094	0.8782	0.9503
12	0.9077	0.0741	0.8867	0.9538

Table 7 illustrates the evaluation results for the (5–128) structure. The TPR performance is best when the latent layer size is 12, and the FPR performance is best when the latent layer size is 6. However, the MCC and accuracy performance are best when the latent layer size is 9.

Table 7. Evaluation results of the (5–128) structure.

Latent Layer Size	TPR	FPR	MCC	Accuracy
3	0.8524	0.0706	0.8906	0.9551
6	0.8776	0.1089	0.8808	0.9512
9	0.9038	0.0541	0.9054	0.9612
12	0.9085	0.0859	0.8864	0.9536

Table 8 lists the evaluation results for the (7–64) structure. The TPR performance is the best when the latent layer size is 3, and the FPR performance is the best when the latent layer size is 9. However, the MCC and accuracy performance are at their best when the latent layer size is 12.

Table 8. Evaluation results of the (7–64) structure.

Latent Layer Size	TPR	FPR	MCC	Accuracy
3	0.9440	0.1356	0.8581	0.9423
6	0.8436	0.0676	0.8926	0.9559
9	0.8253	0.0219	0.8981	0.9581
12	0.8470	0.0319	0.9036	0.9604

Table 9 shows the evaluation results for the (7–128) structure. The TPR performance is the best when the latent layer size is 9, and the FPR, MCC, and accuracy performance are the best when the latent layer size is 6.

Table 9. Evaluation results of the (7–128) structure.

Latent Layer Size	TPR	FPR	MCC	Accuracy
3	0.9367	0.0792	0.8867	0.9538
6	0.8750	0.0518	0.8980	0.9583
9	0.9319	0.0585	0.8979	0.9583
12	0.8947	0.0556	0.8955	0.9573

Table 10 illustrates the best model structures based on MCC scores. In the comparison results, the MCC evaluation metric confirms that the (5–128) structure is the best at classifying the ship operating mode when the latent layer size is 9. Furthermore, Figure 5 shows the MCC value according to the latent layer size of the models used in the comparison experiments.

Table 10. Best models from structures based on MCC score.

Structure	Latent Layer Size	TPR	FPR	MCC	Accuracy
(5–32)	9	0.9541	0.0974	0.8877	0.9541
(5–64)	12	0.9077	0.0741	0.8867	0.9538
(5–128)	9	0.9038	0.0541	0.9054	0.9612
(7–64)	12	0.8470	0.0319	0.9036	0.9604
(7–128)	6	0.8750	0.0518	0.8980	0.9583

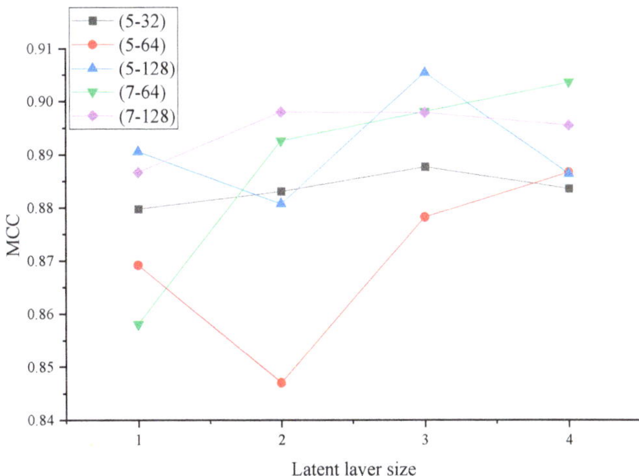

Figure 5. MCC results of structures with latent layer sizes.

Tables 11–15 exhibit the confusion matrices of the best model structures based on MCC scores. The rows of the tables present the classification results, and the columns show the actual classification classes. The diagonal cells of the matrices show the numbers of successful classification results, and the off-diagonal cells show the numbers of misclassified results.

Table 11 presents the confusion matrix of the (5–32) structure with a latent layer size of 9. It can be seen that the classification performance for the Stand By state is the best among the confusion matrix entries compared.

Table 11. Confusion matrix of (5–32) structure with a latent layer size of 9.

	In Port	Stand by	At Sea
In port	991	16	0
Stand by	97	333	10
At sea	0	155	4466
Correct	0.9108	0.6607	0.9977
Incorrect	0.0891	0.3392	0.0022

Table 12 shows the confusion matrix of the (5–64) structure with a latent layer size of 12. The classification performance for the At Sea state is the best among the confusion matrix entries compared.

Table 12. Confusion matrix of (5–64) structure with a latent layer size of 12.

	In Port	Stand by	At Sea
In port	1012	31	0
Stand by	76	305	5
At sea	0	168	4471
Correct	0.9301	0.6051	0.9988
Incorrect	0.0698	0.3948	0.0011

Table 13 provides the confusion matrix of the (5–128) structure with a latent layer size of 9. The classification performance for the In Port and Stand By states is relatively high among the confusion matrix entries compared, whereas the classification performance for the At Sea state is the lowest. However, it can also be seen that the classification model performance for the At Sea state is not especially low, with only 20 misclassified data points.

Table 13. Confusion matrix of (5–128) structure with a latent layer size of 9.

	In Port	Stand by	At Sea
In port	1048	35	0
Stand by	40	329	20
At sea	0	140	4456
Correct	0.9632	0.6527	0.9955
Incorrect	0.0367	0.3472	0.0044

Table 14 shows the confusion matrix of the (7–64) structure with a latent layer size of 12. The classification performance for the In Port state is the best among the confusion matrix entries compared. Conversely, the performance for the Stand By state is the lowest.

Table 14. Confusion matrix of (7–64) structure with a latent layer size of 12.

	In Port	Stand by	At Sea
In port	1060	54	0
Stand by	28	299	7
At sea	0	151	4469
Correct	0.9742	0.5932	0.9984
Incorrect	0.0257	0.4067	0.0015

Table 15 presents the confusion matrix of the (7–128) structure with a latent layer size of 6. This model has the lowest average performance out of the confusion matrix entries compared.

Table 15. Confusion matrix of (7–128) structure with a latent layer size of 6.

	In Port	Stand by	At Sea
In port	1043	44	0
Stand by	45	308	12
At sea	0	152	4464
Correct	0.9586	0.6111	0.9973
Incorrect	0.0413	0.3888	0.0026

5.2. Discussion

Performance by Model Structure and Latent Size

This section discusses the importance of the findings of this study. First, we focused on model structure, which was found to affect the classification performance: the (5–128) structure gave the best results. This less complex model structure (5–128) had a better performance result than more complex model structures (7–64, 7–128).

Second, we investigated the effect of latent size on model performance. We found that the performance of the model increased with latent size up to a certain point and declined after that point. Hence, finding the appropriate latent size can improve performance even when a model's structure is fixed.

Through this study, several limitations of the variational autoencoder (VAE) model were identified. First, the proposed VAE model utilizes only container vessel data. These data may differ from those obtained from other types of ships, such as LNG ships and bulk ships. This pilot study focused on performing comparative experiments using confusion matrices and evaluation metrics to evaluate the ship state classification performance according to changes in the parameters of VAE models using actual ship data. Therefore, the direct discussion that is possible based on the current results is limited. In order to improve the performance of the ship state classification model in the future, a comparative study with deep learning-based classification models using various structures that are currently used in other fields is needed.

Second, we found that the classification performance for the Stand By state was low. The model is affected by the balance of data, and the quantity of data for the Stand By state was very small in this study. This class imbalance problem can be solved in three ways: data-level, algorithm-level, and hybrid methods [37]. Data-level methods apply various data sampling methods that try to create balanced, distributed data for the training dataset. The cost-sensitive approaches focus on diminishing the bias toward major groups [38]. Hybrid methods combine the benefits of the previous two types of methods and minimize their weaknesses to improve classification model performance [39]. By adopting these methods, the problem of imbalanced data can be solved.

In the commercial realm, no algorithm has been presented that can automatically classify current ship states. However, as research on smart ships, reducing ship fuel consumption, and eco-friendly algorithms is conducted, the VAE model proposed in this pilot study can provide ship state-applied ship data to smart ship researchers. In addition, research can be conducted on reducing ship fuel consumption and improving eco-friendliness using the characteristics of the ship's state.

6. Conclusions

Artificial intelligence research using ship data is being actively conducted as the shipping market evolves. However, studies have not been performed on classifying ship operating modes, despite previous studies on using ship data to predict power loads. An SAE has the advantage of being able to perform effective classification by analyzing the features of the input data. Therefore, we conducted a pilot study on deep learning models that can classify ship operating modes using an SAE. Furthermore, experiments were performed to compare the performance according to changes in the structure of the SAE and changes in the size of its latent layer. The key points to be verified through research are as follows:

1. TPR, FPR, accuracy, and MCC were selected as evaluation metrics to perform experiments that compared the performance according to changes in the structure of the SAE and the size of its latent layer. Even though previous studies have not been conducted in this area, performance comparison among the models was possible according to changes in the structure of the SAE and the size of its latent layer.
2. In the results of the SAE model comparison experiments, the (5–128) structure with a latent layer size of 9 showed the best operating mode classification performance.
3. The classification performance for the In Port and At Sea modes used in the experiments was generally excellent; however, the classification performance for the Stand By mode was low. Therefore, more data are needed to improve the prediction performance of the Stand By mode.

Through this pilot study, we found that our VAE-based deep learning model can be used to analyze ship operating modes. Furthermore, a model that can be used as a comparison target group for the classification of the actual operating state of the ship in the future was found. Based on this model structure, it will be possible to develop enhanced models. Although the classification performance for the Stand By mode was limited because of the imbalanced data, it was possible to propose a VAE model structure that can maximize the data classification performance of the In Port and At Sea modes. However, further research is required to address some limitations of this study. First, the issue of handling imbalanced data needs to be studied using real ship data; data-level techniques, algorithm-level methods, and hybrid methodologies can be utilized to find the most appropriate method to improve the classification model. Second, research to compare denoising, sparse, and stacking autoencoder models could be carried out to improve classification performance and establish which autoencoder-based model best interprets the features of real ship data.

Author Contributions: Conceptualization, methodology, and software, J.-Y.K.; project administration, funding acquisition, J.-S.O. All authors have read and agreed to the published version of the manuscript.

Funding: This research was supported by the Korea Institute of Marine Science and Technology Promotion (KIMST), funded by the Korea Coast Guard (20190460).

Institutional Review Board Statement: Not applicable.

Informed Consent Statement: Not applicable.

Data Availability Statement: Not applicable.

Conflicts of Interest: The author declares no conflict of interest.

References

1. Tang, Y.-l.; Shao, N.-n. Design and Research of Integrated Information Platform for Smart Ship. In Proceedings of the 2017 4th International Conference on Transportation Information and Safety (ICTIS), Banff, AB, Canada, 8–10 August 2017; pp. 37–41. Available online: https://ieeexplore.ieee.org/document/8047739 (accessed on 12 January 2023).
2. Smart Shipping: Comprehensive Automation in the Maritime Sector. Available online: https://www.government.nl/topics/maritime-transport-and-seaports/smart-shipping-comprehensive-automation-in-the-maritime-sector (accessed on 12 January 2023).
3. Zeng, X.; Chen, M. A novel big data collection system for ship energy efficiency monitoring and analysis based on BeiDou system. *J. Adv. Transp.* **2021**, *2021*, 9914720. [CrossRef]
4. Reilly, G.; Jorgensen, J. Classification considerations for cyber safety and security in the smart ship era. In Proceedings of the International Smart Ships Technology Conference, London, UK, 26–27 January 2016; pp. 26–27. Available online: https://ww2.eagle.org/content/dam/eagle/Archived-Assets/leadership/articles-archives/ABS-RINA-Cyber-Safety-Security-Ship-Tech.pdf (accessed on 27 January 2023).
5. Pérez Fernández, R.; Benayas Ayuso, A.; Pérez Arribas, F.L. Data Management for Smart Ship or How to Reduce Machine Learning Cost in IoS Applications. 2018. Available online: https://www.researchgate.net/publication/322635826_DATA_MANAGEMENT_FOR_SMART_SHIP_OR_HOW_TO_REDUCE_MACHINE_LEARNING_COST_IN_IoS_APPLICATIONS (accessed on 10 January 2023).
6. Xiao, Y.; Chen, Z.; McNeil, L. Digital empowerment for shipping development: A framework for establishing a smart shipping index system. *Marit. Pol. Manag.* **2022**, *49*, 850–863. [CrossRef]
7. Aslam, S.; Michaelides, M.P.; Herodotou, H. Internet of ships: A survey on architectures, emerging applications, and challenges. *IEEE Internet Things J.* **2020**, *7*, 9714–9727. [CrossRef]
8. Ahn, Y.G.; Kim, T.; Kim, B.R.; Lee, M.K. A study on the development priority of smart shipping items—Focusing on the expert survey. *Sustainability* **2022**, *14*, 6892. [CrossRef]
9. Chen, C.; Chen, X.Q.; Ma, F.; Zeng, X.J.; Wang, J. A knowledge-free path planning approach for smart ships based on reinforcement learning. *Ocean Eng.* **2019**, *189*, 106299. [CrossRef]
10. Ding, Y.; Li, R.; Shen, H.; Li, J.; Cao, L. A novel energy-saving route planning algorithm for marine vehicles. *Appl. Sci.* **2022**, *12*, 5971. [CrossRef]
11. Accetta, A.; Pucci, M. A First Approach for the Energy Management System in DC Micro-Grids with Integrated RES of Smart Ships. In Proceedings of the 2017 IEEE Energy Conversion Congress and Exposition (ECCE), Cincinnati, OH, USA, 1–5 October 2017; pp. 550–557. Available online: https://ieeexplore.ieee.org/document/8095831 (accessed on 19 February 2023).
12. Hasanvand, S.; Rafiei, M.; Gheisarnejad, M.; Khooban, M.H. Reliable power scheduling of an emission-free ship: Multiobjective deep reinforcement learning. *IEEE Trans. Transp. Electr.* **2020**, *6*, 832–843. [CrossRef]
13. Kim, J.Y.; Lee, J.H.; Oh, J.H.; Oh, J.S. A comparative study on energy consumption forecast methods for electric propulsion ship. *J. Mar. Sci. Eng.* **2022**, *10*, 32. [CrossRef]
14. Almotiri, J.; Elleithy, K.; Elleithy, A. Comparison of Autoencoder and Principal Component Analysis Followed by Neural Network for E-learning Using Handwritten Recognition. In Proceedings of the IEEE Long Island Systems, Applications and Technology Conference, Farmingdale, NY, USA, 5 May 2017. Available online: https://ieeexplore.ieee.org/document/8001963 (accessed on 19 February 2023).
15. Yao, R.; Liu, C.; Zhang, L.; Peng, P. Unsupervised Anomaly Detection Using Variational Auto-Encoder based Feature Extraction. In Proceedings of the International Conference on Prognostics and Health Management, San Francisco, CA, USA, 17 June 2019. Available online: https://ieeexplore.ieee.org/abstract/document/8819434 (accessed on 10 February 2023).
16. Ma, S.; Chen, M.; Wu, J.; Wang, Y.; Jia, B.; Jiang, Y. High-voltage circuit breaker fault diagnosis using a hybrid feature transformation approach based on random forest and stacked autoencoder. *IEEE Trans. Ind. Electron.* **2018**, *66*, 9777–9788. [CrossRef]

17. Thirukovalluru, R.; Dixit, S.; Sevakula, R.K.; Verma, N.K.; Salour, A. Generating Feature Sets for Fault Diagnosis using Denoising Stacked Auto-encoder. In Proceedings of the International Conference on Prognostics and Health Management, Ottawa, ON, Canada, 20 June 2016. Available online: https://ieeexplore.ieee.org/stamp/stamp.jsp?tp=&arnumber=7542865 (accessed on 12 February 2023).
18. Schmidhuber, J. Deep learning in neural networks: An overview. *Neural Netw.* **2015**, *61*, 85–117. [CrossRef]
19. Bank, D.; Koenigstein, N.; Giryes, R. APR, 2021. Autoencoders. Available online: https://arxiv.org/pdf/2003.05991.pdf (accessed on 12 February 2023).
20. Ghosh, S.; Laksana, E.; Morency, L.P.; Scherer, S. Representation Learning for Speech Emotion Recognition. In Proceedings of the INTERSPEECH 2016, San Francisco, CA, USA, 8 September 2016. Available online: https://doi.org/10.21437/Interspeech.2016-692 (accessed on 19 February 2023).
21. Ambaw, A.B.; Bari, M.; Doroslovački, M. A case for stacked autoencoder based order recognition of continuous-phase FSK. In Proceedings of the Annual Conference on Information Sciences and Systems (CISS), Baltimore, MD, USA, 22 March 2017. Available online: https://ieeexplore.ieee.org/xpl/conhome/7917217/proceeding (accessed on 19 February 2023).
22. Singh, K.; Malhotra, J. Stacked Autoencoders Based Deep Learning Approach for Automatic Epileptic Seizure Detection. In Proceedings of the (ICSCCC), 2018 First International Conference on Secure Cyber Computing and Communication, Jalandhar, India, 15 December 2018. Available online: https://ieeexplore.ieee.org/document/8703357 (accessed on 2 February 2023).
23. Law, A.; Ghosh, A. Multi-label classification using a cascade of stacked autoencoder and extreme learning machines. *Neurocomputing* **2019**, *358*, 222–234. [CrossRef]
24. Aouedi, O.; Piamrat, K.; Bagadthey, D. A Semi-supervised Stacked Autoencoder Approach for Network Traffic Classification. In Proceedings of the International Conference on Network Protocols, Madrid, Spain, 13 October 2020. Available online: https://ieeexplore.ieee.org/xpl/conhome/9259328/proceeding (accessed on 30 January 2023).
25. Deperlioglu, O. Heart sound classification with signal instant energy and stacked autoencoder network. *Biomed. Signal Process. Control* **2021**, *64*, 102211. [CrossRef]
26. Gokhale, M.; Mohanty, S.; Ojh, A. A stacked autoencoder based gene selection and cancer classification framework. *Biomed. Signal Process. Control* **2022**, *78*, 103999. [CrossRef]
27. Arafa, A.; El-Fishawy, N.; Badawy, M.; Radad, M. RN-Autoencoder: Reduced Noise Autoencoder for classifying imbalanced cancer genomic data. *J. Biol. Eng.* **2023**, *17*, 7. [CrossRef] [PubMed]
28. Baldi, P. Autoencoders, unsupervised learning, and deep architectures. In Proceedings of the ICML Workshop on Unsupervised and Transfer Learning, Irvine, CA, USA, June 2012; pp. 37–49. Available online: http://proceedings.mlr.press/v27/baldi12a/baldi12a.pdf (accessed on 6 February 2023).
29. Bengio, Y.; Lamblin, P.; Popovici, D.; Larochelle, H. Greedy layer-wise training of deep networks. *Adv. Neural. Inf. Process. Syst.* **2006**, *19*.
30. Kim, J.H.; Choi, J.E.; Choi, B.J.; Chung, S.H. Twisted rudder for reducing fuel-oil consumption. *Int. J. Naval Archit. Ocean Eng.* **2014**, *6*, 715–722. [CrossRef]
31. Nguyen, D.H.; Le, M.D.; Ohtsu, K. Ship's optimal autopilot with a multivariate auto-regressive exogenous model. *IFAC Proc. Vol.* **2000**, *33*, 277–282. [CrossRef]
32. Nwankpa, C.; Ijomah, W.; Gachagan, A.; Marshall, S. Activation functions: Comparison of trends in practice and research for deep learning. *arXiv* **2018**, arXiv:1811.03378. Available online: https://arxiv.org/pdf/1811.03378.pdf (accessed on 6 February 2023).
33. Kagalkar, A.; Raghuram, S. Activation Functions: CORDIC Based Implementation of the Softmax Activation Function. In Proceedings of the International Symposium on VLSI Design and Test, Bhubaneswar, India, 23 July 2020. Available online: https://ieeexplore.ieee.org/abstract/document/9190498 (accessed on 30 January 2023).
34. Gorodkin, J. Comparing two K-category assignments by a K-category correlation coefficient. *Comp. Biol. Chem.* **2004**, *28*, 367–374. [CrossRef]
35. Jurman, G.; Riccadonna, S.; Furlanello, C. A Comparison of MCC and CEN Error Measures in Multi-Class Prediction. 2012. Available online: https://journals.plos.org/plosone/article?id=10.1371/journal.pone.0041882 (accessed on 20 January 2023).
36. Kingma, D.P.; Ba, J. Adam: A method for stochastic optimization. *arXiv* **2014**, arXiv:1412.6980. Available online: https://arxiv.org/abs/1412.6980 (accessed on 15 January 2023).
37. Krawczyk, B. Learning from imbalanced data: Open challenges and future directions. *Prog. Artif. Intell.* **2016**, *5*, 221–232. [CrossRef]
38. Zięba, M.; Tomczak, J.M. Boosted SVM with active learning strategy for imbalanced data. *Soft Comput.* **2015**, *19*, 3357–3368. [CrossRef]
39. Woźniak, M. Hybrid classifiers. In *Methods of Data, Knowledge, and Classifier Fusion*; Springer: Berlin/Heidelberg, Germany, 2013; p. 519. [CrossRef]

Article

Characteristics and Causes of Coastal Water Chemistry in Qionghai City, China

Junyi Jiang [1,†], Guowei Fu [1,†], Yu Feng [2,3], Xinchen Gu [4,5,*], Pan Jiang [6,*], Cheng Shen [1] and Zongyi Chen [7]

1 Haikou Marine Geological Survey Center, China Geological Survey, Haikou 571127, China
2 School of Environment, Tsinghua University, Beijing 100084, China
3 Chinese Research Academy of Environmental Sciences, Beijing 100012, China
4 State Key Laboratory of Hydraulic Engineering Simulation and Safety, School of Civil Engineering, Tianjin University, Tianjin 300072, China
5 State Key Laboratory of Simulation and Regulation of Water Cycle in River Basin, China Institute of Water Resources and Hydropower Research, Beijing 100044, China
6 College of Environment and Resources, Southwest University of Science and Technology, Mianyang 621010, China
7 Monitoring Centre for Hydrology and Water Resources Upstream Ganjiang River, Ganzhou 341000, China
* Correspondence: gxc@tju.edu.cn (X.G.); jiangpan@mails.swust.edu.cn (P.J.)
† These authors contributed equally to this work and should be considered co-first authors.

Abstract: The coastal zone area of Qionghai City is one of the important coastal zones in the South China Sea, and its water environment has been affected by human activities such as urbanization and industrialization. In order to protect the water resources and ecological environment of this area, the water chemistry characteristics of the main watersheds and their causes in the coastal zone area of eastern Hainan Island were investigated to provide a scientific basis for environmental protection and sustainable development. In this study, the characteristics and sources of water chemical ion components were analyzed using a Piper trilinear diagram, Gibbs diagram, and correlation analysis with the coastal zone area of Qionghai city as the research object. The results show the following: (1) the dominant cation of water chemistry in the coastal zone of Qionghai City is Na^+ with a mean value of 35.001 $mg \cdot L^{-1}$, and the dominant anion is Cl^- with a mean value of 30.69 $mg \cdot L^{-1}$; (2) the dominant cation content in the coastal zone of Qionghai City is $Na^+ > Ca^{2+} > Mg^{2+} > K^+$, and the dominant anion content is $Cl^- > SO_4{}^{2-} > HCO_3{}^- > CO_3{}^{2-}$; (3) at the five collection sites in the study area, the ion concentrations showed different trends, with the highest ion concentration in the water samples collected from aquaculture ponds, and the main water chemistry type was Na-Cl; the lowest ion concentration was in the water samples collected from the rivers, and the main type of water chemistry was $Ca \cdot Mg\text{-}HCO_3$. The source of water chemistry ions in the study area mainly included seawater, rock weathering, atmospheric precipitation, and evaporation concentration. The results of this study can provide a scientific basis for the development, utilization, and management of local water resources and provide basic data for environmental protection and sustainable development.

Keywords: Qionghai coastal zone; water chemical characteristics; cause analysis; source of ions

Citation: Jiang, J.; Fu, G.; Feng, Y.; Gu, X.; Jiang, P.; Shen, C.; Chen, Z. Characteristics and Causes of Coastal Water Chemistry in Qionghai City, China. *Appl. Sci.* **2023**, *13*, 5579. https://doi.org/10.3390/app13095579

Academic Editor: Mauro Marini

Received: 25 March 2023
Revised: 25 April 2023
Accepted: 28 April 2023
Published: 30 April 2023

1. Introduction

In recent years, water quality studies for the coastal zone area of Qionghai City have gradually increased, including the monitoring and analysis of seawater, river water, lake water, and other water bodies. Some bays and estuaries in the region have water pollution problems, mainly from agricultural, industrial, and urban sewage sources. Water chemistry plays an important role in ecological environmental protection, water resource utilization, water pollution control, and water safety assurance. The water chemistry of watersheds can reflect the effects of rock weathering, atmospheric deposition, and human activities in the watershed [1]. By analyzing the characteristics and spatial and temporal changes in river water chemistry, the basic processes that proceed from water chemistry component formation could be effectively discerned [2]. In addition, the analysis of the water chemistry characteristics of coastal zone water bodies is a very critical part of the evaluation of the quality of water resources in the basin, which has non-negligible value for the scientific development and natural protection of water resources in the basin [3]. The water chemistry composition characteristics of watershed water bodies are mainly influenced by physical geography [4]; however, more and more studies show that the water chemistry composition of watershed waters is also influenced by human activities and has become an important issue affecting human survival and social development [4–6].

For the above-mentioned studies, many results have been achieved both in China and abroad, and research methods tend to be diversified [7,8]. Since the middle of the 20th century, a large amount of theoretical knowledge and a great number of technical means have been used to study the characteristics and evolution of the chemistry in both surface water and groundwater. Among them, there are many mathematical methods to deal with water quality data, generally including cluster analysis, principal component analysis, correlation analysis, factor analysis, etc. [9,10]. Based on the analysis of water quality data and the further use of piper trilinear plots [11], Gibbs [12], the water quality simulation method, the isotope analysis method [13,14], etc., investigations into the ion chemical characteristics of rivers and the influence of major weathering processes in watersheds have been undertaken [15]. Following this, the GIS visualization function appeared as a clear display of water quality results [16]. Gibbs' analysis of the water chemistry components of various global water bodies (precipitation, seawater, lakes, and rivers) suggests that rock weathering, atmospheric precipitation input, and evapotranspiration–crystallization are the three major controlling factors of global surface water chemistry components [12]. Kattan's analysis of the water chemistry of the Euphrates River in Syria showed that the ionic fraction in the river was influenced by the dissolution of rock weathering, water temperature, and evaporation [17]. Different study areas were located in different environmental contexts, including a variety of factors such as geological formations, climate, and human activities. These factors affect the ion concentrations and chemical composition of water bodies, resulting in significant differences in the water chemistry characteristics of different regions. Therefore, when conducting water resources management and environmental protection, the influence of these factors needs to be fully considered when developing management measures that are appropriate for the region [18–20].

Qionghai City, located on the east coast of Hainan Island, is a coastal city with rich coastal ecological resources. The quality of the water environment in the area is important for maintaining the local ecological environment and developing tourism [21]. Water chemistry is the study of the distribution, transformation, and transport patterns of substances such as dissolved inorganic compounds, organic matter, and biological elements in water bodies and their impact on the ecological environment.

Therefore, what type of water chemistry is found in Qionghai City? Additionally, what are the sources of water chemistry ions? In this paper, the main water chemistry indicators, including the total dissolved solids (TDS), pH value, and anion and cation concentration, are analyzed to reveal the chemical composition of the area's water bodies, aiming to provide a scientific basis for environmental protection and sustainable development in the region.

2. Materials and Methods

2.1. Study Area Overview

Qionghai City is located in the center of the eastern Hainan Province [22], located at 110°07′~110°41′ E, 18°59′~19°29′ N, east of the South China Sea, west of Tunchang County and Qiongzhong County, north of Dingan County and Wenchang City, and south of Wanning City, with a total area of 1692 km^2. Qionghai City belongs to the tropical monsoon and marine humid climate zone, which is greatly influenced by monsoons, abundant light, high temperatures, rain, frequent typhoons, four indistinct seasons, and distinct dry and rainy seasons. The average annual rainfall is 2072 mm, the average annual sunshine is 2155 h, the average annual temperature is 24 °C, and there is no frost and snow all year round. There are many rivers in this territory, abundant water resources, and the total water resources of Qionghai City reach 2.216 billion m^3, ranking third in Hainan Province, with available water resources at 8.40 million m^3 (37.9% of the total water resources). In 2020, the regional GDP reached CNY 112.98 billion, an increase of 3.1% year on year. At the same time, Qionghai City completed the preliminary results of the "Qionghai City Coastal Zone Protection and Utilization Comprehensive Plan", which aimed to strengthen the protection of environmental resources and the development and utilization management of the coastal zone.

2.2. Sample Collection and Determination

The water quality of civil wells, reservoirs, aquaculture ponds, rivers, and canals is a key area for water quality in Qionghai City. The specificity of Qionghai City needs to be considered in order to consider the evolution of water chemistry in the coastal zone area of Qionghai City in an integrated way. A total of 177 samples were collected from civil wells, reservoirs, aquaculture ponds, rivers, and canals in Qionghai between May and June 2022. A total of 5 samples were sent for testing, and the distribution of the sampling sites is shown in Figure 1. Water samples were generally collected below 10 cm from the water surface and then filtered through a 0.45 μm filter membrane, and those for cation (Ca^{2+}, Mg^{+}, K^{+}, and Na^{+}) analysis were acidified with nitric acid at a pH < 2, and those for anion analysis (Cl^{-}, $CO_3{}^{2-}$, $HCO_3{}^{-}$, and $SO_4{}^{2}$, the anion samples (Cl^{-}, $CO_3{}^{2-}$, $HCO_3{}^{-}$, and $SO_4{}^{2-}$)) were not added with the reagents; the samples were sealed with paraffin and stored away from light. The samples were sent to the laboratory of the China Geological Survey Haikou Marine Geological Survey Center for the determination of K^{+} and Na^{+} by flame emission spectrometry, Ca^{2+} and Mg^{2+} by disodium ethylenediaminetetraacetate titration, Cl^{-} and $SO_4{}^{2-}$ by ion chromatography, and $HCO_3{}^{-}$ and $CO_3{}^{2-}$ by the hydrochloric acid volumetric method. The test results met the quality requirements.

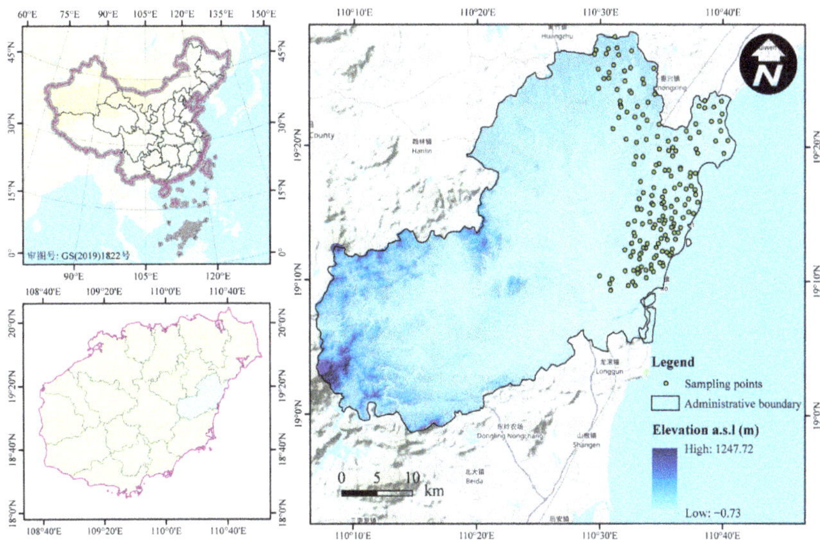

Figure 1. Schematic diagram of the distribution of sampling points of water bodies along Qionghai's coasts.

2.3. Analytical Methods and Quality Assurance

In this paper, the water chemistry data were counted by Excel and mathematical statistics, Pearson correlation analysis, Piper's trilinear diagram, and Gibbs diagram were used to analyze the water chemistry characteristics and control factors in combination with hydrogeological conditions in the study area. Among them, Piper trilinear plots and Gibbs plots were developed using Origin 2020, and inter-ion correlations were analyzed using SPSS 22 statistics.

The Pearson correlation coefficient, also known as the product difference correlation coefficient, is a statistical indicator that expresses the degree and direction of linear correlation between the two variables. The correlation coefficient of a sample is denoted by the symbol ***r*** and is calculated by the following formula [23,24]:

$$r = \frac{\Sigma(X-\overline{X})(Y-\overline{Y})}{\sqrt{\Sigma(X-\overline{X})^2\Sigma(Y-\overline{Y})^2}} = \frac{l_{XY}}{\sqrt{l_{XX}l_{YY}}} \tag{1}$$

$$l_{XX} = (X-\overline{X})^2 = \sum X^2 - \frac{(\sum X)^2}{n} \tag{2}$$

$$l_{YY} = (Y-\overline{Y})^2 = \sum Y^2 - \frac{(\sum Y)^2}{n} \tag{3}$$

$$l_{XY} = \Sigma(X-\overline{X})(Y-\overline{Y}) = \sum XY - \frac{(\sum X)(\sum Y)}{n} \tag{4}$$

where l_{YY} denotes the off-average sum of squares for X; l_{YY} denotes the off-average sum of squares for Y; l_{XY} denotes the off-average sum of squares for X and Y; Pearson's correlation coefficient is a dimensionless statistical indicator with a range of $-1 \le r \le 1$. A correlation coefficient less than 0 is a negative correlation, greater than 0 is a positive correlation, and equal to 0 indicates no correlation.

The larger the absolute value of the correlation coefficient, the closer the correlation between the two variables. To determine whether there is a linear relationship, the researcher needs to look at the scatter plot of the two variables; if the scatter plot is roughly a straight line, it indicates a linear relationship, and if it is not a straight line, there is no linear relationship.

The cluster analysis K-means algorithm, also known as the K-means clustering algorithm, is a widely used clustering algorithm that is the basis for other clustering algorithms [25,26]. Assuming that the input samples are $S = X_1, X_2, \cdots, X_m$, the algorithm steps are as follows: (1) select the initial k category centers $\mu_1, \mu_2, \ldots \mu_k$; (2) for each sample X_i, mark it as the closest category to the category center (distance can be calculated using Euclidean distance); (3) update each category center to the mean of all samples belonging to that category; (4) repeat the last two steps until the centroid is unchanged or a predetermined number of iterations is reached, and the algorithm terminates.

The objective function of the K-means algorithm is the within-cluster sum of squared errors (SSE), also known as cluster inertia.

$$SSE = \sum_{i=1}^{n}\sum_{j=1}^{k} w^{(i,j)} \|x^i - \mu^j\|^2 \tag{5}$$

where $w^{(i,j)} = 1$ if x^i belongs to cluster j. Otherwise, it is 0.

3. Results

3.1. Basic Physical and Chemical Properties of Water Bodies

The minimum, maximum, mean, and standard deviation of each indicator of the analyzed water samples in terms of the results are shown in Table 1. The content of the dominant cations in the water in the study area showed $Na^+ > Ca^{2+} > Mg^{2+} > K^+$ and the average concentration of the dominant cation Na^+ was 34.432 mg·L^{-1}. The dominant anion content showed $Cl^- > SO_4^{2-} > HCO_3^- > CO_3^{2-}$, and the average concentrations of the dominant anions Cl^-, SO_4^{2-}, and HCO_3^- were 30.69 mg·L^{-1}, 18.49 mg·L^{-1}, and

15.40 mg·L^{-1}, respectively. The mean equivalent concentration of Cl^- accounted for 46% of the total anions, and the mean equivalent concentration of SO_4^{2-} accounted for 28% of the total anions. The coefficient of variation could characterize the stability of ionic components, and the larger the coefficient of variation, the more susceptible the ionic component was to the influence of the external environment.

Table 1. Analysis results of major ion components in the Qionghai coastal basin.

Ions	Maximum Value (mg·L^{-1})	Minimum Value (mg·L^{-1})	Average Value (mg·L^{-1})	Standard Deviation (mg·L^{-1})	Coefficient of Variation (%)
Ca^{2+}	86.47	10.61	34.43	17.72	52.00
Mg^{2+}	66.84	2.14	22.47	13.04	58.00
Na^+	90.90	20.20	35.00	13.39	38.00
K^+	11.85	0.45	3.75	2.81	75.00
Cl^-	107.00	4.01	30.69	25.68	84.00
SO_4^{2-}	69.30	0.34	18.49	15.05	81.00
CO_3^{2-}	5.56	0.12	1.34	1.05	78.00
HCO_3^-	55.03	1.34	15.36	9.90	64.00
TDS	852.00	12.00	203.60	150.44	74.00
Total hardness	3.51	0.14	0.95	0.66	70.00
pH	9.08	2.41	6.93	0.98	14.00

3.2. Spatial Variation Characteristics of Major Ion Components

The trend of all types of ion concentrations is given in Figure 2. The trend of all ion concentrations is the trend of water chemical ions. Figure 2 shows the temporal trends of the water chemical ion fraction content in the coastal zone area of Qionghai City, and it was found that the overall trend showed the ion concentration to gradually decrease. The trends of different ions also showed their different characteristics, and ion concentrations were affected by many disturbing factors due to the different environments and conditions in which different collection sites were located.

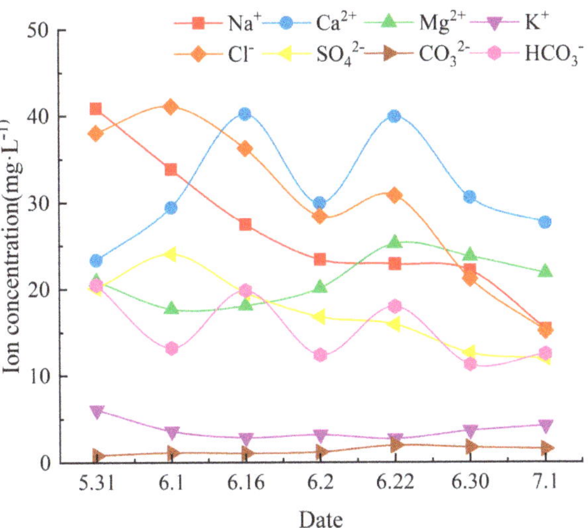

Figure 2. Spatial variation characteristics of water chemical ion component content in Qionghai coastal water.

The spatial distribution map shows (Figure 3) that the ionic concentration content of the cation Na^+ was higher at 40.90 mg·L^{-1}, while the anion Cl^- had the highest ionic concentration content of 41.10 mg·L^{-1}. Among the selected collection sites, the water column ion concentration was highest in the aquaculture ponds and lowest in the rivers. In general, the trend of ion concentration was the result of the combined influence of many factors.

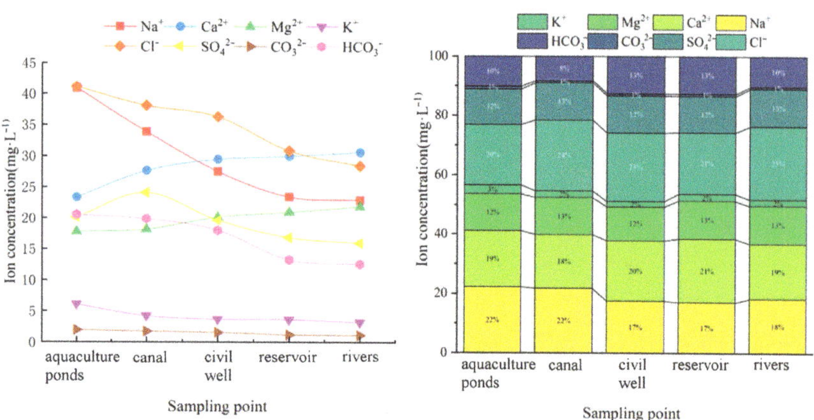

Figure 3. Spatially variable characteristics of the average concentration of major ions on the Wanning coast.

Through the mathematical and statistical analysis of the water chemistry indexes of water bodies in the coastal zone of Qionghai City, we could obtain a preliminary understanding of the enrichment degree and the changing pattern of the water chemistry of water bodies in the study area [27]. Figure 4 shows that the average pH value of water bodies in the area was 6.93, with a coefficient of variation of 14%. In addition, the concentration of total dissolved substance (TDS) values showed a gradual decrease and varied between 12 and 852 mg·L^{-1}, with an average value of 203.6 mg·L^{-1}.

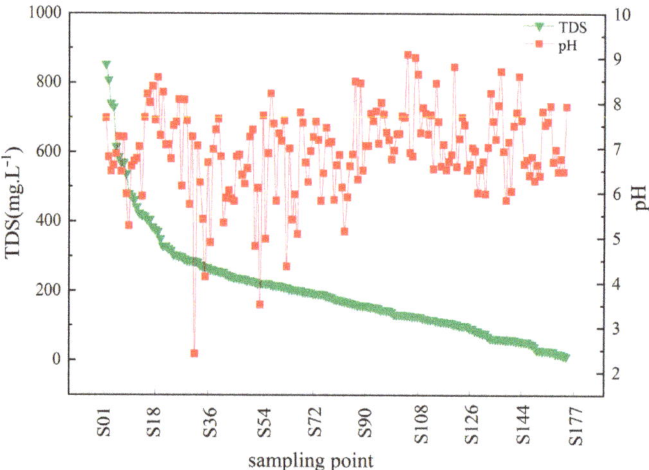

Figure 4. Trends of chemical TDS and pH in Qionghai coastal water (TDS: total dissolved solids; pH: Pondus Hydrogenii).

3.3. Water Chemistry Type

The Piper trilinear diagram is a graphical method used to describe the chemical components of water. By analyzing the distribution and combination type of ion content in the Piper trilinear map, the water chemistry type and water quality characteristics of the surface water could be reflected [28,29]. The Piper trilinear diagram of the major ions in the surface water of the coastal zone area of Qionghai City is shown in Figure 5. From the figure, the main water chemistry type of surface water in the aquaculture ponds was the Na-Cl type, which accounted for the largest proportion; this was followed by the Ca-Mg-Cl type water, while the SO_4 type water accounted for a smaller proportion. The main water chemistry type of the canal was the Na^-Ca^-Cl type, the main water chemistry type of the civil well was the Ca·Mg-Cl type, the main water chemistry type of the reservoir was the Ca-Mg-SO_4 type, and the main water chemistry type of the river was the Ca·Mg-HCO_3 type. The surface water in the area of the aquaculture ponds was mainly influenced by seawater, which had a high content of Na and Cl, while the content of other ions was relatively low, so the ion content in the surface water could also be dominated by Na^+ and Cl^-. In addition, the high content of Na also reflected the high salinity of groundwater in the area, and there could be some degree of seawater intrusion. Finally, we also drew an ilr-ion plot (Figure 5f) of all five types [30].

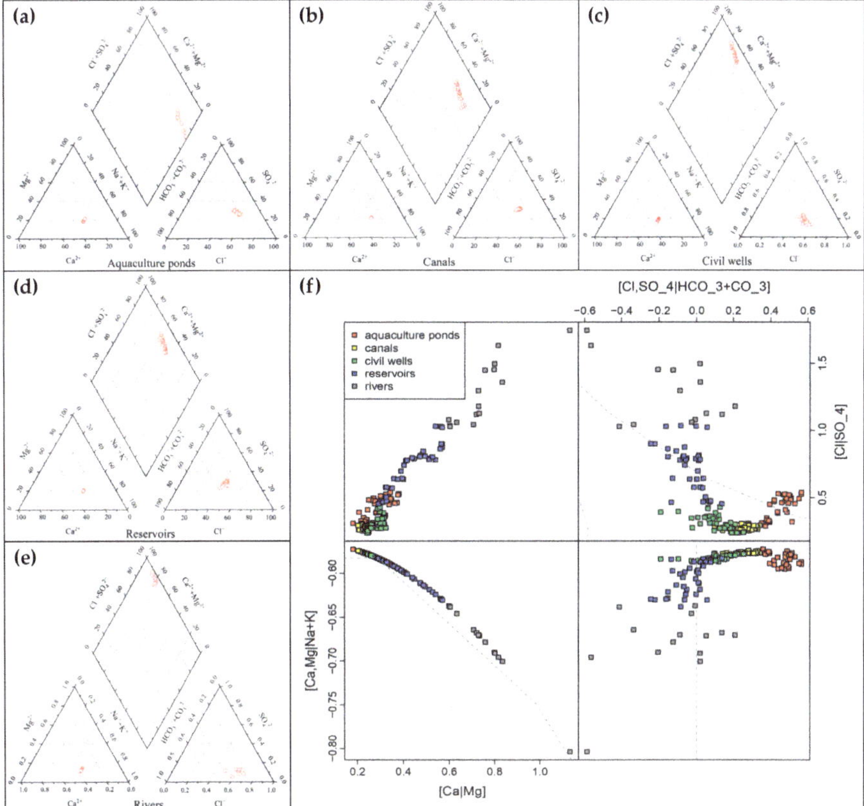

Figure 5. Major ions of water chemistry in Qionghai: (**a**) Piper trilinear diagram of the aquaculture ponds; (**b**) Piper trilinear diagram of the canals; (**c**) Piper trilinear diagram of the civil wells; (**d**) Piper trilinear diagram of the reservoirs; (**e**) Piper trilinear diagram of the civil rivers; (**f**) Ilr-ion plot of all 5 types.

3.4. Analysis of Major Ion Sources and Control Factors

The Gibbs diagram is a graphical method that can describe the source of water chemical components. By analyzing the distribution of each ion in the Gibbs diagram, it is possible to qualitatively determine the influence of factors such as atmospheric rainfall, rock weathering, and evaporation concentration on the source of ions in the surface water [31]. In the Gibbs plot of the coastal zone area of Qionghai City (Figure 6), it can be seen that the cation $Na^+/(Na^+ + Ca^{2+})$ was mainly between 8% and 60%, and the anion $Cl^-/(Cl^- + HCO_3^-)$ was mainly between 9% and 50%. This indicates that the main sources of ions in surface water in the region are atmospheric precipitation and seawater, while the contribution of rock weathering is small. In this region, the influence of seawater was high, so the proportion of $Na^+/(Na^+ + Ca^{2+})$ was relatively high, while the influence of evaporation concentration was relatively small, so the proportion of $Cl^-/(Cl^- + HCO_3^-)$ was also not high.

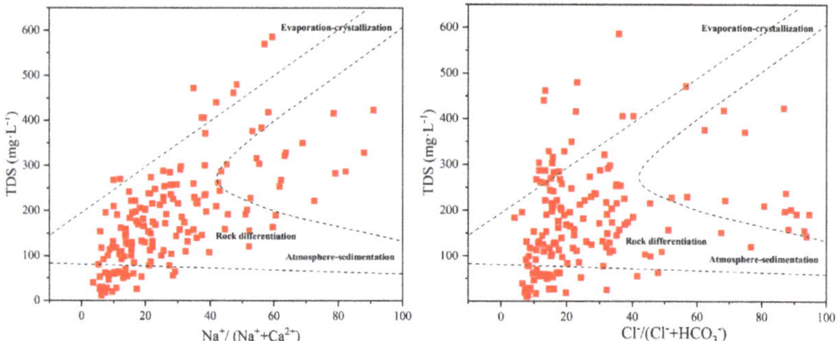

Figure 6. Qionghai Coastal Water Quality Gibbs Chart.

According to the results of the Pearson correlation coefficient analysis in Table 2, it can be seen that there was a significant correlation ($p < 0.05$) between TDS and Na^+, Ca^{2+}, Cl^-, and SO_4^{2-} in the surface water of the coastal zone of Qionghai City, indicating that there is a significant linear relationship between the content of these ions in surface water and the content of the total dissolved solids (TDS). Thus, both Na^+ and Cl^- are major contributors of ions to the water of this region. In addition, there was a significant correlation between Na^+ and Cl^-, suggesting that these two ions have the same source and may be influenced by seawater. In addition, Table 2 showed a significance of 0.14 between Ca^{2+} and HCO_3^-, probably because Ca^{2+} was derived from rock weathering processes, while HCO_3^- was mainly derived from atmospheric precipitation and evaporative concentration. It can be further speculated that the sources of surface water ions in the coastal zone of Qionghai City are mainly seawater, rock weathering, atmospheric precipitation, and evaporation concentration, and other factors [32–36].

Table 2. Correlations between conventional indicators.

	Ca^{2+}	Mg^{2+}	Na^+	K^+	Cl^-	SO_4^{2-}	HCO_3^-	CO_3^{2-}	TDS
Ca^{2+}	1								
Mg^{2+}	0.526 **	1							
Na^+	0.887 **	0.536 *	1						
K^+	0.125	0.524 *	0.128	1					
Cl^-	0.791 **	0.619 *	0.844 **	0.116	1				
SO_4^{2-}	0.689	0.199	0.743 **	0.147	0.674 **	1			
HCO_3^-	0.140	0.635 *	0.535 **	0.141	0.450 *	0.507 **	1		
CO_3^{2-}	0.296	0.133	0.372	0.117	0.161	0.136	0.140	1	
TDS	0.712 **	0.569 *	0.897 **	0.327	0.830 **	0.730 **	0.568 **	0.177	1

*. Correlation is significant at the 0.05 level (two-tailed); **. Correlation is significant at the 0.01 level (two-tailed).

4. Discussion

This study provides an in-depth analysis of the water chemistry characteristics of the main watersheds in the coastal zone area of Qionghai City and their causes. Using Piper's trilinear diagram, Gibbs diagram, and correlation analysis, this study drew significant conclusions about the water chemistry characteristics and sources in this region. The results show that the characteristics of water chemistry ion fractions in this region were similar to many other coastal zone regions.

Firstly, the dominant cation was Na^+, and the dominant anion was Cl^-, which is consistent with the characteristics of being located in the coastal zone area. Secondly, the content of dominant cations showed $Na^+ > Ca^{2+} > Mg^{2+} > K^+$, while the content of dominant anions was $Cl^- > SO_4^{2-} > HCO_3^- > CO_3^{2-}$, and these results indicate that the influence of seawater was very significant in this region. In addition, the ion concentrations showed different trends in the water samples from different collection points, mainly because of the different sources of water and environmental factors.

For example, the highest ion concentration in the water samples collected from the farming ponds, and the main water chemistry type, was Na^-Cl, which might be related to the development of the surrounding farming industry, seawater backflow, and other factors. Therefore, measures must be taken to limit the scale of the farming industry and reduce the phenomenon of seawater backflow to reduce the risk of water pollution. On the contrary, the water samples collected in the river had the lowest ion concentration and the main water chemistry type was the $Ca{\cdot}Mg\text{-}HCO_3$ type; this is because water bodies in the area may be affected by factors such as rock weathering and atmospheric precipitation.

Yang et al. [37] studied the water chemistry characteristics of the southwest coastal zone area of Hainan Province and found that Na^+ and Cl^- were the main water chemistry ions and that water bodies in the area were influenced by human activities, with agricultural activities contributing more to NO_3^- and SO_4^{2-} ions in the water bodies. In addition, Xi et al. [38] studied the water chemistry of groundwater in the east coastal zone area of Hainan Island and found that Ca^{2+} and Mg^{2+} were the main cations, HCO_3^- and Cl^- were the main anions, and the water chemistry of the area was influenced by different hydrogeological conditions and human activities. These findings are consistent with the results of this paper.

The results of this study also indicate that the sources of water chemical ions in the region resulted from various factors such as seawater, rock weathering, atmospheric precipitation, and evapotranspiration concentration effects. This conclusion provides a scientific basis for the development, utilization, and management of local water resources, which could help to protect water resources and the ecological environment of the region and promote the sustainable development of the area.

However, due to the limited scope of this study, other possible factors were not explored in depth. Our future study will further expand this topic, collecting more water quality samples in Qionghai to explore the details and characteristics and possible spatial and temporal variability of the sources of water chemistry ions in the region.

5. Conclusions

Based on the analysis of the chemical composition of surface water in the coastal areas of Qionghai, this study found significant variations in the ion concentrations and chemical types among different sampling sites. The dominant cations followed the order of $Na^+ > Ca^{2+} > Mg^{2+} > K^+$, while the dominant anions were $Cl^- > SO_4^{2-} > HCO^{3-} > CO_3^{2-}$. Specifically, the water chemistry in aquaculture farms was dominated by the Na^-Cl type, while the main chemical types in canals, wells, reservoirs, and rivers were $Na{\cdot}Ca\text{-}Cl$, $Ca{\cdot}Mg\text{-}Cl$, $Ca\text{-}Mg\text{-}SO_4$, and $Ca{\cdot}Mg^-HCO_3$, respectively.

Furthermore, significant correlations were observed between TDS and the major ions Na^+, Ca^{2+}, Cl^-, SO_4^{2-}, and HCO_3^-, indicating their contribution to the overall water chemistry of the region. The correlation between Na^+ and Cl^- suggests that they have a common source that is possibly influenced by seawater. Additionally, the correlation between Ca^{2+} and HCO_3^- may be related to different geological processes. Human activities and natural factors can influence water chemistry characteristics, as demonstrated in the coastal areas of Qionghai. Thus, long-term monitoring is needed in future studies to ensure the protection of natural water quality. Moreover, surface water and groundwater are often closely related, and further research is required to investigate their interactions in the coastal areas of Qionghai.

Author Contributions: Conceptualization, J.J. and X.G.; data curation, G.F.; methodology, P.J.; project administration, J.J. and Z.C.; resources, Z.C.; software, Y.F.; supervision, X.G. and C.S.; validation, J.J. and C.S.; writing—original draft, J.J. and G.F.; writing—review and editing, J.J. and X.G. All authors have read and agreed to the published version of the manuscript.

Funding: This research was funded by the Comprehensive Survey of Natural Resources in Haichengwen Coastal Zone: DD20230414, and the Natural Science Foundation of Jiangxi, China, grant number 20224BAB203034.

Institutional Review Board Statement: Not applicable.

Informed Consent Statement: Not applicable.

Data Availability Statement: Publicly available datasets were analyzed in this study. These data can be found here: https://www.gscloud.cn (accessed on 28 February 2023). The data are not publicly available as the project is not yet completed.

Acknowledgments: This study used data sampled from field experiments, which were obtained and provided to us by Chunwu Cao. The format of the manuscript was revised by Yanwei Song. The graphical production (visualization) of the manuscript was carried out by Yang Wang. Zhiguo Chen also provided some corrections to the manuscript. We would like to express our gratitude to them. Likewise, we are grateful to the editors and reviewers who revised the manuscript for processing.

Conflicts of Interest: The authors declare no conflict of interest.

References

1. Heydarizad, M.; Gimeno, L.; Amiri, S.; Minaei, M.; Mohammadabadi, H.G. A Comprehensive Overview of the Hydrochemical Characteristics of Precipitation across the Middle East. *Water* **2022**, *14*, 2657. [CrossRef]
2. Zhang, Z. The Study of Hydrochemical Characteristicsand lon Sources of Precipitation and Riverwater in Mountainous Areas in the UpperReaches of the Shiyang River. Master's Thesis, Northwest Normal University, Lanzhou, China, 2021.
3. Huang, Q.B.; Qin, X.Q.; Liu, P.Y.; Lan, F.N.; Zhang, L.K.; Su, C.T. Major lonic Features and Their Controlling Factors in the Upper-Middle Reaches of Wujiang River. *Environ. Sci.* **2016**, *37*, 1779–1787. [CrossRef]
4. Wang, M.; Yang, L.; Li, J.; Liang, Q. Hydrochemical Characteristics and Controlling Factors of Surface Water in Upper Nujiang River, Qinghai-Tibet Plateau. *Minerals* **2022**, *12*, 490. [CrossRef]
5. Guo, Y.W.; Tian, F.Q.; Hu, H.C.; Liu, Y.P.; Zhao, S.H. Characteristics and Significance of Stable lsotopes and Hydrochemistry inSurface Water and Groundwater in Nanxiaohegou Basin. *Environ. Sci.* **2020**, *41*, 682–690. [CrossRef]
6. Shen, B.B.; Wu, J.L.; Zhan, S.; Jin, M.; Saparov, A.S.; Abuduwaili, J. Spatial variations and controls on the hydrochemistry of surface waters across the Ili-Balkhash Basin, arid Central Asia. *J. Hydrol.* **2021**, *600*, 126565. [CrossRef]
7. Lima, V.P.; de Lima, R.A.F.; Joner, F.; Siddique, I.; Raes, N.; Ter Steege, H. Climate change threatens native potential agroforestry plant species in Brazil. *Sci. Rep.* **2022**, *12*, 2267. [CrossRef]
8. Ye, H.; Gan, J.; Li, G.; Hu, J.; Su, L.; Wu, J.; Wang, S. Temporal and Spatial Distributions and Controlling Factors of Hydrochemistry at WuyishanNational Park Water Body. *Fujan J. Agric. Sci.* **2022**, *37*, 1362–1370. [CrossRef]
9. Zhang, J.; Zhu, B. Hydrochemical characteristics and influencing factors in Northern Xinjiang: Research progress and overview. *Geogr. Res.* **2022**, *41*, 1437–1458.
10. Jiang, L.G.; Yao, Z.J.; Liu, Z.F.; Wang, R.; Wu, S.S. Hydrochemistry and its controlling factors of rivers in the source region of the Yangtze River on the Tibetan Plateau. *J. Geochem. Explor.* **2015**, *155*, 76–83. [CrossRef]
11. Mahloch, J.L. Graphical interpretation of water quality data. *Water Air Soil Pollut.* **1974**, *3*, 217–236. [CrossRef]
12. Gibbs, R.J. Mechanisms Controlling World Water Chemistry. *Science* **1970**, *170*, 1088–1090. [CrossRef]
13. Jia, Y.F.; Guo, H.M.; Xi, B.D.; Jiang, Y.H.; Zhang, Z.; Yuan, R.X.; Yi, W.X.; Xue, X.L. Sources of groundwater salinity and potential impact on arsenic mobility in the western Hetao Basin, Inner Mongolia. *Sci. Total Environ.* **2017**, *601*, 691–702. [CrossRef] [PubMed]
14. Liu, F.; Wang, S.; Wang, L.S.; Shi, L.M.; Song, X.F.; Yeh, T.C.J.; Zhen, P.N. Coupling hydrochemistry and stable isotopes to identify the major factors affecting groundwater geochemical evolution in the Heilongdong Spring Basin, North China. *J. Geochem. Explor.* **2019**, *205*, 106352. [CrossRef]

15. Zou, J.; Liu, F.; Zhang, K. Hydrochemical characteristics and formation mechanism of shallow groundwater in typicawater-receiving areas of the South-to-North Water Diversion Project. *China Environ. Sci.* **2022**, *42*, 2260–2268. [CrossRef]
16. Xiaobo, L.; Hng, L.; Baoping, Y.; Xincun, Z. Hydrochemical characteristics and formation mechanism of water source area in Jiuxian County, Tai'an City. *J. Environ. Eng. Technol.* **2023**, *15*, 1–13.
17. Kattan, Z. Chemical and isotopic characteristics of the Euphrates River water, Syria: Factors controlling its geochemistry. *Environ. Earth Sci.* **2015**, *73*, 4763–4778. [CrossRef]
18. Okosa, I.; Ndukwu, M.C.; Horsfall, I.T.; Igbojionu, D.O. The combined effect of water management and environmental control on the planting of two varieties of garden egg in a partially shaded greenhouse: An energy and yield indicator analysis. *Energy Nexus* **2022**, *7*, 100132. [CrossRef]
19. Papas, M. Supporting Sustainable Water Management: Insights from Australia's Reform Journey and Future Directions for the Murray–Darling Basin. *Water* **2018**, *10*, 1649. [CrossRef]
20. Zhang, P.; Cui, X.; Gao, Y.; Mo, J.; Chen, D. Research on Water Environment Situation and Regulation Progress of Urban Lakes. *IOP Conf. Ser. Earth Environ. Sci.* **2021**, *826*, 012020.
21. Xiumei, H. Study on Coupling Development of RuralTourism and Local Urbanization in Oionghai. Master's Thesis, Fujian Normal University, Fujian, China, 2017.
22. Cheng, Y.; Zhai, M.; Wang, Y.; Zhang, J. Development Model and Driving Forces of New Urbanization in Hainan Province: Qionghai City as a Case. *Sci. Geogr. Sin.* **2019**, *39*, 1902–1909. [CrossRef]
23. Li, Z.; Yang, Y.; Li, L.; Wang, D. A weighted Pearson correlation coefficient based multi-fault comprehensive diagnosis for battery circuits. *J. Energy Storage* **2023**, *60*, 106584. [CrossRef]
24. Pawan; Rohtash, D. Electroencephalogram channel selection based on pearson correlation coefficient for motor imagery-brain-computer interface. *Meas. Sens.* **2023**, *25*, 100616. [CrossRef]
25. Li, X.; Meng, X.; Ji, X.; Zhou, J.; Pan, C.; Gao, N. Zoning technology for the management of ecological and clean small-watersheds via k-means clustering and entropy-weighted TOPSIS: A case study in Beijing. *J. Clean. Prod.* **2023**, *397*, 136449. [CrossRef]
26. Izzati, M.T.N.; Aini, A.M.N.; Shahnorbanun, S. Identification of Student Behavioral Patterns in Higher Education Using K-Means Clustering and Support Vector Machine. *Appl. Sci.* **2023**, *13*, 3267.
27. Guo, X.; Wang, X.; Shi, X.; Yu, H.; Huirong, Z.; Fang, Z. Hydrochemical characteristics and sources of chemical constituents in groundwater in Hunchun River Basin, Northeast China. *Arab. J. Geosci.* **2022**, *15*, 694. [CrossRef]
28. Wei, H.; Wu, J.K.; Shen, Y.P.; Zhang, W.; Liu, S.W.; Zhou, J.X. Hydrochemical Characteristics of Snow Meltwater and River Water During Snow-meltinoPeriod in the Headwaters of the Ertis River, Xinjiang. *Environ. Sci.* **2016**, *37*, 1345–1352. [CrossRef]
29. Liu, F.; Li, Z.; Hao, J.; Liang, P.; Wang, F.; Zhang, H. Study on the hydrochemical and stable isotope characteristics at theheadwaters of the Irtysh River in spring. *Arid. Land Geogr.* **2020**, *42*, 234–242.
30. Shelton, J.L.; Engle, M.A.; Buccianti, A.; Blondes, M.S. The isometric log-ratio (ilr)-ion plot: A proposed alternative to the Piper diagram. *J. Geochem. Explor.* **2018**, *190*, 130–141. [CrossRef]
31. Wen, Y.; Qiu, J.; Cheng, S.; Xu, C.; Gao, X. Hydrochemical Evolution Mechanisms of Shallow Groundwater and Its Quality Assessment in the Estuarine Coastal Zone: A Case Study of Qidong, China. *Int. J. Environ. Res. Public Health* **2020**, *17*, 3382. [CrossRef]
32. Zhang, Y.; Xu, M.; Li, X.; Qi, J.; Zhang, Q.; Guo, J.; Yu, L.; Zhao, R. Hydrochemical Characteristics and Multivariate Statistical Analysis of Natural Water System: A Case Study in Kangding County, Southwestern China. *Water* **2018**, *10*, 80. [CrossRef]
33. Xiao, J.; Jin, Z.; Zhang, F. Geochemical controls on fluoride concentrations in natural waters from the middle Loess Plateau, China. *J. Geochem. Explor.* **2015**, *159*, 252–261. [CrossRef]
34. Cui, Y.H.; Wang, J.; Liu, Y.C.; Hao, S.; Gao, X. Hydro-chemical Characteristics and Ion Origin Analysis of Surface Groundwater at the ShengjinLake and Yangtze River Interface. *Environ. Sci.* **2021**, *42*, 3223–3231. [CrossRef]
35. Yongjun, Z.; Yuhui, Y.; Yicheng, H.; Xiancheng, F.; Jingyan, Y. Temporal and spatial variation characteristics of Hydrochemistry and irrigation adaptabilityevaluation in Kashi River Basin, Xinjiang. *Arid. Land Geogr.* **2022**, *13*, 1–15.
36. An, S.K.; Jiang, C.L.; Zhang, W.X.; Chen, X.; Zheng, L.G. Influencing factors of the hydrochemical characteristics of surface water and shallow groundwater in the subsidence area of the Huainan Coalfield. *Arab. J. Geosci.* **2020**, *13*, 191. [CrossRef]
37. Yang, K.; Liu, W.-Q.; Xu, X.-Y.; Chen, G.-Q.; Liu, Y.-J.; Fu, T.-F.; Wang, C.-J.; Fu, Y.-X. Evaluation of seawater intrusion in typical coastal zones of Hainan Province. *Mar. Sci.* **2019**, *43*, 57–63.
38. Xi, L.; Chen, K.; Huang, X.; Gan, H.; Xia, Z.; Tan, X. Hydrogeochemistry and origin of groundwater in the south coast of Hainan. *Geol. Bull. China* **2021**, *40*, 350–363.

growth and physiological properties, mainly in terms of a reduced growth rate, altered leaf morphology, reduced biomass [10–12], and increased consumption of non-structural carbohydrates [13]. However, there is some variation in the response of different species to changes in the light environment due to factors such as the growth, habitat and life history of seagrasses [14].

Enhalus acoroides is a large, long-lived seagrass with the ability to reproduce both sexually and asexually. It is widely distributed in the Indo-West Pacific, and in China, it is found only on Hainan Island, mainly in Xincun Bay and Li'an Lagoon in Lingshui [15]. The fishery is well developed in Li'an Lagoon, and the discharge of aquaculture sewage and domestic wastewater, among other pollutants, has a significant impact on the health status of seagrass beds [16]. In the past 10–20 years, the distribution area and plant density of *E. acoroides* have decreased dramatically, and few studies have been reported regarding the effects of external environmental changes on *E. acoroides* [17]. In this study, we examined *E. acoroides*, simulated an environment reflecting the reduced light transmission of water by building a shade shelter in situ, explored differences in the response of *E. acoroides* to changes in the light environment, analyzed the degradation mechanism of *E. acoroides*, and provided a theoretical basis for determining a reasonable conservation strategy.

2. Materials and Methods

2.1. Study Area

Li'an Lagoon is located in the southeastern part of Lingshui County, with an area of approximately 9 km^2, and it is connected to the outer sea only by a tidal branch of approximately 60 m [16]. The distribution area of seagrasses in the lagoon is approximately 1.42 km^2, and the species include *E. acoroides*, *Thalassia hemprichii*, *Halophila ovalis* and *Cymodocea rotundata* [18]. However, due to long-term human interference, the seagrass meadow of Li'an Lagoon is gradually developing into an ecosystem, with *E. acoroides* as the single dominant species. Therefore, *E. acoroides* was selected as the research object in this experiment.

2.2. Experimental Design and Sample Collection

The experiment was divided into two phases. The first phase consisted of three and six months of shading, and the second phase consisted of three months of recovering light. *E. acoroides*, located in the intertidal zone with a patchy distribution and basically uniform growth (only *E. acoroides* was distributed in the area where shading was performed), were selected in the study area on December 2020, and the light environment with reduced water transmittance was simulated by building a shading shelter in situ. The size of the shading shed was 2 m $\times$ 2 m, and the height of the shed was slightly higher than the plant height of *E. acoroides* in a completely floating state. The shading material was a black shading net with uniform density, and the shading intensity was manipulated by changing the number of overlapped shading nets and measuring the light intensity directly below the shading net in situ using an underwater irradiance meter (ZDS-10W-2D). The light measurements were carried out in full sunlight at noon, and the results of the measurements are shown in Table 1.

Based on the light measurements, the shading rates of the two shading treatments were approximated as 60% (single shading) and 90% (double shading), which were labelled as moderate shading (MS) and high shading (HS), respectively, while the full-light treatment (no shading) was used as the control (CK). Three sample plots were selected in the study area, labeled ZG1, ZG2, ZG3. Three light treatments were set up in each sample plot, with three sets of replicates for each treatment, for a total of 27 sample squares (Figure 1). Due to tidal changes, the shading canopy is sometimes exposed to the water, and sometimes completely submerged. To reduce the experimental error caused by wave interference or the attachment of other marine organisms, we visited the sample site regularly every month to replace the shading net.

Table 1. The measurement of light intensity in situ.

Site	Control/lx	Single/lx	Intensity	Control/lx	Double/lx	Intensity
	76,000	28,300	62.76	78,400	6580	91.61
ZG1	78,700	27,800	64.68	76,900	7000	90.90
	75,300	29,400	60.96	76,200	7800	89.76
	56,500	18,500	67.26	51,800	2530	95.12
ZG2	59,200	20,500	65.37	52,400	3310	93.68
	58,300	22,700	61.06	55,600	2890	94.80
	36,900	14,500	60.70	25,800	2280	91.16
ZG3	28,500	12,400	56.49	26,900	2560	90.48
	35,200	16,700	52.56	26,200	1680	93.59
Average	——	——	61.32	——	——	92.34

Note: ZG1, ZG2, ZG3 are three sample plots in study area. Single, the light intensity of the single-layer shading net; Double, the light intensity of the double-layer shading net.

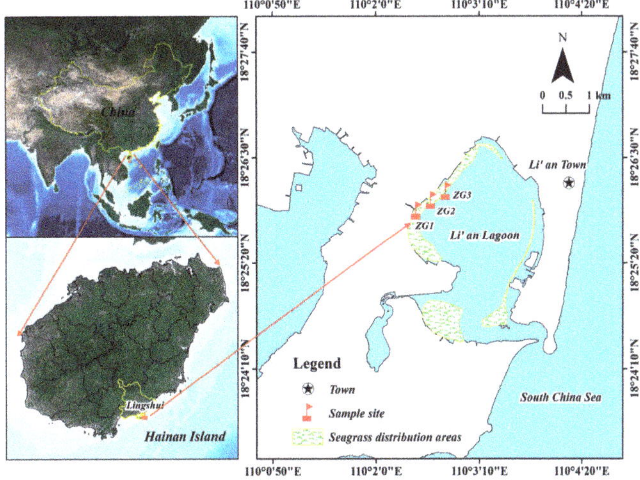

Figure 1. Study area.

Samples were collected at low tide when the *E. acoroides* was exposed to the surface. The seagrasses were collected in the test plots using 25 cm × 25 cm sampling frames directly below the shading shelters. The *E. acoroides* were dug out from the sampling frames together with the rhizomes, and the sediments were collected using PVC pipes at a depth of 30 cm with intervals of 0–5, 5–10, 10–20 and 20–30 cm.

2.3. Measurement of Indicators

The leaf length and leaf width of *E. acoroides* in all sampling frames were measured with a measuring tape, and the number of leaves was subsequently counted. Plant density was calculated by taking the number of *E. acoroides* in each sampling frame and dividing it by the area of the sampling frame (0.0625 m^2) to calculate the plant density of *E. acoroides* under different light treatments (plants/m^2). Seagrasses were carefully retrieved, and subsequently dried to constant weight at 60 °C for the determination of biomass [12].

Three *E. acoroides* were randomly selected from each sample frame as the sample plants to be tested, and the middle part of their leaves was taken for the determination of the chlorophyll content. The fresh leaves were cut into small pieces of about 0.20 cm, mixed well and weighed to 0.20 g. The leaves were then put into a stoppered graduated test tube with 80% acetone solution for chlorophyll extraction. When the leaf tissue had all turned white, it indicated that the chlorophyll had been extracted cleanly [19]. After that, the absorbance values of the extracts at the corresponding wavelengths were measured using a UV spectrophotometer (UV-5200), and the content of chlorophyll and chlorophyll a/b value (Chl a/Chl b) were calculated accordingly [20].

Sediment samples were stored in desiccators prior to analysis. After being naturally air dried, the sediment samples were ground and homogenized with a mortar and pestle and subsequently passed through a 100-mesh sieve for the determination of the sediment organic carbon content. The sediment organic carbon content was determined via the potassium dichromate–sulfuric acid oxidation method. The organic carbon of the sediment was oxidized with a potassium dichromate–sulfuric acid solution under oil bath heating, and the remaining potassium dichromate was titrated with ferrous sulfate to calculate the organic carbon content from the amount of potassium dichromate consumed [21].

2.4. Statistical Analysis

Data were processed and analyzed using Excel 2019 and SPSS 26.0. The variability of the morphological characteristics, including the density, biomass and photosynthetic pigment content of the *E. acoroides*, and the sediment carbon content under different light treatments, were analyzed via one-way analysis of variance (ANOVA) and least significant difference (LSD) tests ($\alpha = 0.05$). The data in the graphs are the "mean ± standard error". Plots were generated using Origin 9.8.

3. Results

3.1. Morphology

When light was reduced, the leaf length and leaf width of *E. acoroides* decreased significantly ($p < 0.05$), and the decrease was higher in the 6-month shading treatment than in the 3-month shading treatment (Table 2). With the increase in shading intensity, the number of leaves of *E. acoroides* showed a decreasing trend. After the restoration of light, the leaf length of the plants in the shading treatment was still significantly smaller than that of the plants in the full-light treatment ($p < 0.05$), but the leaf width and leaf number of the plants in the moderate shading treatment increased after light restoration compared with the shading period; the leaf length and leaf width of the plants in the high shading treatment did not tend to increase throughout the restoration period, indicating that the greater the shading intensity was, the weaker the ability of *E. acoroides* to restore normal growth was, and the effect caused by high shading still occurred.

Table 2. Effects of different light treatments on morphological characteristics of *E. acoroides*.

Period	Treatments	Leaf Length/cm	Leaf Width/cm	Leaf Number
S3	CK	34.98 ± 0.96 a	1.65 ± 0.02 a	4.34 ± 0.09 a
	MS	29.78 ± 1.17 b	1.47 ± 0.02 b	3.60 ± 0.11 b
	HS	25.25 ± 1.16 c	1.38 ± 0.02 c	3.35 ± 0.11 b
S6	CK	54.98 ± 1.55 a	1.84 ± 0.02 a	6.63 ± 0.08 a
	MS	22.79 ± 1.80 b	1.49 ± 0.03 b	5.00 ± 0.12 b
	HS	21.47 ± 1.40 b	1.40 ± 0.02 c	4.84 ± 0.12 b
R3	CK	24.31 ± 1.26 a	1.75 ± 0.02 a	5.41 ± 0.14 a
	MS	19.12 ± 0.87 b	1.50 ± 0.03 b	4.23 ± 0.12 b
	HS	16.30 ± 0.91 b	1.35 ± 0.03 c	3.58 ± 0.12 c

Note: S3, shading for 3 months; S6, shading for 6 months; R3, recovery for 3 months. CK–Control; MS–Moderate shading; HS–High shading. Different letters indicate significant differences at the 0.05 level. The same below.

3.2. Density

When the shading period was 3 months, the shoot density of *E. acoroides* decreased but was not significantly different from that of the control ($p > 0.05$) (Figure 2). After 6 months of shading treatment, the shoot density in the moderate shading and high shading treatments was significantly lower than that in the full-light treatment ($p < 0.05$), decreasing by 55.36% and 59.82%, respectively, compared with that of the control. After the restoration of light, the shoot density increased in all shading treatments but did not return to the same level as the control.

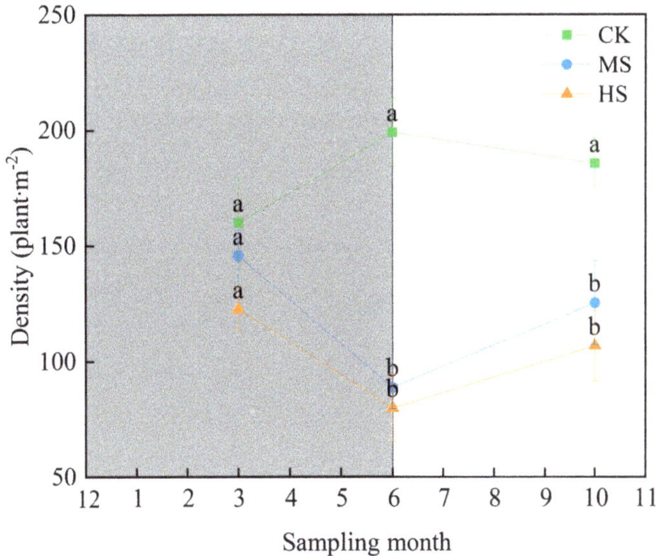

Figure 2. Effects of different light treatments on the density of *E. acoroides*. Note: The shaded part represents shading, and the blank part represents light recovery. Different letters in the same sampling period indicate significant differences at the 0.05 level.

3.3. Biomass

Three months of shading treatment caused a significant reduction in the biomass of *E. acoroides* compared to that in the full-light treatment ($p < 0.05$) (Figure 3). With an increasing shading treatment time, the biomass of the moderate shading and heavy shading groups decreased by 53.36% and 54.10%, respectively, compared to that of the full-light treatment group. The biomass remained lower during the period of light restoration.

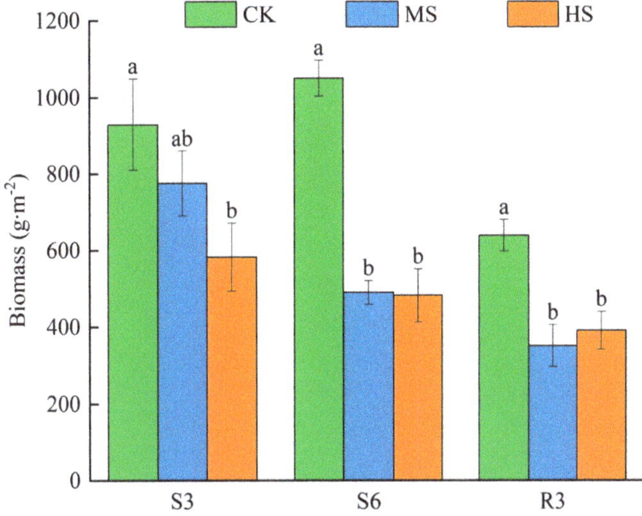

Figure 3. Effects of different light treatments on the biomass of *E. acoroides*. Note: Different letters in the same sampling period indicate significant differences at the 0.05 level.

3.4. Chlorophyll

The maximum chlorophyll a and chlorophyll b contents in *E. acoroides* occurred in the control group throughout the shading period. As the shading intensity increased, the chlorophyll a/b values showed a tendency to decrease (Table 3). When the shading duration reached 6 months, the chlorophyll content of the shading treatment was significantly lower than that of the control ($p < 0.05$). After the restoration of light, the total chlorophyll content increased significantly compared to that in the shading period ($p < 0.05$), and the chlorophyll a/b values did not differ significantly from those in the control ($p > 0.05$).

Table 3. Effects of different light treatments on the chlorophyll content of *E. acoroides*.

Period	Treatments	Photosynthetic Pigments (mg·g^{-1})			
		Chl a	Chl b	Total Chl	Chl a/Chl b
S3	CK	0.142 ± 0.016 a	0.062 ± 0.007 a	0.204 ± 0.022 a	2.419 ± 0.206 a
	MS	0.108 ± 0.010 a	0.056 ± 0.004 a	0.164 ± 0.014 a	1.899 ± 0.094 b
	HS	0.125 ± 0.014 a	0.049 ± 0.004 a	0.175 ± 0.018 a	2.495 ± 0.153 a
S6	CK	0.207 ± 0.012 a	0.084 ± 0.005 a	0.291 ± 0.018 a	2.484 ± 0.036 a
	MS	0.087 ± 0.008 b	0.041 ± 0.004 b	0.128 ± 0.011 b	2.222 ± 0.083 b
	HS	0.082 ± 0.007 b	0.039 ± 0.004 b	0.121 ± 0.011 b	2.196 ± 0.093 b
R3	CK	0.301 ± 0.024 a	0.142 ± 0.019 a	0.434 ± 0.032 a	2.156 ± 0.341 a
	MS	0.229 ± 0.019 b	0.102 ± 0.014 a	0.330 ± 0.022 b	2.233 ± 0.205 a
	HS	0.230 ± 0.018 b	0.100 ± 0.016 a	0.334 ± 0.018 b	2.539 ± 0.475 a

Note: Chl a—chlorophyll a content; Chl b—chlorophyll b content; Total Chl—Total chlorophyll content; Chl a/Chl b—chlorophyll a/b values. Different letters indicate significant differences at the 0.05 level.

3.5. Sediment Carbon

The sediment carbon content was higher in the control than in the shading treatment (Figure 4). There was no significant difference in the sediment carbon content at different depths in the control ($p > 0.05$). When the light was reduced, the sediment carbon content at 0–5 cm in the high shading treatment was significantly lower than that in the control at the same depth (34.36%). With increasing depth, the sediment carbon content under moderate shading decreased significantly ($p < 0.05$).

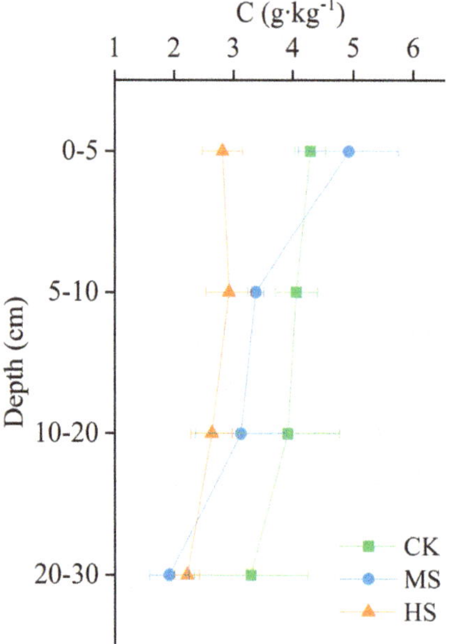

Figure 4. Effects of shading on sediment carbon content in seagrass beds.

4. Discussion

4.1. Response of Seagrass Growth to Shading

Changes in the leaf morphology of seagrass can be used to visualize its response to changes in the light environment [4]. Studies have shown that light reduction significantly reduces the leaf length and width of *Zostera marina* [22]. At 64% and 75% shading intensities, the number of leaves of *T. hemprichii* was significantly lower than that in the full-light treatment [23]. In the present study, the leaf length and width of *E. acoroides* decreased with increasing shading intensity, and the number of leaves also tended to decrease. When light is insufficient, seagrasses are able to persist in this stressful environment by altering their leaf morphology or maintaining a small number of leaves to effectively reduce their respiratory demand in the low-light environment. In contrast, an increase in the leaf length can enhance the capture of light by seagrasses in low-light environments [24]. The reason for this difference could be due to the higher shading intensity and longer shading period in this study. In addition, the shoot density of *E. acoroides* did not change significantly at 3 months of shading treatment, which may be related to the time of shading initiation. We started the shading treatment of *E. acoroides* in winter, when the growth rate of *E. acoroides* was low. Indeed, carbohydrates were accumulated by seagrasses in the spring and summer when the sunlight was sufficient to meet their carbon demand in the low-light environment in the early period. The shoot density of *E. acoroides* decreased significantly with the increasing duration of the shading treatment, which was in summer. Seagrasses exposed to high temperatures may require more light to maintain a carbon balance and are thus more susceptible to light reduction [25].

4.2. Response of Seagrass Chlorophyll Content to Shading

Photosynthetic pigments play an irreplaceable role in photosynthesis in plants [26]. Studies have shown that the regulation of the photosynthetic structure in response to low light usually includes an increase in the chlorophyll content and a decrease in the chlorophyll a/b ratio [27,28]. The increase in chlorophyll content can significantly enhance the absorption, transfer and conversion of energy from weak light in the optical system to improve the utilization efficiency of light energy [29]. Different results were obtained in our study. When the duration of the shading treatment was 3 months, the chlorophyll content of *E. acoroides* was not significantly different from that of the control, indicating that this seagrass was tolerant to short-term light stress; as the duration of shading increased, the chlorophyll content decreased sharply. This shows that a long-term light intensity that is too low is detrimental to chlorophyll synthesis, which is similar to observations made in previous studies [30]. After shading, the chlorophyll a/b values of *E. acoroides* were lower than those in the full-light treatment. Light reduction usually decreases the proportion of red light absorption in seagrass, and chlorophyll b is more capable of absorbing blue light than chlorophyll a. The reduction in chlorophyll a/b values in low-light environments indicates that the seagrass has an enhanced ability to utilize blue light, thus improving light capture in low-light environments [31].

4.3. Effect of Shading on the Carbon Storage of Seagrass Beds

Seagrass accumulates organic matter via photosynthesis, and light reduction decreases its photosynthetic rate [32], which in turn affects productivity [33] and even causes biomass loss [34]. In the present study, as the shading intensity increased, the amount of effective light that *E. acoroides* could receive decreased, thus inhibiting its growth and biomass accumulation. The loss of aboveground biomass reflects both the growth and development of seagrasses in low-light environments and suggests that seagrasses are able to avoid being shaded by their own leaves by reducing their leaf area [35]. The loss of leaves reduces photosynthesis in seagrasses to some extent, and the reduction in oxygen delivery to the rhizome may cause anaerobic conditions in the subsurface tissues, thus causing a reduction in the biomass of the subsurface [36].

The seagrass canopy capture of organic matter from the surrounding habitat is the main source of sediment carbon, and the underground biomass usually has a high lignin content and relatively low decomposition rate, so the reduction in the seagrass biomass disrupts its CO_2 uptake process [37]. There is a distinct lack of research on the effects of reduced light on the sediment carbon storage of seagrass beds, but existing studies suggest that the degradation and loss of seagrass beds may result in the loss and release of sediment carbon [38]. For example, the large-scale loss of *Posidonia oceanica* has been shown to lead to an 11–25% reduction in sediment carbon stocks [39]. In *Thalassia testudinum*, shading treatments have also been shown to result in a loss of up to 47% of sediment carbon stocks in the surface layer alone [40]. In the present study, the sediment carbon was lower in the shading treatment than in the control, and it decreased significantly with increasing depth. This may be due to the loss of aboveground biomass caused by reduced light, resulting in a decrease in its efficient trapping of allochthonous carbon and reducing the deposition of aboveground tissues to the detritus layer [37]. In addition, a reduction in root biomass would decrease the flux of the root exudation of the dissolved organic carbon into the sediment [41]. Therefore, biomass loss due to light reduction would further lead to a reduction in sediment carbon stocks. However, *Zostera nigricaulis* showed no significant change in the carbon content of sediment during up to two years of intense disturbance [42]. Possible reasons for this difference in results include the contribution of seagrass itself to the sediment carbon pool [43], the hydrological conditions of the study area [44], and the influence of the composition and activity of the microbial community on sediment mineralization [45].

4.4. Response of Seagrass to Light Restoration

After the restoration of light, the chlorophyll contents of *E. acoroides* increased compared to those in the shading period, providing a favorable material basis for its photosynthesis under suitable light conditions. However, the physiological recovery was not reflected by the leaf morphology, shoot density or biomass. During the period of light restoration, the leaf and biomass of *E. acoroides* remained reduced. Similar differences were observed in the growth and physiological responses of *Amphibolis griffithii* and *Z. marina* to restored light [13,46]. This may be related to the seasonal growth of seagrass. In addition, the growth of *E. acoroides* during the period of light restoration may be limited by typhoons. However, the more important reason is that long-term light stress may have a devastating effect on the growth of seagrass and that recovery would be difficult in the short term even when the stress is removed. Future increases in the photosynthetic rate may be effective in alleviating the damage caused by the pre-light-stressed environment, but long-term light reduction will affect photosynthesis and photomorphic construction, eventually leading to death. The sediment carbon of *Posidonia australis* also increased significantly after restoration compared to the disturbed period [47]. This shows that a suitable growth environment is conducive to promoting the accumulation of organic matter in seagrasses, thereby increasing their CO_2 sequestration capacity.

5. Conclusions

In this study, the effects of shading and the restoration of light on the growth and carbon storage capacity of *E. acoroides* were investigated through in situ experiments over a period of 10 months. The results show that the growth of *E. acoroides* was significantly negatively affected by the low-light environment, which also caused a significant loss of shoot density. Short-term low-light conditions had no significant effect on the chlorophyll content of *E. acoroides*, but a prolonged reduction in light was detrimental to its chlorophyll synthesis. In addition, long-term light reduction not only causes the loss of seagrass biomass, but also may lead to the carbon that is stored being unable to have a long-term stable existence in its sediment due to the loss of seagrass shelter. After the restoration of light, the chlorophyll content of *E. acoroides* increased compared to this content during the shading period, but its leaf morphology, shoot density and biomass did not return to the level of the full-light treatment, and the effects caused by shading remained.

The degradation and disappearance of seagrass beds and the loss of their carbon storage capacity are issues of concern. Our study highlights that the carbon storage capacity of seagrass beds may eventually decrease in environments with reduced light exposure over time, at least at a 60–90% shading intensity. This implies that we need to promote the conservation and restoration of seagrass ecosystems through the enhanced management of water quality. To provide better management measures, we can learn as much as possible about the light compensation points of *E. acoroides* by conducting experiments with wider shading intensity ranges in the future.

Author Contributions: Conceptualization, X.Z.; investigation, X.Z. and M.F.; writing—original draft preparation, M.F.; writing—review and editing, X.Z., Y.W., Y.S. and G.F.; project administration, G.F.; funding acquisition, G.F. and X.Z. All authors have read and agreed to the published version of the manuscript.

Funding: This research was funded by the Comprehensive Survey of Natural Resources in "Hai Cheng Wen" Coastal Zone (Grant No. DD20230414), the Comprehensive Survey of Natural Resources in Huizhou-Shanwei Coastal Zone (Grant No. DD20230415) and the E-Talent project of Hainan University (KYQD2R1968).

Institutional Review Board Statement: Not applicable.

Informed Consent Statement: Not applicable.

Data Availability Statement: The data are unavailable due to privacy.

Acknowledgments: We deeply thank Liguo Liao for the assistance in setting up the experiment and all the help in the sampling process. We express our thanks to Shiquan Chen for the always enlightening and fruitful discussions. We would also like to thank the editors and reviewers for their comments and suggestions on the manuscript.

Conflicts of Interest: The authors declare no conflict of interest.

References

1. Tang, J.W.; Ye, S.F.; Chen, X.C.; Yang, H.L.; Sun, X.H.; Wang, F.M.; Wen, Q.; Chen, S.B. Coastal blue carbon: Concept, study method, and the application to ecological restoration. *Sci. China Earth Sci.* **2018**, *48*, 661–670. [CrossRef]
2. Ramesh, R.; Banerjee, K.; Paneerselvam, A.; Raghuraman, R.; Purvaja, R.; Lakshmi, A. Importance of Seagrass Management for Effective Mitigation of Climate Change. *Coast. Manag.* **2019**, *14*, 283–299.
3. Ontoria, Y.; Gonzalez-Guedes, E.; Sanmarti, N.; Bernardeau-Estellerb, J.; Ruiz, J.M.; Romerol, J.; Perez, M. Interactive effects of global warming and eutrophication on a fast-growing Mediterranean seagrass. *Mar. Environ. Res.* **2019**, *145*, 27–38. [CrossRef] [PubMed]
4. Browne, N.K.; Yaakub, S.M.; Tay, J.; Todd, P.A. Recreating the shading effects of ship wake induced turbidity to test acclimation responses in the seagrass *Thalass. Hemprichii. Estuar. Coast. Shelf Sci.* **2017**, *199*, 87–95. [CrossRef]
5. Statton, J.; Mcmahon, K.; Lavery, P.; Kendrick, G.A. Determining light stress responses for a tropical multi-species seagrass assemblage. *Mar. Pollut. Bull.* **2018**, *128*, 508–518. [CrossRef]
6. Duarte, C.M. Seagrass depth limits. *Aquat. Bot.* **1991**, *40*, 363–377. [CrossRef]
7. Fraser, M.W.; Short, J.; Kendrick, G.; McLeana, D.; Keesinga, J.; Byrne, M.; Caley, M.J.; Clarke, D.; Davis, A.R.; Erftemeijer, P.L.A.; et al. Effects of dredging on critical ecological processes for marine invertebrates, seagrasses and macroalgae, and the potential for management with environmental windows using Western Australia as a case study. *Ecol. Indic.* **2017**, *78*, 229–242. [CrossRef]
8. Eriander, L.; Laas, K.; Bergström, P.; Gipperth, L.; Moksnes, P. The effects of small-scale coastal development on the eelgrass (*Zostera marina* L.) distribution along the Swedish west coast-ecological impact and legal challenges. *Ocean. Coast. Manag.* **2017**, *148*, 182–194. [CrossRef]
9. Lapointe, B.E.; Herren, L.W.; Brewton, R.A.; Alderman, P.K. Nutrient over-enrichment and light limitation of seagrass communities in the Indian River Lagoon, an urbanized subtropical estuary. *Sci. Total Environ.* **2020**, *699*, 134068. [CrossRef]
10. Benham, C.F.; Beavis, S.G.; Hendryr, A.; Jackson, E.L. Growth effects of shading and sedimentation in two tropical seagrass species: Implications for port management and impact assessment. *Mar. Pollut. Bull.* **2016**, *109*, 461–470. [CrossRef]
11. Bertelli, C.M.; Creed, J.C.; Nuuttila, H.K.; Unsworth, R.K.F. The response of the seagrass *Halodule wrightii* Ascherson to environmental stressors. *Estuar. Coast. Shelf Sci.* **2020**, *238*, 106693. [CrossRef]
12. Collier, C.J.; Waycott, M.; Ospina, A.G. Responses of four Indo-West Pacific seagrass species to shading. *Mar. Pollut. Bull.* **2012**, *65*, 342–354. [CrossRef] [PubMed]

13. Wong, M.C.; Griffiths, G.; Vercaemer, B. Seasonal response and recovery of eelgrass (*Zostera marina*) to short-term reductions in light availability. *Estuaries Coasts J. Coast. Estuar. Res. Fed.* **2019**, *43*, 120–134. [CrossRef]
14. Kilminster, K.; McMahon, K.; Waycott, M.; Kendrick, G.A.; Scanes, P.; Mckenzie, L.; O'Brien, K.R.; Lyons, M.; Ferguson, A.; Maxwell, P. Unravelling complexity in seagrass systems for management: Australia as a microcosm. *Sci. Total Environ.* **2015**, *534*, 97–109. [CrossRef]
15. Yu, S.; Liu, S.L.; Jiang, K.; Zhang, J.P.; Jiang, Z.J.; Wu, Y.C.; Huang, C.; Zhao, C.Y.; Huang, X.P.; Trevathan-Tackett, S.M. Population genetic structure of the threatened tropical seagrass *Enhalus acoroides* in Hainan Island, China. *Aquat. Bot.* **2018**, *150*, 64–70. [CrossRef]
16. Chen, S.Q.; Pang, Q.Z.; Cai, Z.F.; Wu, Z.J.; Shen, J.; Wang, D.R.; Chen, H.Y. Analysis of distribution characteristics, health status, and influencing factors of seagrass bed in Li'an Lagoon, Hainan Island. *Mar. Sci.* **2020**, *44*, 57–64.
17. Chen, S.Q.; Wang, D.R.; Wu, Z.J.; Zhang, G.C.; Li, Y.C.; Tu, Z.G.; Yao, H.J.; Cai, Z.F. Discussion of the change trend of the seagrass beds in the east coast of Hainan Island in nearly a decade. *Mar. Environ. Sci.* **2015**, *34*, 48–53.
18. Cai, Z.F.; Chen, S.Q.; Wu, Z.J.; Liang, D.M.; Yin, F.; Tong, Y.H.; Wang, D.R. Distribution Differences and Environmental Effects of Seagrasses Between Bays and Lagoons of Hainan Island. *Trans. Oceanol. Limnol.* **2017**, *3*, 74–84. [CrossRef]
19. Moore, B.R.H. *Laboratory Guide for Elementary Plant Physiology*; Burgess Publishing: Minneapolis, MN, USA, 1957; pp. 73–74.
20. Gao, J.F. *Experimental Guidance for Plant Physiology*; Higher Education Press: Beijing, China, 2006; pp. 74–77.
21. Bao, S.D. *Agrochemical Analysis of Soil*, 3rd ed.; China Agriculture Press: Beijing, China, 2000; pp. 285–292.
22. Bertelli, C.M.; Unsworth, R. Light Stress Responses by the Eelgrass, *Zostera marina* (L.). *Front. Environ. Sci.* **2018**, *6*, 39–51. [CrossRef]
23. Deyanova, D.; Gullstrom, M.; Lyimo, L.D.; Dahl, M.; Hamisi, M.I.; Mtolera, M.S.P.; Björk, M. Contribution of seagrass plants to CO_2 capture in a tropical seagrass meadow under experimental disturbance. *PLoS ONE* **2017**, *12*, e0181386. [CrossRef]
24. Suykerbuyk, W.; Govers, L.L.; van Oven, W.G.; Giesen, K.; Giesen, W.B.J.T.; de Jong, D.J.; Bouma, T.J.; van Katwijk, M.M. Living in the intertidal; desiccation and shading reduce seagrass growth, but high salinity or population of origin have no additional effect. *PeerJ* **2018**, *6*, e5234. [CrossRef] [PubMed]
25. Kim, Y.K.; Kim, S.H.; Lee, K.S. Seasonal growth responses of the seagrass *Zostera marina* under severely diminished light conditions. *Estuaries Coasts* **2015**, *38*, 558–568. [CrossRef]
26. Enríquez, S. Light absorption efficiency and the package effect in the leaves of the seagrass *Thalassia testudinum*. *Mar. Ecol. Prog.* **2005**, *289*, 141–150. [CrossRef]
27. Longstaff, B.J.; Loneragan, N.R.; O'donohue, M.J.; Dennison, W.C. Effects of light deprivation on the survival and recovery of the seagrass *Halophila ovalis* (R.Br.) Hook. *J. Exp. Mar. Biol. Ecol.* **1999**, *234*, 1–27. [CrossRef]
28. Silva, J.; Barrote, I.; Costa, M.M.; Albano, S.; Santos, R. Physiological responses of *Zostera marina* and *Cymodocea nodosa* to light-limitation stress. *PLoS ONE* **2013**, *8*, e81058. [CrossRef] [PubMed]
29. Fokeera-Wahedally, S.; Bhikajee, M. The effects of in situ shading on the growth of a seagrass, *Syringodium isoetifolium*. *Estuar. Coast. Shelf Sci.* **2005**, *64*, 149–155. [CrossRef]
30. Wong, M.C.; Vercaemer, B.M.; Griffiths, G. Response and recovery of eelgrass (*Zostera marina*) to chronic and episodic light disturbance. *Estuaries Coasts* **2020**, *44*, 312–324. [CrossRef]
31. Longstaff, B.J.; Dennison, W.C. Seagrass survival during pulsed turbidity events: The effects of light deprivation on the seagrasses *Halodule pinifolia* and *Halophila ovalis*. *Aquat. Bot.* **1999**, *65*, 105–121. [CrossRef]
32. Bite, J.S.; Campbell, S.J.; Mckenzie, L.J.; Coles, R.G. Chlorophyll fluorescence measures of seagrasses *Halophila ovalis* and *Zostera capricorni* reveal differences in response to experimental shading. *Mar. Biol.* **2007**, *152*, 405–414. [CrossRef]
33. Lee, K.S.; Sang, R.P.; Kim, Y.K. Effects of irradiance, temperature, and nutrients on growth dynamics of seagrasses: A review. *J. Exp. Mar. Biol. Ecol.* **2007**, *350*, 144–175. [CrossRef]
34. Serrano, O.; Mateo, M.; Renom, P. Seasonal response of *Posidonia oceanica* to light disturbances. *Mar. Ecol. Prog. Ser.* **2011**, *423*, 29–38. [CrossRef]
35. Enríquez, S.; Pantoja-Reyes, N.I. Form-function analysis of the effect of canopy morphology on leaf self-shading in the seagrass *Thalassia testudinum*. *Oecologia* **2005**, *145*, 235–243. [CrossRef] [PubMed]
36. Enríquez, S.; Marbà, N.; Duarte, C.M.; van Tussenbroek, B.I.; Reyes-Zavala, G. Effects of seagrass *Thalassia testudinum* on sediment redox. *Mar. Ecol. Prog.* **2001**, *219*, 149–158. [CrossRef]
37. Dahl, M.; Deyanova, D.; Lyimo, L.D.; Näslund, J.; Samuelsson, G.S.; Mtolera, M.S.P.; Björk, M.; Gullström, M. Effects of shading and simulated grazing on carbon sequestration in a tropical seagrass meadow. *J. Ecol.* **2016**, *104*, 654–664. [CrossRef]
38. Liu, S.L.; Jiang, Z.J.; Wu, Y.C.; Zhang, J.P.; Zhao, C.Y.; Huang, X.P. Mechanisms of sediment carbon sequestration in seagrass meadows and its responses to eutrophication (in Chinese). *Chin. Sci. Bull.* **2017**, *62*, 3309–3320. [CrossRef]
39. Marbà, N.; Díaz-Almela, E.; Duarte, C.M. Mediterranean seagrass (*Posidonia oceanica*) loss between 1842 and 2009. *Biol. Conserv.* **2014**, *176*, 183–190. [CrossRef]
40. Trevathan-Tackett, S.M.; Wessel, C.; Cebrian, J.; Ralph, P.J.; Masqué, P.; Macreadie, P.I. Effects of small-scale, shading-induced seagrass loss on blue carbon storage: Implications for management of degraded seagrass ecosystems. *J. Appl. Ecol.* **2018**, *55*, 1351–1359. [CrossRef]

41. Premarathne, C.; Jiang, Z.J.; He, J.; Fang, Y.; Chen, Q.M.; Cui, L.J.; Wu, Y.C.; Liu, S.L.; Zhao, C.Y.; Vijerathna, P.; et al. Low Light Availability Reduces the Subsurface Sediment Carbon Content in *Halophila beccarii* From the South China Sea. *Front. Plant Sci.* **2021**, *12*, 664060. [CrossRef]
42. Macreadie, P.I.; York, P.H.; Sherman, C.; Keough, M.J.; Ross, D.J.; Ricart, A.M.; Smith, T.M. No detectable impact of small-scale disturbances on "blue carbon" within seagrass beds. *Mar. Biol.* **2014**, *161*, 2939–2944. [CrossRef]
43. Pendleton, L.; Donato, D.C.; Murray, B.C.; Crooks, S.; Jenkins, W.A.; Sifleet, S.; Craft, C.; Fourqurean, J.W.; Kauffman, B.; Marbà, N.; et al. Estimating global "blue carbon" emissions from conversion and degradation of vegetated coastal ecosystems. *PLoS ONE* **2012**, *7*, e43542. [CrossRef]
44. Oakes, J.M.; Eyre, B.D. Transformation and fate of microphytobenthos carbon in subtropical, intertidal sediments: Potential for long-term carbon retention revealed by ^{13}C-labeling. *Biogeosciences* **2014**, *11*, 1927–1940. [CrossRef]
45. Martin, B.C.; Gleeson, D.; Statton, J.; Siebers, A.R.; Grierson, P.; Ryan, M.H.; Kendrick, G.A. Low light availability alters root exudation and reduces putative beneficial microorganisms in seagrass roots. *Front. Microbiol.* **2018**, *8*, 2667. [CrossRef] [PubMed]
46. Mcmahon, K.; Lavery, P.S.; Mulligan, M. Recovery from the impact of light reduction on the seagrass *Amphibolis griffithii*, insights for dredging management. *Mar. Pollut. Bull.* **2011**, *62*, 270–283. [CrossRef] [PubMed]
47. Macreadie, P.I.; Trevathan-Tackett, S.M.; Skilbeck, C.G.; Sanderman, J.; Curlevski, N.; Jacobsen, G.; Seymour, J.R. Losses and recovery of organic carbon from a seagrass ecosystem following disturbance. *Proc. R. Soc. B Biol. Sci.* **2015**, *282*, 20151537. [CrossRef] [PubMed]

Article

Numerical Study on the Green-Water Loads and Structural Responses of Ship Bow Structures Caused by Freak Waves

Chengzhe Zhang [1], Weiyi Zhang [2], Hao Qin [1,*], Yunwu Han [3,*], Enjin Zhao [1], Lin Mu [4] and Haoran Zhang [5]

1 Hubei Key Laboratory of Marine Geological Resources, College of Marine Science and Technology, China University of Geosciences, Wuhan 430074, China
2 Marine Design & Research Institute of China, Shanghai 200023, China
3 Shandong Maritime Safety Administration Logistics Management Center, Qingdao 266002, China
4 College of Life Sciences and Oceanography, Shenzhen University, Shenzhen 518060, China
5 State Key Laboratory of Ocean Engineering, Shanghai Jiao Tong University, Shanghai 200240, China
* Correspondence: qinhao@cug.edu.cn (H.Q.); yhan36824@gmail.com (Y.H.)

Abstract: In recent decades, freak waves, characterized by their unusual high amplitude, sharp crest, and concentrated energy, have attracted researchers' attention due to their potential threat to marine structures. Green-water loads caused by freak waves can be significant and may lead to local damage to the ship structures. Therefore, this paper focuses on the study of green-water loads and examines the structural responses of ship bow structures under the influence of the green-water loads caused by freak waves. Firstly, a three-dimensional numerical wave tank is established in which the superposition model is used to generate freak waves. Validations on the freak-wave generation, ship motion response and the wave loading are carried out to verify the present solvers. The simulation on the interaction between the freak wave and the ship are conducted to obtain the interaction process and green-water loads. Secondly, a finite element (FEM) model of the ship bow is built, on which the green-water loads are applied to calculate the structural responses. Finally, the displacement and stress of the deck and breakwater structures are analyzed. It is found that green water events caused by freak waves can generate enormous impact forces on the bow deck and breakwater, resulting in severe structural responses and even possible damage to the structures. The local strength of structures under freak waves needs to be considered in practical engineering applications.

Keywords: freak wave; green-water loads; finite element method; structural response

Citation: Zhang, C.; Zhang, W.; Qin, H.; Han, Y.; Zhao, E.; Mu, L.; Zhang, H. Numerical Study on the Green-Water Loads and Structural Responses of Ship Bow Structures Caused by Freak Waves. *Appl. Sci.* **2023**, *13*, 6791. https://doi.org/10.3390/app13116791

Academic Editors: Koji Murai and Inwon Lee

Received: 26 April 2023
Revised: 25 May 2023
Accepted: 31 May 2023
Published: 2 June 2023

1. Introduction

A ship is one main way of the transportation of goods in international trade. However, during the transportation process, ships may encounter various adverse sea conditions that endanger their safety. Freak waves, also known as rogue waves, which are rare and unpredictable waves characterized by an unusual high amplitude, sharp crest, and concentrated energy [1–3], are one of the most dangerous. The interaction between ships and freak waves can result in complex nonlinear phenomena including slamming and green water on deck, which may lead to severe damages on the ship structures, e.g., the 'World Glory' accident [4], the 'SS El Faro' accident [5] and many other destructions [6,7]. Therefore, investigation on the impact loads and structural response of the ship structures is crucial for ship safety.

In recent decades, many studies have been conducted to investigate the effects of freak waves on marine structures in order to understand the damage mechanisms caused by these waves. Some scholars have studied the wave force induced by freak waves. For example, Sparboom et al. [8] examined the impact of freak waves on cylinders in a large-scale physical model test and studied the influence of tilt angles. They found that the maximum impact pressure of freak waves on cylinders occurred when the incoming wave direction was opposite to the cylinder's tilt direction. Corte and Grilli [9] utilized a wave superposition model to simulate freak waves and investigated the impact loads of freak waves on vertical rigid cylindrical piles. Kim et al. [10] investigated the impact pressure of Draupner freak waves and fifth-order Stokes waves on cylinders and found that the impact load of the freak wave was 2.8 times that of regular waves of the same scale. Li et al. [11] used an energy-focusing model in the laboratory to simulate freak waves and studied the influence of wave parameters such as amplitude, period, and spectral width on the impact load of freak waves on cylinders. Bunnik et al. [12] conducted numerical simulations of fixed offshore platforms subjected to freak-wave impacts using a numerical wave tank. They predicted the impact loads on offshore platforms caused by freak waves and designed model experiments based on simulated conditions. The experimental results were in good agreement with those of the numerical simulations. In the academic circle, focused wave groups are frequently used to generate the freak waves, and as such, some scholars have researched the interaction between focused wave groups and marine/coastal structures. Gao et al. [13] utilized OpenFOAM to examine the transient fluid resonance phenomenon occurring in a narrow gap between two adjacent boxes that were excited by incident focused waves with varying spectral peak periods and focused wave amplitudes. Liu et al. [14] investigated the transient fluid motion within a narrow gap formed by two fixed boxes under the action of an incident focused wave group, using a two-dimensional viscous flow numerical wave flume. In the study of Gao et al. [15], based on the Morlet wavelet transform and discrete Fourier transform techniques, the capability of focused transient wave groups to trigger the harbor resonance phenomenon was revealed for the first time. Some scholars have investigated the motion response and longitudinal strength of marine structures under freak waves. For example, Liu et al. [16] described how to evaluate the strength of a ship based on its nonlinear vertical bending moment (VBM), and investigated the influence of freak-wave height and speed on VBMs and deformation. Soares et al. [17] investigated the VBM in the midship induced by freak waves and explored the influence of the position in space where the freak waves were generated on the variation of the maximum bending moment. Qin et al. [18] studied the impact of freak waves and second-order Stokes waves on a fixed elastic deck using a strong coupled fluid–structure interaction method and discussed the hydroelastic responses. Luo et al. [19] conducted physical experiments to investigate the impact of a freak wave on a tension-leg platform, which suggested that a high-crest freak wave could induce violent motions of a floating platform and snap loads in tethers. Rudman et al. [20] utilized the smoothed particle hydrodynamic (SPH) method to simulate the effect of freak waves on a semi-submersible platform and compared and analyzed the motion response characteristics of the semi-submersible platform under two mooring systems: a tension leg platform (TLP) and taut spread mooring (TSM). Focusing on the local wave-structure interaction phenomena, freak waves with huge wave heights usually lead to significant green water events. Therefore, some other scholars have examined the green-water loads caused by freak waves. Hu et al. [21] found asymmetric and irregular characteristics in green water caused by freak waves when compared to regular waves. Qin et al. [22] studied the impact of green-water loads caused by Peregrine breather-type freak waves on a flat deck, and proposed an empirical formula for the quick prediction of nonlinear freak-wave forces on such deck structures. Zhang et al. [23] studied the wave height, deck pressure, and superstructure pressure of a fixed FPSO under the influence of freak waves. Liu et al. [24] investigated the mechanism and impact form of green water events under the influence of freak waves, and analyzed the ship's motion response, water variation and pressure. Wang et al. [25] discussed the relationship between ship motion response and green water events under the influence of freak waves, and the effect of ship speed on wave-ship interaction, revealing the impact of freak-wave peaks and sequences on roll, heave, and impact pressure.

The serious wave loads during green water events may lead to severe structural responses and even local damages of the ship bow, especially to the deck and breakwater structures. However, previous research focused on the structural analysis during bottom slamming, bow flare slamming and stern slamming. The structural responses and local damage of the deck and breakwater structures were rarely examined. For example, Maki et al. [26] used computational fluid dynamics (CFD) and FEM methods to study the water entry slamming of the ship's body and predict the structural hydroelastic response. Yang et al. [27] established a three-dimensional FEM model of a 1700 TEU container ship's bow structure and studied the dynamic response of bow flare slamming under impact loads, using a one-fold-plate-thickness deformation criterion to determine whether the structure buckled. Ren et al. [28] studied the dynamic response of bow flare structures under impact loads using LS-DYNA software. Xie et al. [29] proposed an uncoupled CFD and FEM method to study the local dynamic response of bow flare on a 21,000 TEU container ship under extreme impact loads. Kim et al. [30] studied the slamming response of the stern of an LNG ship using numerical simulation. Qin et al. [31,32] studied the impact of green-water loads caused by freak waves on the deck and deck-house structures, which further obtained the displacement of the structures using a strong coupled fluid–structure interaction method. However, the structures they used were fixed deck models simplified as plates and beams.

From these studies, it is seen that the interaction between freak waves and marine structures has been widely studied in the aspects of wave force, motion response, longitudinal strength and green water events. Nevertheless, the green-water loads of a ship with free motions induced by freak waves and the corresponding structural analysis of the deck and breakwater structures have rarely been examined. Therefore, this paper provides a detailed study on the green-water loads induced by the freak wave and the corresponding structural responses of the deck and breakwater structures of a ship bow. To achieve this, firstly, a three-dimensional numerical wave tank is established in which the superposition model is used to generate freak waves. Validations on the freak-wave generation, ship motion response and wave loading are carried out to verify the present issues. The simulation of the interaction between the freak wave and the ship are conducted to obtain the interaction process and green-water loads. Secondly, a finite element model of the ship bow is built, on which the green-water loads are applied to calculate the structural responses. Finally, the displacement and stress of the deck and breakwater structures are analyzed, while conclusions on the safety of the ship bow is drawn. The novelty of the study lies in that a comprehensive investigation of the ship's safety under freak-wave conditions are conducted, including the motion response, green-water impact loading and the structural response. More importantly, the structural responses and damages of the deck and breakwater structures during the green-water event induced by the realistic scale freak wave are examined for the first time, which provides a meaningful reference value for the practical engineering design of ship deck and breakwater structures.

2. Numerical Implementation

2.1. Freak Wave Based on Superposition Model

The superposition model, commonly employed as a freak-wave model, has been shown to be effective in generating freak waves experimentally and numerically. In the present study, a two-wave-train superposition model is adopted, which combines a random wave train and a convergent wave train to form the wave surface elevation. By adjusting the two trains of the waves' energy ratio and the phase of the linear cosine component waves of the convergent wave train, a freak wave is produced at a particular time and location [33]. The wave surface elevation of the two-wave-train superposition method can be expressed as:

$$\zeta = \sum_{1i=1}^{N_1} a_{1i}\cos(o_{1i}x - w_{1i}t + \theta_{1i}) + \sum_{2i=1}^{N_2} a_{2i}\cos[o_{2i}(x - x_c) - w_{2i}(t - t_c)] \tag{1}$$

In Equation (1), the first item creates the background random waves, while the second item generates the freak-wave peaks. N_1 and N_2 represent the respective quantities of component waves present in the two wave trains. o_{1i} and o_{2i} represent the wave numbers of individual component waves within the two wave trains. w_{1i} and w_{2i} denote the circular frequency of each component wave within the two wave trains. θ_{1i} determines the phase of each component wave in the random wave train, while $o_{2i}t_c - o_{2i}x_c$ determines the phase of each component wave in the convergent wave train. t and x are the time and coordinate, and x_c and t_c are the position and focusing time when the freak-wave crest is formed. a_{1i} and a_{2i} are the amplitudes of the individual component waves within the two wave trains.

2.2. Fluid Solver

To simulate the interaction between freak waves and the ship, a numerical wave tank is required. This study utilizes a three-dimensional fluid solver to model this interaction. The incompressible N-S equations and a VOF method were used by the solver to reconstruct free surfaces. The fluid was assumed to be viscous, Newtonian, and incompressible, with governing equations consisting of continuity, momentum conservation, and volume transportation equations, written as follows:

$$\frac{\partial \boldsymbol{u}^f}{\partial x} = 0 \tag{2}$$

$$\frac{\partial \boldsymbol{u}^f}{\partial t} + \boldsymbol{u}^f \frac{\partial \boldsymbol{u}^f}{\partial x} = \frac{1}{\rho^f}\frac{\partial \boldsymbol{\sigma}^f}{\partial x} + \boldsymbol{f}^f + \boldsymbol{v}\frac{\partial^2 \boldsymbol{u}^f}{\partial x^2} \tag{3}$$

$$\frac{\partial F}{\partial t} + \frac{\partial \boldsymbol{u}^f F}{\partial x} = 0 \tag{4}$$

where t and x denote the time and coordinate. $\boldsymbol{u}^f$, ρ^f, $\boldsymbol{f}^f$, $\boldsymbol{\sigma}^f$, v and F are the velocity, density, body force, Cauchy stress tensor of the fluid, kinetic viscosity coefficient and the transportation volume of the fluid. The standard $k - \omega$ model [34] is applied for the solution of the governing equations.

Omitting the surface tension, boundary conditions at the free surface are given as follows:

$$\frac{\partial u_n}{\partial \tau} + \frac{\partial u_\tau}{\partial n} = 0 \tag{5}$$

$$-p + 2\mu \frac{\partial u_n}{\partial n} = -p_0 \tag{6}$$

where u_n and u_τ are the normal component and the tangential component of the velocity vector on the boundary, respectively, and p, p_0 and μ are the liquid pressure, air pressure and dynamic viscosity coefficient.

The governing equations are solved using the Semi-Implicit Method for Pressure Linked Equations (SIMPLE) algorithm [35]. The solution of ship motion response is obtained by using the rigid-body six-degree-of-freedom motion equation and employing the FAVOR method to handle the dynamic boundary (Wang et al. [25], Patankar and Spalding [36]; Hirt and Sicilian [37]).

2.3. Structure Solver

According to Newton's second law, the rate of change of structural momentum is equal to the external force load acting on the structure. Therefore, the momentum equation for the structure can be written as:

$$\rho^s \ddot{\boldsymbol{z}} = \nabla \cdot \boldsymbol{\sigma}^s + \boldsymbol{f}^s \tag{7}$$

where ρ^s is the structural density, $\boldsymbol{z}$ is the displacement, $\boldsymbol{\sigma}^s$ is the first Piola–Kirchhoff stress tensor, and $\boldsymbol{f}^s$ is the volume force acting on the structure. The expression for $\boldsymbol{\sigma}^s$ is given by the following equation:

$$\sigma^s = \mu^s\left[\nabla \mathbf{z} + (\nabla \mathbf{z})^T\right] + \lambda^s(\nabla \cdot \mathbf{z})\boldsymbol{I} \tag{8}$$

where μ^s and λ^s are the Lame Constants, $\boldsymbol{I}$ is the identity tensor.

The structural responses of the deck and breakwater structures are calculated by discretizing the momentum conservation equation for the structure, using FEM for spatial discretization and the $Newmark - \beta$ method [38] for temporal discretization. The discrete form of the structural dynamic equation is:

$$\boldsymbol{M} \cdot \ddot{z} + \boldsymbol{C} \cdot \dot{z} + \boldsymbol{K} \cdot z = f \tag{9}$$

where $\boldsymbol{M}$ is the mass, $\boldsymbol{C}$ is the damping and $\boldsymbol{K}$ is stiffness matrix matrices of the structure. f is the force vector.

Time can be discretized using the $Newmark - \beta$ method [38], which involves the calculation of displacement, velocity, and acceleration within a specific time interval.

$$\dot{z}^{n+1} = \dot{z}^n + \left[(1-\beta)\ddot{z}^{n+1} + \beta \cdot \ddot{z}^{n+1}\right]\Delta t \tag{10}$$

$$z^{n+1} = z^n + \dot{z}^n \Delta t + \left[\left(\frac{1}{2} - \gamma\right)\ddot{z}^n + \gamma \cdot \ddot{z}^{n+1}\right]\Delta t^2 \tag{11}$$

where Δt is time step. β and γ are parameters that determine the accuracy and stability of the equation. Using Nastran as the structural solver in this paper, which is a widely used finite element analysis software, various dynamic loads can be defined in the time domain, and its powerful analytical capabilities have been extensively applied to structural dynamics analysis [39].

The structural boundary conditions need to satisfy the continuity of velocity and surface force at the boundary, as shown below:

$$\boldsymbol{u}^s = \boldsymbol{u}^s_{boundary} \tag{12}$$

$$\sigma^s \boldsymbol{n}^s = \boldsymbol{T}_{boundary} \tag{13}$$

here, $\boldsymbol{u}^s$ is the velocity of the structure, $\boldsymbol{u}^s_{boundary}$ and $\boldsymbol{T}_{boundary}$ represent the velocity and surface force vector, respectively, given in the outer domain of the structure at its boundary, and $\boldsymbol{n}^s$ is the unit normal vector at the structural boundary.

3. Numerical Validations

3.1. Validation on the Ship Motion Responses

The container ship model chosen for this validation is developed at the Institute of Ship Technology, Ocean Engineering and Transport Systems (Duisburg-Essen, Germany) as a benchmark for numerical methods by Moctar et al. [40]. The model has been widely used by scholars [41–43], including Mei et al. [43] who conducted physical experiments and numerical simulations to investigate the motion of this ship model in regular waves. Figure 1 presents side views of this ship; Table 1 lists its principal particulars. The regular wave is defined by a height of 0.06235 m and a period of 1.38 s.

Figure 1. Side views of the container ship model.

Table 1. Main particulars of the model.

Designation	Scale
Length (m)	3.984 ± 0.001
Breadth (m)	0.572 ± 0.001
Draft (m)	0.163 ± 0.001
Displacement (kg)	245.155
Block coefficient (-)	0.661
Froude number (-)	0.139

Figure 2 illustrates the comparison between the heave and pitch motions of the container ship obtained using the present numerical method and experimental and numerical measurements conducted by Mei et al. [43]. Due to the reflection of the beach and wave maker [44], as well as the shallow-water squatting phenomenon, there are some discrepancies between the numerical simulation results and the experimental results. However, good consistency was found when comparing them with the numerical results. The simulation results obtained from the present method were in strong agreement with the numerical simulations conducted by Mei et al. [43]. This indicates that both the present solver and meshing have the ability to produce precise simulations of a ship's heave and pitch motions in waves.

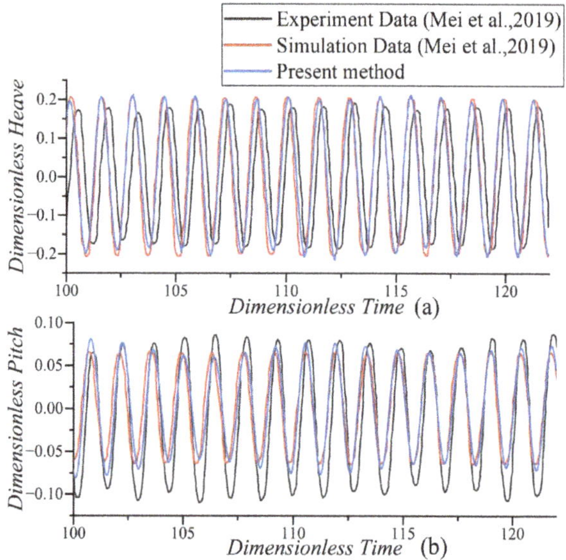

Figure 2. Comparisons of heave and pitch time histories from present method and the method of Mei et al. [43]: (**a**) Heave; (**b**) pitch.

3.2. Validation of the Green Water Loading

Rosetti et al. [45] conducted research on the green-water loads of various regular waves using a fixed FPSO model in a wave tank at the University of Sao Paulo. The experimental investigation was conducted in the wave flume of the Department of Naval Architecture and Ocean Engineering, University of São Paulo. As shown in Figure 3, the FPSO model employed in the experiment had a slightly inclined deck to facilitate water drainage after green-water events. A wave probe (WM01) was located on the upstream surface of the deck; the load cell used for measuring the impact force was located on a 20 mm × 20 mm plate. The regular wave height was 0.088 m, the wave period was 0.97 s, and the wavelength was 1.46 m.

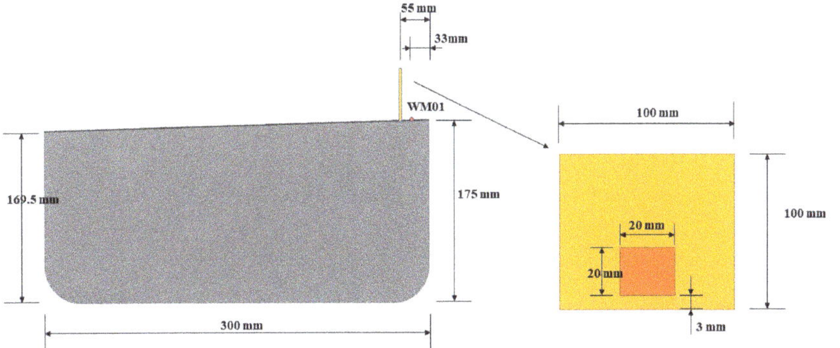

Figure 3. The model setup in the experiment.

Figure 4 illustrates the comparison between the water elevations detected by WM01 and impact forces among the results obtained from the present solver, and the experimental data by Rosetti et al. [45]. They noted that CFD simulations tend to slightly overestimate the maximum value, which could result from monitoring errors during testing, but such results remain satisfactory. These findings indicate that the present simulation methodology is highly effective in accurately simulating green-water events and loads.

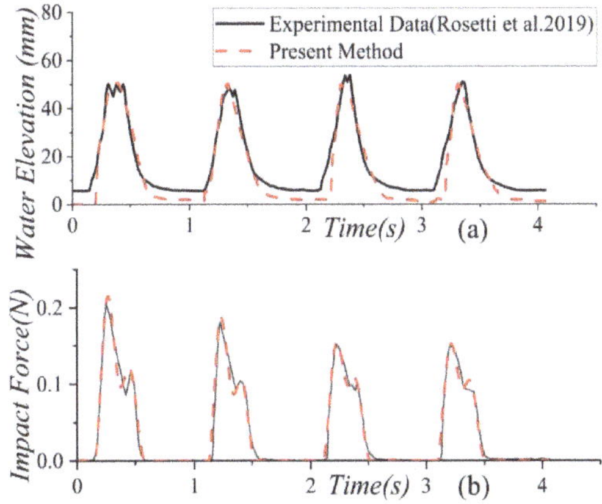

Figure 4. The comparison between experimental data [45] and present method: (**a**) Water elevation; (**b**) impact force.

4. Wave-Ship Interaction and Green-Water Loads

4.1. Numerical Wave Tank

A three-dimensional numerical wave tank was constructed to simulate the interaction between container ships and freak waves. As depicted in Figure 5, the wave tank features a length of 1000 m, width of 300 m, and water depth of 180 m. A container ship model, shown in Figure 6, was situated 150 m from the left end of the tank. It should be mentioned that an on-deck breakwater structure was considered in this ship model, which is usually used for on-deck water blocking. A Piston wave-maker was located at the left boundary to generate freak waves, while a wave absorber zone was positioned near the right boundary to eliminate wave reflections based on the sponge layer relaxation method proposed by Mayer et al. [46]. As the present study focused solely on the head wave condition, the ship was limited to a rigid container ship model with heave and pitch motions, which is similar to the condition of Zhao and Hu [47].

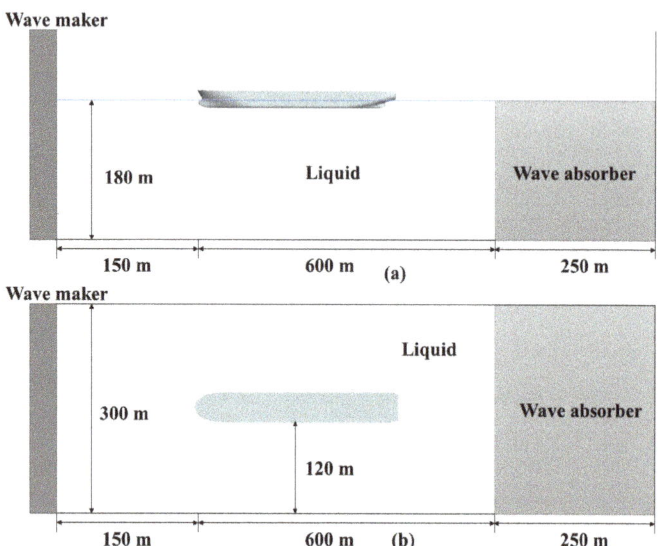

Figure 5. Layout of the numerical wave tank: (**a**) Front view, (**b**) top view. The plot is not to scale.

Figure 6. A sketch of the container ship model.

4.2. Generation of the Freak Wave

The target freak wave was similar to the one recorded at Yura fishery in the Japanese sea area by Mori et al. [48]. The measured freak wave had a height of 11.2 m, crest height of 7.4 m, and significant wave height of 4.6 m, occurring in water with a depth of 43 m. Considering that the container ship model used in present study possesses dimensions of 398 m in length, 57 m in width, and 34 m in height, the freak wave was scaled with a scaling factor of 1:3.3. A wave surface elevation monitoring point was set 150 m away from the wave-making boundary to record the wave surface elevation at that location within the numerical wave tank. Subsequently, the free surface elevation of the freak wave was transformed into the displacement of the Piston. As mentioned by Cui et al. [49], the displacement $S(t)$ of the Piston wave-maker for freak waves can be expressed as:

$$S(t) = \frac{\sum_{1i=1}^{N_1} a_{1i}\cos(o_{1i}x - w_{1i}t + \theta_{1i})}{W_{1i}} + \frac{\sum_{2i=1}^{N_2} a_{2i}\cos[o_{2i}(x - x_c) - w_{2i}(t - t_c)]}{W_{2i}} \tag{14}$$

$$W_i = \frac{4\sinh^2(o_i d)}{\sinh^2(o_i d) + 2o_i d} \tag{15}$$

where W_i is the transfer function associated with the ith component of the propagating wave, d is the water depth, while the other variables have the same definitions as in Equation (1). Additionally, the theoretical value of the surface elevation at this location was calculated using Equation (1) as a reference for comparison with the numerical simulation results.

Figure 7a displays the free surface elevations of freak waves generated using three different grid groups: coarse, medium, and fine. The coarse grid had a size of 1 m × 1 m × 1 m, the medium grid had a size of 0.5 m × 0.5 m × 0.5 m, and the fine grid had a size of 0.25 m × 0.25 m × 0.25 m. Figure 7b gives a comparison between the theoretical and medium-grid results. As can be observed, the numerical simulation results are in excellent agreement with the theoretical values. Despite some small differences in the random waves used as background waves, the heights of the freak-wave crest and peak are highly consistent.

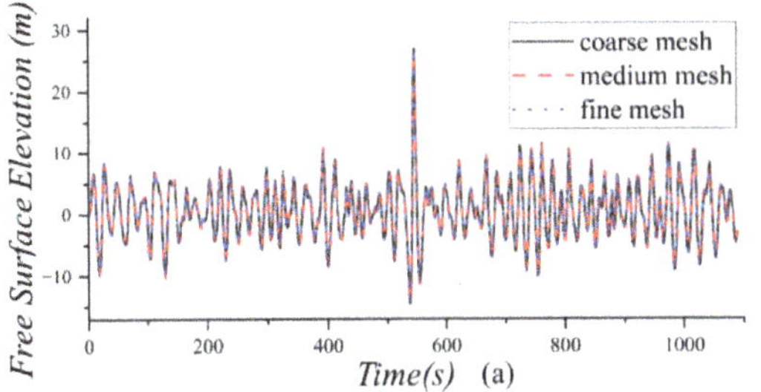

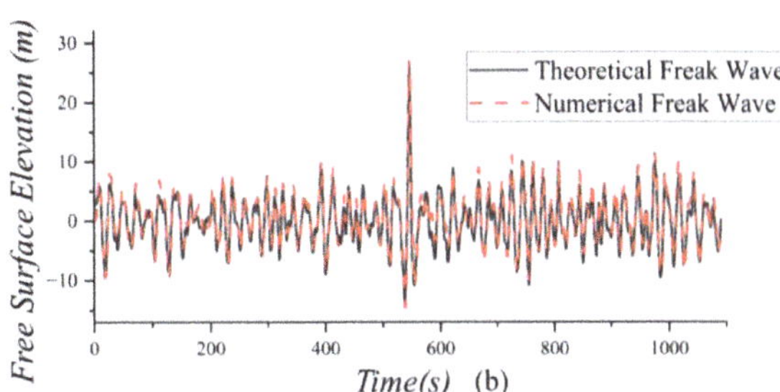

Figure 7. Verification of the grid independence and method feasibility of the freak wave: (**a**) Grid independence; (**b**) method feasibility.

Klinting and Sand [50] put forward α, β_1, β_2 and μ to describe the features of freak waves. These characteristic parameters are defined as follows: (1) $\alpha = H_f/H_a$; (2) $\beta_1 = H_f/H_{f-1}$; (3) $\beta_2 = H_f/H_{f+1}$; (4) $\mu = \eta_f/H_f$. Here, H_f is the freak wave's height, H_{f+1} and H_{f-1} are the wave heights of adjacent waves before and after the freak wave, H_a is the significant wave height of the wave train, and η_f is the peak height corresponding to H_f. As shown in Table 2, it can be observed that the characteristic parameters of the generated freak wave largely agree with those of the measured freak wave, indicating the effectiveness of the numerical wave tank and freak-wave model in generating such waves.

Table 2. Comparison of parameters between the theoretical result, numerical result and the measured freak wave in Yura fishery.

Parameters	α	β_1	β_2	μ
Measured by Mori et al. [48]	2.42	4.03	2.02	0.67
Theoretical result	2.46	3.65	1.96	0.70
Numerical result	2.48	3.70	1.84	0.71

4.3. Measurement of the Impact Pressures

In order to obtain the green-water loads induced by the target freak wave, pressure monitoring points were installed on both the deck and breakwater structures of the ship bow. As illustrated in Figures 8 and 9, D100 to D811 were set to obtain the pressures on the deck, while B100 to B805 were set to obtain the pressures on the breakwater. Upon obtaining the pressure time histories during the green water event, the green-water loads could be approximated and loaded to the FEM model of the ship bow structures.

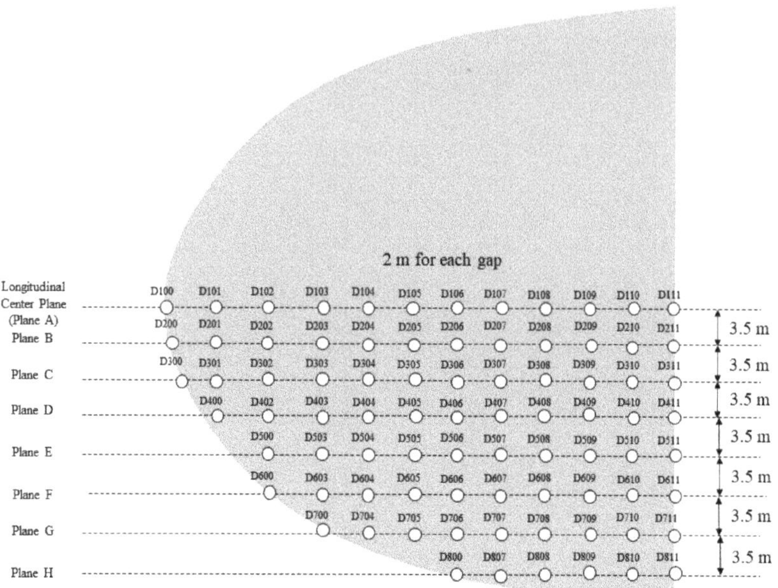

Figure 8. A sketch of pressure monitoring points on the deck.

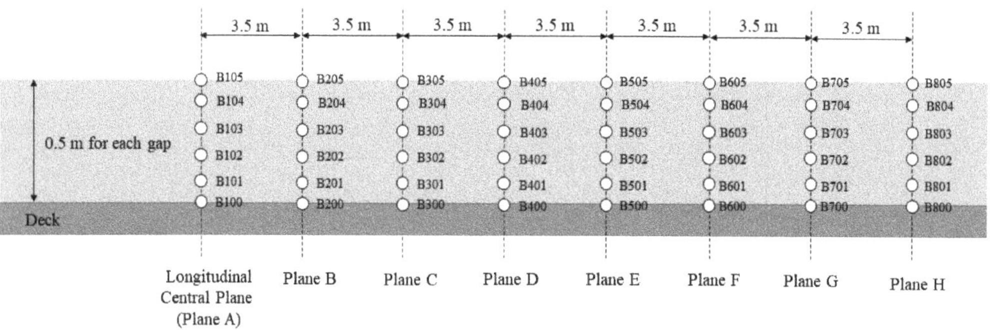

Figure 9. A sketch of pressure monitoring points on the breakwater.

4.4. Ship Motion Responses and Green-Water Loads

4.4.1. Ship Motion Responses Induced by the Freak Wave

To ensure result accuracy, grid independence testing must be conducted beforehand. Given the complexity of wave-ship interactions, it is preferable to select larger computational domains while maintaining a fine simulation of the flow field, thereby effectively reducing the impact of wave reflection caused by the boundary. As such, the present study utilized nested mesh methods, dividing the mesh region into two parts: background and free surface regions, with the free surface regions being denser. Through these methods, computational speed was maximized while maintaining result accuracy. Three types of meshes were tested, including coarse mesh (background region mesh size of 2 m × 2 m × 2 m, free surface region mesh size of 1 m × 1 m × 1 m), medium mesh (background region mesh size of 1 m × 1 m × 1 m, free surface region mesh size of 0.5 m × 0.5 m × 0.5 m), and fine mesh (background region mesh size of 0.5 m × 0.5 m × 0.5 m, free surface region mesh size of 0.25 m × 0.25 m × 0.25 m). Simulation time steps were determined according to the Courant number criterion (Anderson and Wendt [37]), with values of 0.1 s, 0.05 s, and

0.025 s used for the coarse, medium, and fine meshes, respectively. As shown in Figure 10, the results of the grid-independence tests indicate that the medium mesh size provides a balance between simulation accuracy and computational speed.

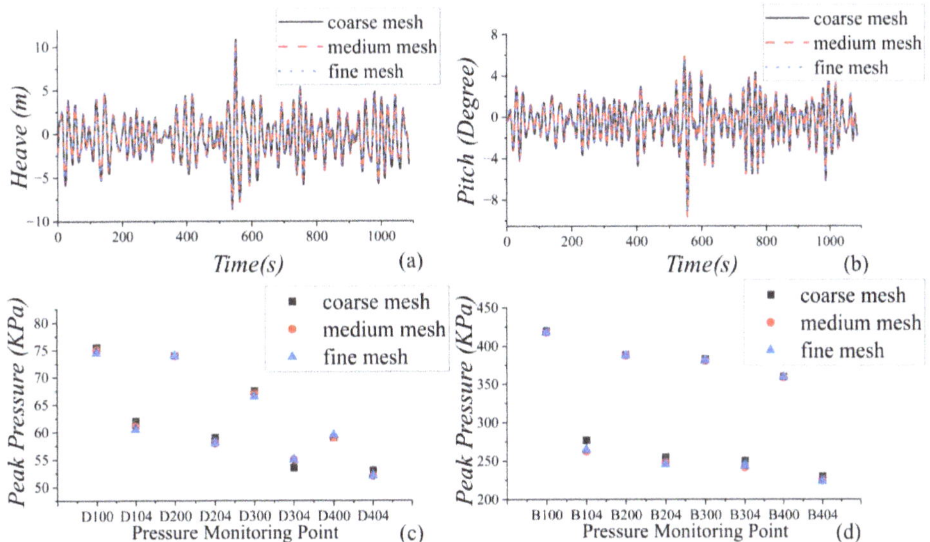

Figure 10. Results of grid independence test: (**a**) Heave; (**b**) pitch; (**c**) peak pressure of deck; (**d**) peak pressure of breakwater.

Since present study focuses on the interaction between freak waves and the ship, the following content examines only the period from when a freak wave appears until the interaction with the ship concludes. Specifically, this refers to the time interval between 530 s and 565 s in Figure 7 (hereinafter, 0 s corresponds to 530 s in Figure 7).

As shown in Figure 11, at 13.4 s, the freak wave comes in touch with the ship bow, causing the wave to roll over due to the large bow flare. Subsequently, a substantial amount of water surges over the deck at an extremely high speed. During this process, the ship bow is lifted by the freak wave. At 15.7 s, the on-deck water slams the breakwater and rapidly moves upwards along its surface. At 16.6 s, the water climbs to the highest point, with some of it overtopping the breakwater. Following this, the water drains off the deck, marking the end of the first green water impact. At this moment, the ship bow is greatly lifted, with increased heave motion reaching the maximum value, and the pitch motion also increasing to the first peak value. Additionally, then, the freak wave moves towards the middle of the ship, and the amplitude of the heave and pitch motions decreases as the ship gradually returns to its normal floating state. As the freak wave approaches the stern of the ship, the stern is gradually lifted, causing the ship bow to sink into the water. At 27.0 s, the bow is submerged below the free surface. At 29.20 s, a small amount of water surges over the deck, leading to the second green water impact but far less than the first. At this moment, the absolute value of the pitch motion reaches its maximum. As the freak wave propagates across the ship, the interaction between the freak wave and the ship ends at about 34.0 s.

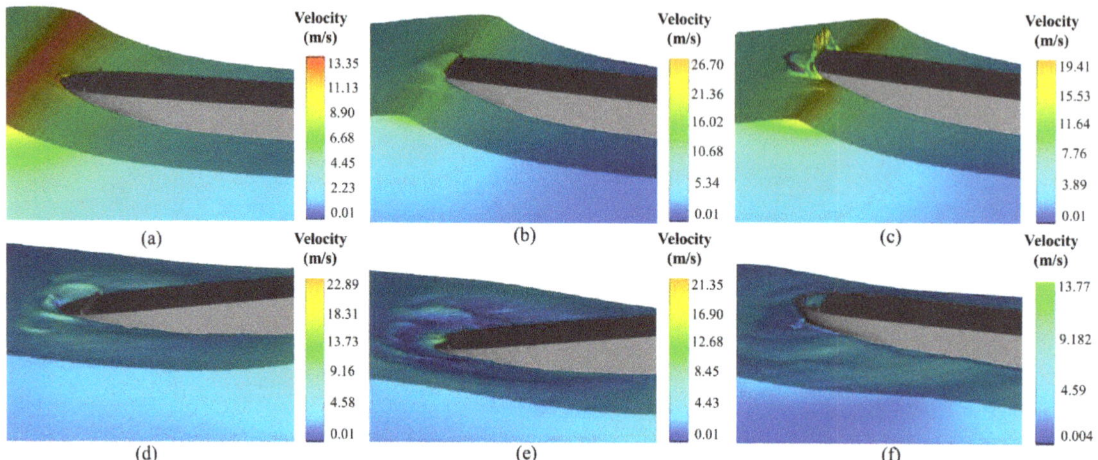

Figure 11. Three-dimensional snapshots of interaction between freak waves and container ships: (**a**) t = 13.4 s, (**b**) t = 15.7 s, (**c**) t = 16.6 s, (**d**) t = 27.0 s, (**e**) t = 29.2 s, (**f**) t = 34.0 s.

4.4.2. Green-Water Loads Induced by the Freak Wave

As the amount of water on the deck during the second green-water impact was significantly less than the first impact, the impact pressure generated from the second impact on the deck was also much smaller than that of the first. Therefore, the following sections only study the impact pressure generated by the first green water impact.

Figures 12 and 13 show the selected time histories of the impact pressure on the deck and breakwater caused by the freak wave. For the deck, the impact pressure starts at 13.8 s and gradually returns to zero after 21.5 s. The maximum impact pressure on the deck is located near the breakwater. The impact pressure at the front of the deck increases slowly, while the area of the deck near the breakwater reaches its peak value at a faster rate. These phenomena may be due to the gradual reduction in water thickness as the water flows over the deck, resulting in a decrease in pressure. However, the velocity of the water flow also increases, and when the water reaches the area of the deck near the breakwater, the flow velocity is the highest. At the same time, a large amount of water accumulates near the breakwater, leading to a large impact pressure in a short period of time. At around 19.6 s, a small impact pressure appears again due to the falling water. Compared with other literature, such as Zhang et al. [23], the impact pressure generated from the falling water was relatively smaller here. This is because other models used in previous studies applied higher superstructures rather than the lower breakwater in the present study. Due to the lower height of the breakwater, most water overtops over the breakwater and only a small portion of the water falls and impacts the deck, resulting in a smaller impact pressure from the falling water.

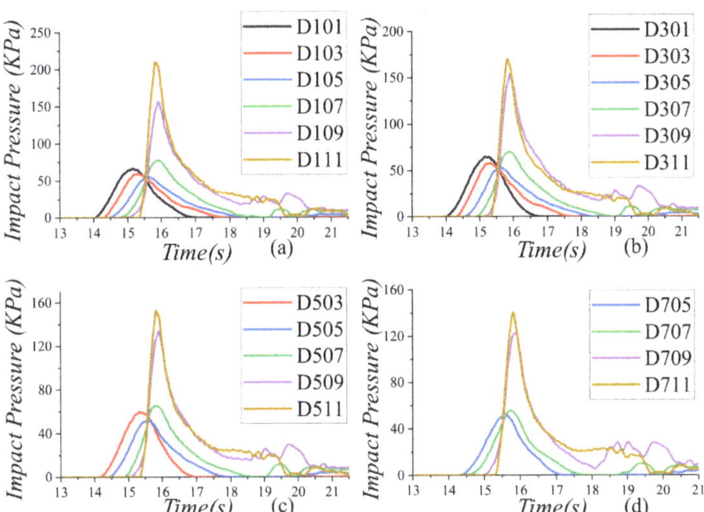

Figure 12. Impact pressures on the deck during the crest-induced green-water event: (**a**) In plane A; (**b**) in plane c; (**c**) in plane E; (**d**) in plane G.

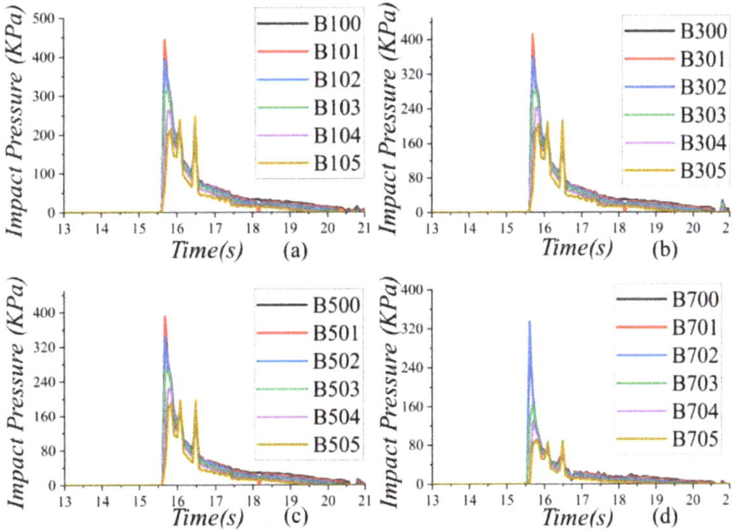

Figure 13. Impact pressures on the breakwater during the crest-induced green-water event: (**a**) In plane A; (**b**) in plane c; (**c**) in plane E; (**d**) in plane G.

The breakwater starts to experience impact pressure at 15.5 s and gradually returns to zero at 20.5 s. The maximum impact pressure on the breakwater occurs near the bottom close to the deck. For the breakwater, the impact pressure it experiences is greater than that of the deck, generating a large pressure peak during the impact moment before decreasing at a slightly slower rate. This phenomenon may be due to the fact that when the water flow impacts the surface of the breakwater, the flow velocity reaches its maximum, carrying a huge amount of energy to impact the surface of the breakwater. At this moment, the breakwater mainly experiences dynamic pressure, reaching the peak pressure in an instant. After the peak pressure, two small peaks appear on the surface of the breakwater due to the water overturning under the influence of gravity, resulting in another impact on the breakwater. It is worth noting that the breakwater was impacted by green-water loads before the deck area near it, as has been elaborated by Gomez-Gesteira et al. [51] and Hu et al. [52] in their articles, and will not be reiterated here.

5. Structural Responses of the Deck and Breakwater

5.1. Modelling of the Ship Bow Structures

To investigate the structural response of the deck and breakwater structures under freak waves, it is essential to establish a finite element model of the ship. The dimensions of the container ship were identical to those detailed in Section 4.1. To reduce computational costs, the bow section was selected for the model. The slamming areas subjected to green-water wave loads lack a clear definition among classification societies across various countries. Zhang et al. [23] and Liu et al. [24] conducted research on the slamming load characteristics from the bow to the superstructure area. According to the guidelines of DNV [53], BV [54], GL [55], and KR [56], the impact zone of the bow flare and bottom is situated at a range of 0.1 L from the FP (forward perpendicular) [29,57]. Based on this information, the truncated region extending 0.1 L from the FP is selected as the slamming region to construct an FEM model of the ship's truncated bow, which is illustrated in Figure 14.

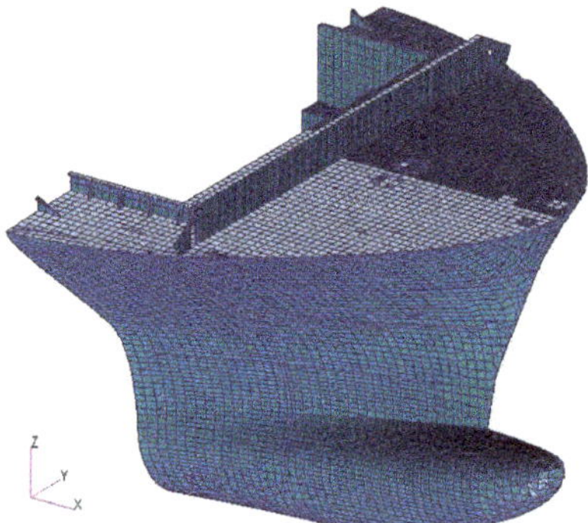

Figure 14. Three-dimensional finite element model of ship bow.

The finite element model employs three types of elements: plate elements are primarily utilized to simulate the shell structures of various components, including the deck, side shells, longitudinal bulkheads, top transverse bulkheads, deck beams, transverse bulkheads, panels, transverse webs, hatch coaming and its panel, and elbow plates. The majority of plate elements are quadrilateral elements, with a minimal number of triangular elements used for component connections and arc transitions. Beam elements are mainly used to simulate large and continuous longitudinal members, such as stringers, stiffeners, and supporting materials, while accounting for beam cross-sections and eccentricities based on actual conditions. Rod elements are mainly used to simulate small-sized reinforcements, such as openings in panels and discontinuous stiffeners. With a total count of 44,476, the deck and breakwater were set to a grid size of 0.7 m, which is fine enough for finite element analysis [29,57–59]. In addition, mesh convergence testing were conducted. Mesh sizes ranging from 0.3 m to 1.0 m were chosen, and Figure 15 shows the variation in the peak von Mises stress at Deck D111 and Breakwater B105 (the positions of D111 and B105 are shown in Figures 8 and 9) under different mesh sizes. It can be seen that the results converge when the mesh size is less than 0.7 m, and as the size is further reduced, the peak stress difference is less than 2%. Excessively small mesh sizes would significantly increase computational time, which means that a mesh size of 0.7 m can be considered appropriate for simulating the impact problem in this paper. The time step used for finite element analysis was 0.0045 s.

The computer used for the paper is equipped with two AMD EPYC 7402 2.8 GHz CPUs and 128 GB of memory. The time required for mesh independence verification was 624 min, and the average convergence time required per time step was 30 s when performing finite element analysis with a mesh size of 0.7 m.

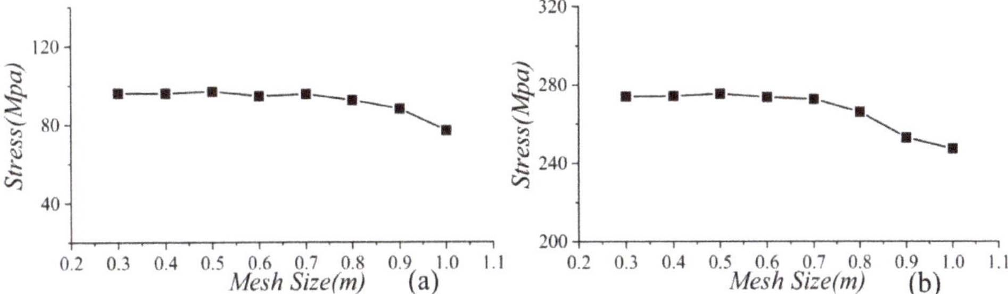

Figure 15. Analysis of mesh convergence: (**a**) The Mises stress at the deck D111, and (**b**) the Mises stress at the breakwater B105.

Reasonable boundary conditions are critical for calculating the dynamic response of structures subjected to slamming loads. Xie et al. [29] used a six-degree-of-freedom constraint to determine the truncation position. Yang et al. [58] indicated that only the aft-side of the model was fixed in six degrees of freedom to ensure that all forces could be balanced with minimal boundary effects on the six brackets. The approach described above was fixed in this article to constrain the ship bow, as illustrated in Figure 16.

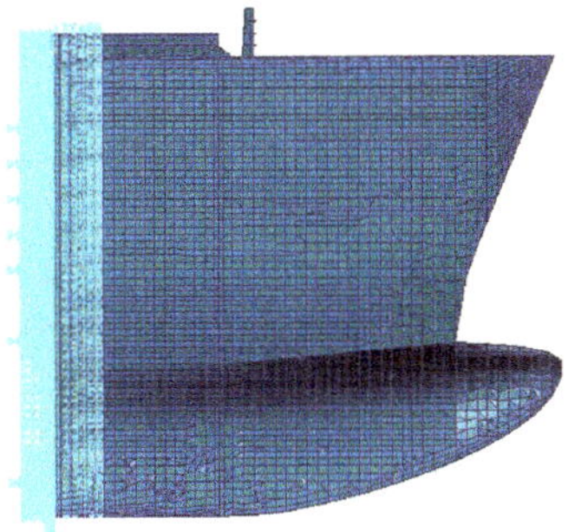

Figure 16. Sketch of boundary condition.

Parameters of the finite element model are as follows: elastic modulus E of 2.06×10^5 MPa, Poisson's ratio v of 0.3, and material density ρ of 7.8×10^3 kg/m^3. When subjected to slamming pressure, the constitutive behavior of the material differs from that observed under static loading conditions. To account for this, the dynamic constitutive model Cowper–Symonds model [60] commonly used in structural impact analysis was introduced.

$$\sigma_{dy} = \sigma_y \left[1 + \left(\frac{\dot{\varepsilon}}{D} \right)^{\frac{1}{e}} \right] \tag{16}$$

where $\dot{\varepsilon}$ represents the plastic strain rate, while D and e are coefficients that must be determined based on test data for various materials. Paik et al. [61] determined a set of Cowper–Symonds coefficients for high-tensile steel, determining D = 3200 s^{-1} and e = 5.

5.2. Loading on the Ship Bow Structures

In this study, the slamming load was first obtained in Section 4.4.2. Thereafter, the impact loads were applied to the FEM model and the resulting structural response was calculated. In problems where structural deformations are significant and impact the fluid field, a coupled fluid–structure interaction method is necessary. However, given that the deformations of the structure can be considered to have a negligible effect on the fluid field, an uncoupled CFD-FEA method was employed in the present study. This approach involved independent utilization of the wave loads simulation and the finite element analysis. For the sake of result accuracy, impact pressure values obtained from the wave load simulation were collected over the hull surface grids every 0.05 s. Subsequently, the time series data were directly applied to the finite element nodes in the same location. In the following sections, the impact pressure obtained from 13.75 to 18.25 s in Section 4.4.2 were applied to the finite element model for structural response calculation. The loading method at different times is shown in Figure 17. The structural solver Nastran [39] was applied for the structural response calculation.

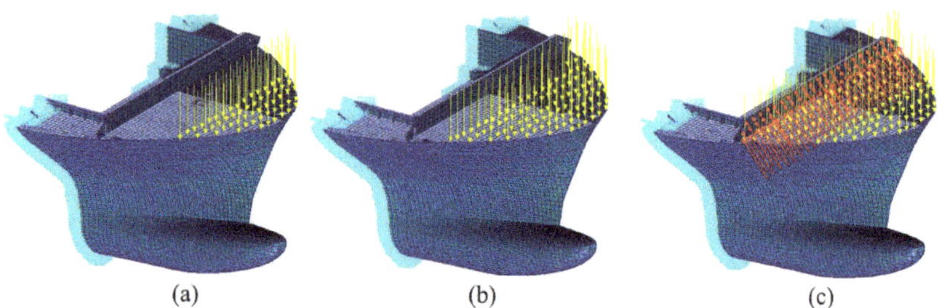

Figure 17. Sketch of external load in finite element model of ship bow: (**a**) t = 14.5 s, (**b**) t = 15 s, (**c**) t = 15.7 s.

5.3. Structural Response Analysis

5.3.1. Structure Responses of the Deck and Breakwater

This paper focuses on the structural strength assessment under slamming loads, and the evaluation process mainly involves verifying the yield strength of the structure. For plate elements, the Mises stress σ_{vm} is:

$$\sigma_{vm} = \sqrt{\sigma_x^2 - \sigma_x\sigma_y + \sigma_y^2 + 3\tau_{xy}^2} \tag{17}$$

where σ_x and σ_y are the normal stresses in the x and y directions of the element, and τ_{xy} is the shear stress of the element.

The yield factor λ_y is used to determine whether there is a yielding phenomenon in the ship's structure, and the specific judgment formula is as follows [29]:

$$\text{Deck}: \ \lambda_y = \frac{k\sigma_{vm}}{225} \tag{18}$$

$$\text{Breakwater}: \ \lambda_y = \frac{k\sigma_{axial}}{235} \tag{19}$$

where k is the material coefficient, which for the high tensile steel used in this model is 0.92 [62].

Using the dynamic analysis method, structural responses of the deck and breakwater structures were obtained. Figure 18 shows the contour plots of total displacement at different moments of the ship bow. During 14.0 to 15.5 s, the green water gradually flows from the deck front to the breakwater, causing the increase of the displacement. The area of the deck affected by green-water loads becomes larger over time. At 15.5 s, most of the deck is subjected to the impact of the green-water loads, but the water has not yet struck the breakwater. At 15.7 s, the impact load is applied to the breakwater, which results in a significant displacement of the breakwater. The maximum displacement occurs above the centerline near B105, and the value of the maximum displacement reaches as high as 1 m. This indicates that the area is likely to have been damaged or even fractured due to the impact. Over time, the green water gradually flows out of the bow, and the deck displacement caused by the impact of green-water loads gradually decreases to zero. It can be preliminarily inferred that the deck remains undamaged during the green water event. However, residual displacement of different degrees still exists in the breakwater area, indicating that the breakwater is likely to be damaged under the green-water loads.

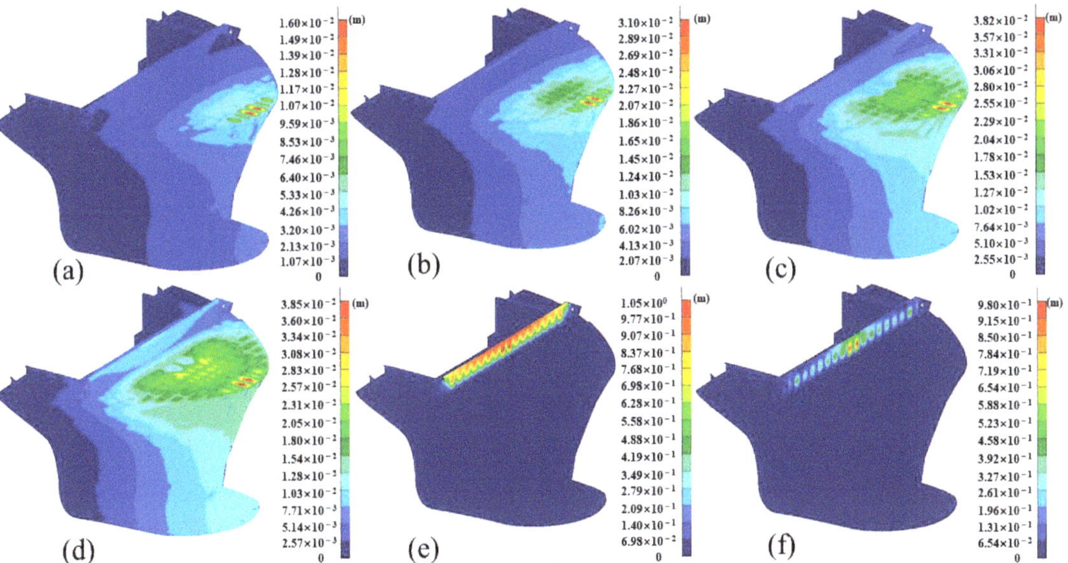

Figure 18. Deformation with different time instants: (**a**) t = 14 s, (**b**) t = 14.5 s, (**c**) t = 15 s, (**d**) t = 15.5 s, (**e**) t = 15.7 s, (**f**) t = 18.25 s.

Figure 19 shows the contour plots of the Von Mises stress at different moments of the ship bow. To facilitate the visualization of the stress distribution beneath the deck, half of the deck plate was hidden in the Mises stress contour plot. Similar to the displacement contour plots, during 14.0 to 15.5 s, the green water event happens as time increases, resulting in larger impact forces and affected deck areas over time. At 15.7 s, the green water impacts the breakwater with a sudden and enormous force. Further investigation reveals that the point of maximum stress response occurs at the stiffeners of the breakwater B101. The stress response value exceeds the material's yield limit, suggesting that this area may have been damaged. Subsequently, the impact force gradually decreases, but residual stresses still exist in the breakwater area. It is noteworthy that the stress response of the deck near the breakwater is greater than that of the forward deck, and the stress response near the longitudinal centerline of the deck and breakwater is greater than that on both sides. Therefore, particular attention should be paid to the above regions when evaluating the safety of the ship's bow.

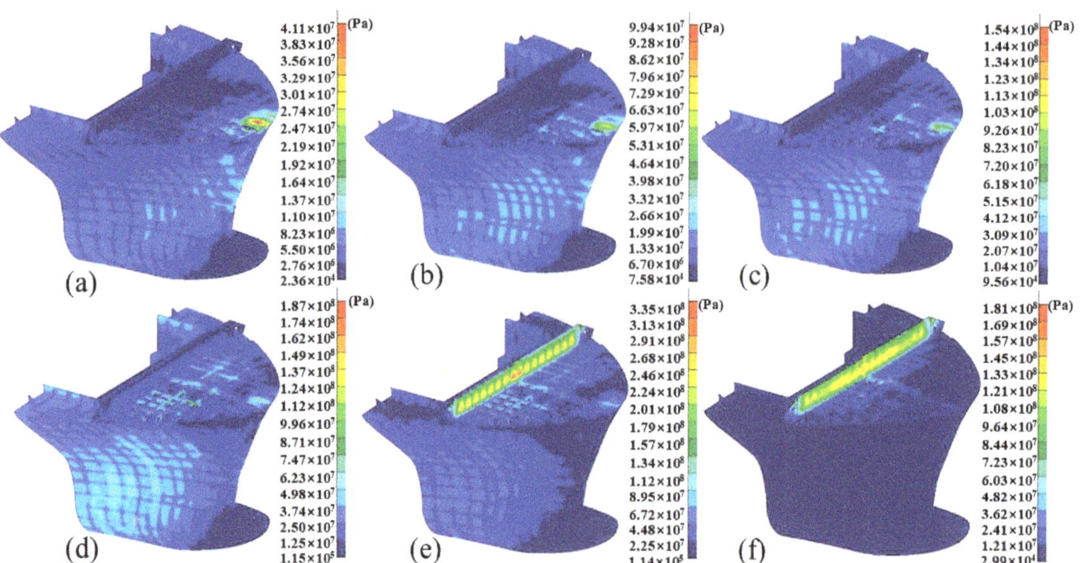

Figure 19. Von Mises stress with different time instants: (**a**) t = 14 s, (**b**) t = 14.5 s, (**c**) t = 15 s, (**d**) t = 15.5 s, (**e**) t = 15.7 s, (**f**) t = 18.25 s.

Based on the previous description, the stress response increases as the deck approaches the position of the breakwater. Figure 20 presents the time histories of the stress response and total displacement at locations D108, D110, and D111 on the deck (the positions of D108, D110, and D111 are shown in Figure 8). These three locations experience a greater impact load compared to other areas of the deck, resulting in a correspondingly greater stress response and displacement. In terms of numerical patterns, compared with the results obtained in Section 4.4.2 for the historical impact load, it can be seen that the stress response results of the deck exhibit synchronous variations with the impact load applied to the deck, and the shape of the curves and the time of occurrence of the maximum values are basically the same, with a relatively short duration of the maximum values. However, after the maximum peak disappears, there is still a slight fluctuation in stress, which may be due to the influence of other positions after the first impact ends, and a slight fluctuation in stress response occurs subsequently. The total displacement at the deck also exhibits synchronous changes with the impact load applied to the deck.

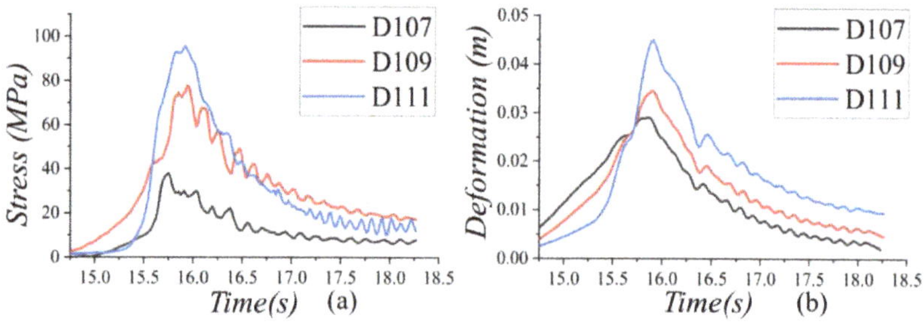

Figure 20. Results of Von Mises stress and total displacement in deck: (**a**) The Von Mises stress at the deck, and (**b**) the total displacement at the deck.

The stress response in the region near the longitudinal centerline of the breakwater was also significant and requires particular attention. The contour plots in Figures 18 and 19 show that the breakwater suffered severe damage due to the impact of the green water. The stress response peak at the breakwater near the longitudinal centerline exceeds its yield limit and may have undergone fracture deformation. Other areas of the breakwater also experience varying degrees of damage, with some residual stress and displacement remaining even after the reduction in the impact loads. When further analyzing the failure mechanism of the breakwater, it was found that the maximum stress response occurs at the bottom of the stiffener behind the breakwater near the longitudinal centerline (i.e., the stiffener located behind breakwater B101), reaching a peak of 315.5 MPa. The areas with the largest displacement were located in the upper part of the breakwater, with the maximum displacement occurring near the longitudinal centerline of the upper part of the breakwater (at breakwater B105). Figure 21 shows the stress response and displacement time histories of the areas with the maximum stress response and displacement. The stress response of the stiffener behind the breakwater was greater than that of the breakwater itself, indicating that the stiffener helped to absorb a considerable amount of the impact force, improving the overall strength and bearing capacity of the structure. However, due to the excessive impact load, the stress response of the stiffener behind breakwater B101 exceeds its yield limit, and the stiffener may fracture under the impact of the green water load. The breakwater near the longitudinal centerline experiences significant displacement as it may have lost support. The upper part of the breakwater is less fixed compared to the lower part, resulting in the largest displacement at that location. The contour plot results at 18.25 s show that in addition to the area near the centerline of the breakwater, where residual stress and displacement exceeds the limit, other parts of the breakwater also exhibit residual stress and displacement. This indicates that other areas of the breakwater also suffered damage under the impact of green water.

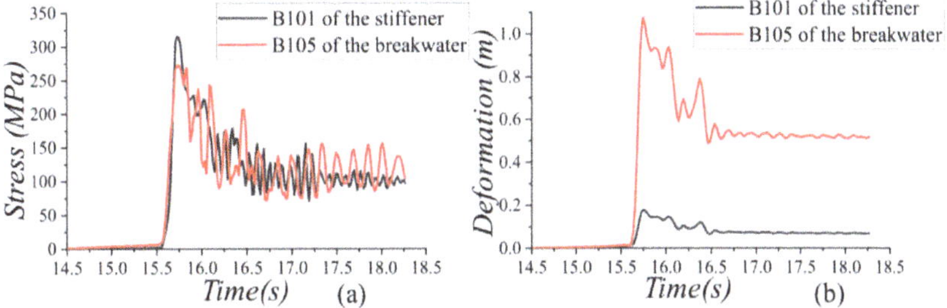

Figure 21. Results of Von Mises stress and displacement in breakwater and stiffener: (**a**) The Mises stress, (**b**) the displacement.

Through comparing the stress response with the yield limit, it was found that the stiffeners near the longitudinal centerline of the breakwater may be damaged during green-water events. However, residual stresses and displacements are still present in other areas of the breakwater after the water drains off the deck, indicating potential damage to these areas as well. Therefore, the safety evaluation of the ship's bow structure cannot solely rely on comparing peak stress responses with the yield limit. In Table 3, the peak stress responses of four key areas on the ship's bow, namely the deck, deck stiffener, breakwater, and breakwater stiffener (excluding the areas that may be damaged due to exceeding the yield limit), are compared against allowable stress. The standard for allowable stress is taken from the Rules for Classification of Sea-going Steel Ships [62] published by the China Classification Society (CCS). The maximum stress responses of the deck and deck stiffener are within the yield limit, indicating that the dynamic response evaluation of the deck and deck stiffener is qualified. However, the dynamic response evaluation of the breakwater and breakwater stiffener is unqualified. For these areas, increasing the thickness of the breakwater and the number of breakwater stiffeners to increase their strength and avoid structural damage is recommended.

Table 3. Safety evaluation results of the ship's bow.

Part	Von Mises Stress (MPa)	Allowable Stress (MPa)	λy	Safety Assessment Results
Deck	95.7	244.5	39.1%	qualified
Deck stiffener	155.4	244.5	63.6%	qualified
Breakwater	264.3	255.4	103.5%	unqualified
Breakwater stiffener	293.0	255.4	114.7%	unqualified

5.3.2. Comparison of Dynamic and Static Analysis Methods

Based on the previous evaluation, the green-water loads caused by the freak wave are likely to cause damage to the ship bow structures. However, the results calculated in Section 5.3.1 are based on the dynamic analysis. Generally, equivalent static-analysis methods are computationally efficient and widely applicable in engineering, but their results are generally conservative, which can easily cause material waste (Yang Bin et al. [57]). The concept of equivalent static analysis is to consolidate the peak values of each component at different times into a single moment, which results in more conservative outcomes, and makes it suitable for the preliminary analysis of ship structures. For ships with high structural requirements, dynamic response analysis is needed (Xie et al. [29]). In the present study, structural responses using dynamic and static analyses at deck location D111 and breakwater stiffener location B101 are compared in Table 4.

Table 4. Comparison of structural response between equivalent static method and dynamic analysis method.

Part	Von Mises Stress (MPa)	
	Equivalent Static Analysis Method	Dynamic Analysis Method
D111 of the deck	232.9	95.7
B101 of the stiffener	735.0	315.5

Both for the undamaged deck and the breakwater stiffeners that may be damaged, the results obtained from the equivalent static analysis method are significantly higher than those from the dynamic analysis method. As analyzed in Section 5.3.1, the impact load varies in space and time, and the peak time at different locations is also different. The equivalent static analysis method concentrates the peak loads of all components into one moment, resulting in a conservative estimate. The dynamic analysis method can reflect the temporal and spatial differences of the impact load on the ship, and can consider the structural dynamic response and nonlinear behavior at different times, thus providing more realistic and accurate results.

5.3.3. Comparison of Linear and Nonlinear Analysis Methods

The finite element analysis method used in this paper was a nonlinear dynamic analysis; however, there is another commonly used finite element analysis method, namely, linear dynamic analysis (Wang et al. [63]). Linear analysis has a broader range of applications and can solve various dynamic-response problems. Additionally, it has a simpler calculation process and faster computational speed. However, linear analysis involves a series of assumptions, such as assuming that materials remain in an elastic state, stress–strain relationships are always proportional, loads are proportional to structural responses, and structures undergo only small displacements and deformations. Nonlinear dynamic analysis, on the other hand, can consider factors such as large deformations, large displacements, and material nonlinearity under load, resulting in the capability to simulate more complex structural responses, more accurately reflect real-world conditions, and provide more reliable design results. To compare the differences between linear and nonlinear dynamic analysis, this paper compared the structural response results obtained by the two methods. Figure 22 presents a comparison of the stress response and total displacement results obtained at deck location D111 and breakwater stiffener location B101 using nonlinear dynamic response analysis and linear dynamic analysis methods.

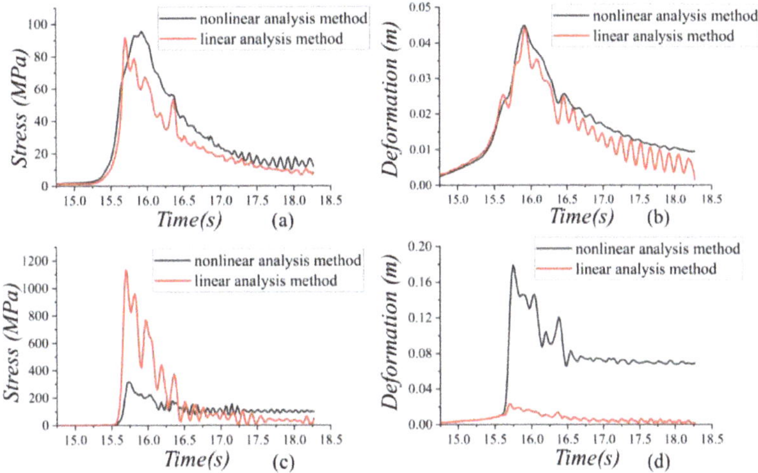

Figure 22. Comparison of nonlinear dynamic analysis and linear dynamic analysis methods: (**a**) The Mises stress at the deck D111, (**b**) the displacement at the deck D111, (**c**) the Mises stress at B101 of the stiffener, and (**d**) the displacement at B101 of the stiffener.

Under the same mesh and time step, for places where the stress response does not exceed the material yield limit, such as the deck, the results of nonlinear dynamic analysis method and linear dynamic analysis method were basically consistent. Figure 22a,b show the analytical results at deck D111. However, for places where the stress response exceeds the material yield limit, such as the breakwater stiffener, there was a significant difference between the results of the nonlinear dynamic analysis method and the linear dynamic analysis method. As shown in Figure 22c,d, in the nonlinear dynamic analysis method, when the impact load causes the failure of the breakwater stiffener and enters the plastic stage, the structure undergoes significant displacement, and the rate of increase in the stress response of the breakwater stiffener begins to slow down. After the impact load decreases, the stress response of the breakwater stiffener does not immediately decrease but decreases slowly. However, the results of the linear dynamic analysis method show a significant stress response and small displacement. This indicates that the nonlinear dynamic analysis method can accurately simulate plastic behavior or instantaneous damage that cannot be described by the linear dynamic analysis method.

Therefore, in the assessment of the slamming strength of the ship hull, if the slamming load is small and does not cause the structure of the ship to exceed its yield limit, the linear dynamic analysis method can be used to simplify the calculation process. However, if the stress response obtained from the linear analysis method is significantly higher than the material's yield limit, then the nonlinear dynamic analysis method should be used to obtain more accurate structural stress-response and displacement results. The nonlinear dynamics analysis method is more complex compared to the linear dynamics analysis method, but the results are generally more accurate. It is more suitable for analyzing ships with higher structural requirements.

6. Conclusions

The purpose of this paper was to study the green-water loads and structural responses of ship bow structures induced by freak waves. To achieve this, a three-dimensional numerical wave tank was established to generate freak waves and to obtain the wave-ship interaction and green-water loads. A finite element model of the ship bow was built, on which the green-water loads were applied to obtain the displacement and stress of the deck and breakwater structures. The main conclusions are as follows:

(1) It can be concluded from numerical simulation results that freak waves have a crucial impact on ship motion response. It should be noted that wave crests and troughs can cause significant heave and pitch motion. Therefore, it is necessary to consider the influence of extreme sea conditions on ship stability in the design and verification of ships.
(2) Freak waves have significant impacts on green-water loads. The green-water event caused by freak waves generates a huge impact force on the deck and breakwater of the ship bow, with a greater slamming load on the breakwater than on the deck. The tremendous impact loads may pose a threat to the safety of personnel on the deck and cause damage to the equipment on deck. Therefore, in engineering applications, it is necessary to pay attention to checking the local strength of the bow deck and the breakwater to ensure the safety of the ship and personnel.
(3) Freak waves are able to cause severe structural responses in the deck and breakwater structures of the ship bow. Under the impact of green water caused by the freak wave, the breakwater and breakwater stiffener suffered relatively severe damage. Therefore, in the design of ships, it is necessary to strengthen the structure of the breakwater to prevent damage to the breakwater caused by green-water events resulting from freak waves. Increasing the thickness of the breakwater and the number of breakwater stiffeners is recommended to guarantee local strength and avoid structural damage.
(4) Equivalent static analysis methods are computationally efficient and widely applicable in engineering, but tend to produce conservative results, which may result in material waste. As such, they are best suited for the preliminary design stage of ship hulls. However, this study focuses on the high-speed water impact caused by freak waves, which is a transient dynamic problem. The equivalent static analysis methods cannot reflect the temporal and spatial differences in the green water impact phenomenon. Therefore, the dynamic analysis methods are more suitable for the analysis of such problems.
(5) The linear dynamic analysis method can capture the spatial and temporal characteristics of slamming loads, but fails to account for geometric and material nonlinearity. It is best suited for scenarios where the slamming load is small and does not result in material failure. However, these methods cannot accurately reflect the state of structural damage or accurately describe the areas of the structure that have been damaged. The damaged areas determined by these methods are often overestimated. Therefore, if the stress response of the linear analysis methods exceeds the yield limit of the material, nonlinear dynamic analysis methods should be used. Nonlinear dynamic analysis methods can not only more accurately display the damaged areas and states of the structure, but also obtain more precise stress responses and displacement results of the structure.

This study may contribute to the better understanding of ship safety design under freak-wave conditions. The green-water loads and corresponding structural analysis may be of reference value from an engineering point of view. Limitations of this study include the lack of physical experiment verification and the assumption of the uncoupled fluid–structure interaction calculation, which could be further investigated.

Author Contributions: Methodology, H.Q. and C.Z.; validation, H.Q., C.Z. and W.Z.; investigation, C.Z. and H.Q.; data curation, C.Z. and H.Q.; writing—original draft, C.Z. and H.Q.; formal analysis, C.Z., H.Q., Y.H. and E.Z.; funding acquisition, H.Q., E.Z. and L.M.; project administration, H.Q. and L.M.; writing—review and editing, H.Q., W.Z., Y.H. and H.Z. All authors have read and agreed to the published version of the manuscript.

Funding: This work was supported by the National Natural Science Foundation of China (Grant No. 52101332, 52001286), National Key Research and Development Program of China (Grant No. 2021YFC3101800), Shenzhen Science and Technology Program (Grant No. KCXFZ20211020164015024), and Shenzhen Fundamental Research Program (Grant No. JCYJ20200109110220482).

Institutional Review Board Statement: Not applicable.

Informed Consent Statement: Not applicable.

Data Availability Statement: The data are unavailable due to privacy.

Acknowledgments: All authors would like to express gratitude to the anonymous reviewers.

Conflicts of Interest: The authors declare no conflict of interest.

References

1. Akhmediev, N.; Ankiewicz, A.; Taki, M. Waves That Appear from Nowhere and Disappear without a Trace. *Phys. Lett. A* **2009**, *373*, 675–678. [CrossRef]
2. Didenkulova, E. Catalogue of Rogue Waves Occurred in the World Ocean from 2011 to 2018 Reported by Mass Media Sources. *Ocean Coast. Manag.* **2020**, *188*, 105076. [CrossRef]
3. Didenkulova, I.; Didenkulova, E.; Didenkulov, O. Freak Wave Accidents in 2019–2021. In Proceedings of the OCEANS 2022—Chennai, Chennai, India, 21–24 February 2022; pp. 1–7.
4. Lavrenov, I.V. The Wave Energy Concentration at the Agulhas Current off South Africa. *Nat. Hazards* **1998**, *17*, 117–127. [CrossRef]
5. Fedele, F.; Brennan, J.; Ponce de León, S.; Dudley, J.; Dias, F. Real World Ocean Rogue Waves Explained without the Modulational Instability. *Sci. Rep.* **2016**, *6*, 27715. [CrossRef] [PubMed]
6. Kharif, C.; Pelinovsky, E. Physical Mechanisms of the Rogue Wave Phenomenon. *Eur. J. Mech.—B/Fluids* **2003**, *22*, 603–634. [CrossRef]
7. Dysthe, K.; Krogstad, H.E.; Müller, P. Oceanic Rogue Waves. *Annu. Rev. Fluid Mech.* **2008**, *40*, 287–310. [CrossRef]
8. Sparboom, U.; Wienke, J.; Oumeraci, H. Laboratory "Freak Wave" Generation for the Study of Extreme Wave Loads on Piles. In Proceedings of the Ocean Wave Measurement and Analysis (2001), Reston, VA, USA, 29 March 2002; pp. 1248–1257.
9. Corte, C.; Grilli, S.T. Numerical modeling of extreme wave slamming on cylindrical offshore support structures. In Proceedings of the Sixteenth International Offshore and Polar Engineering Conference, San Francisco, CA, USA, 28 May–2 June 2006.
10. Kim, N.; Kim, C.H. Investigation of a dynamic property of Draupner freak wave. *Int. J. Offshore Polar Eng.* **2003**, *13*, 38–42.
11. Li, J.; Wang, Z.; Liu, S. Experimental Study of Interactions between Multi-Directional Focused Wave and Vertical Circular Cylinder, Part II: Wave Force. *Coast. Eng.* **2014**, *83*, 233–242. [CrossRef]
12. Bunnik, T.; Veldman, A.; Wellens, P. Prediction of extreme wave loads in focused wave groups. In Proceedings of the Eighteenth International Offshore and Polar Engineering Conference, Vancouver, BC, Canada, 6–11 July 2008.
13. Gao, J.; Lyu, J.; Wang, J.; Zhang, J.; Liu, Q.; Zang, J.; Zou, T. Study on Transient Gap Resonance with Consideration of the Motion of Floating Body. *China Ocean Eng.* **2022**, *36*, 994–1006. [CrossRef]
14. Liu, J.; Gao, J.; Shi, H.; Zang, J.; Liu, Q. Investigations on the Second-Order Transient Gap Resonance Induced by Focused Wave Groups. *Ocean Eng.* **2022**, *263*, 112430. [CrossRef]
15. Gao, J.; Ma, X.; Zang, J.; Dong, G.; Ma, X.; Zhu, Y.; Zhou, L. Numerical Investigation of Harbor Oscillations Induced by Focused Transient Wave Groups. *Coast. Eng.* **2020**, *158*, 103670. [CrossRef]
16. Liu, W.; Suzuki, K.; Shibanuma, K. Nonlinear Dynamic Response and Structural Evaluation of Container Ship in Large Freak Waves. *J. Offshore Mech. Arct. Eng.* **2015**, *137*, 011601. [CrossRef]
17. Guedes Soares, C.; Fonseca, N.; Pascoal, R.; Clauss, G.F.; Schmittner, C.E.; Hennig, J. Analysis of Design Wave Loads on an FPSO Accounting for Abnormal Waves. *J. Offshore Mech. Arct. Eng.* **2006**, *128*, 241–247. [CrossRef]
18. Qin, H.; Tang, W.; Xue, H.; Hu, Z. Numerical Study of Nonlinear Freak Wave Impact underneath a Fixed Horizontal Deck in 2-D Space. *Appl. Ocean Res.* **2017**, *64*, 155–168. [CrossRef]
19. Luo, M.; Koh, C.G.; Lee, W.X.; Lin, P.; Reeve, D.E. Experimental Study of Freak Wave Impacts on a Tension-Leg Platform. *Mar. Struct.* **2020**, *74*, 102821. [CrossRef]
20. Rudman, M.; Cleary, P.; Leontini, J.; Sinnott, M.; Prakash, M. Rogue Wave Impact on a Semi-Submersible Offshore Platform. In Proceedings of the ASME 2008 27th International Conference on Offshore Mechanics and Arctic Engineering, Estoril, Portugal, 15–20 June 2008; pp. 887–894.
21. Hu, Z.; Tang, W.; Xue, H.; Zhang, X.; Wang, K. Numerical Study of Rogue Wave Overtopping with a Fully-Coupled Fluid-Structure Interaction Model. *Ocean Eng.* **2017**, *137*, 48–58. [CrossRef]
22. Qin, H.; Wang, J.; Zhao, E.; Mu, L. Numerical Study of Rogue Wave Forces on Flat Decks Based on the Peregrine Breather Solution under Finite Water Depth. *J. Fluids Struct.* **2022**, *114*, 103744. [CrossRef]
23. Zhang, H.; Tang, W.; Yuan, Y.; Xue, H.; Qin, H. The Three-Dimensional Green-Water Event Study on a Fixed Simplified Wall-Sided Ship under Freak Waves. *Ocean Eng.* **2022**, *251*, 111096. [CrossRef]

24. Liu, D.; Li, F.; Liang, X. Numerical Study on Green Water and Slamming Loads of Ship Advancing in Freaking Wave. *Ocean Eng.* **2022**, *261*, 111768. [CrossRef]
25. Wang, J.; Qin, H.; Hu, Z.; Mu, L. Three-Dimensional Study on the Interaction between a Container Ship and Freak Waves in Beam Sea. *Int. J. Naval Archit. Ocean Eng.* **2023**, *15*, 100509. [CrossRef]
26. Maki, K.J.; Lee, D.; Troesch, A.W.; Vlahopoulos, N. Hydroelastic Impact of a Wedge-Shaped Body. *Ocean Eng.* **2011**, *38*, 621–629. [CrossRef]
27. Yang, S.H.; Chien, H.L.; Chou, C.M.; Tseng, K.C.; Lee, Y.J. A study on the dynamic buckling strength of containership's bow structures subjected to bow flare impact force. In Proceedings of the 3rd International Conference on Marine Structures, Hamburg, Germany, 3–5 April 2011; pp. 249–255.
28. Ren, H.; Yu, P.; Wang, Q.; Li, H. Dynamic Response of the Bow Flare Structure Under Slamming Loads. In Proceedings of the ASME 2015 34th International Conference on Ocean, Offshore and Arctic Engineering, St. John's, NL, Canada, 31 May–5 June 2015.
29. Xie, H.; Liu, X.; Zhang, Z.; Ren, H.; Liu, F. Methodology of Evaluating Local Dynamic Response of a Hull Structure Subjected to Slamming Loads in Extreme Sea. *Ocean Eng.* **2021**, *239*, 109763. [CrossRef]
30. Kim, K.; Shin, Y.S.; Wang, S. A Stern Slamming Analysis Using Three-Dimensional CFD Simulation. In Proceedings of the ASME 2008 27th International Conference on Offshore Mechanics and Arctic Engineering, Estoril, Portugal, 15–20 June 2008; pp. 929–934.
31. Qin, H.; Tang, W.; Hu, Z.; Guo, J. Structural Response of Deck Structures on the Green Water Event Caused by Freak Waves. *J. Fluids Struct.* **2017**, *68*, 322–338. [CrossRef]
32. Qin, H.; Tang, W.; Xue, H.; Hu, Z.; Guo, J. Numerical Study of Wave Impact on the Deck-House Caused by Freak Waves. *Ocean Eng.* **2017**, *133*, 151–169. [CrossRef]
33. Hu, Z.; Tang, W.; Xue, H. A Probability-Based Superposition Model of Freak Wave Simulation. *Appl. Ocean Res.* **2014**, *47*, 284–290. [CrossRef]
34. Wilcox, D.C. Reassessment of the Scale-Determining Equation for Advanced Turbulence Models. *AIAA J.* **1988**, *26*, 1299–1310. [CrossRef]
35. Patankar, S.V.; Spalding, D.B. A Calculation Procedure for Heat, Mass and Momentum Transfer in Three-Dimensional Parabolic Flows. In *Numerical Prediction of Flow, Heat Transfer, Turbulence and Combustion*; Elsevier: Amsterdam, The Netherlands, 1983; pp. 54–73.
36. Patankar, S.V.; Spalding, D.B. A Finite-Difference Procedure for Solving the Equations of the Two-Dimensional Boundary Layer. *Int. J. Heat Mass Transf.* **1967**, *10*, 1389–1411. [CrossRef]
37. Hirt, C.W.; Sicilian, J.M. A Porosity Technique for the Definition of Obstacles in Rectangular Cell Meshes. In Proceedings of the 4th International Conference on Numerical Ship Hydrodynamics, Washington, DC, USA, 24–27 September 1985.
38. Newmark, N.M. A Method of Computation for Structural Dynamics. *J. Eng. Mech. Divis.* **1959**, *85*, 67–94. [CrossRef]
39. MSC. *Patran User's Guide*; MSC Software Corporation: Newport Beach, CA, USA, 2018.
40. el Moctar, O.; Shigunov, V.; Zorn, T. Duisburg Test Case: Post-Panamax Container Ship for Benchmarking. *Ship Technol. Res.* **2012**, *59*, 50–64. [CrossRef]
41. Lyu, W.; el Moctar, O. Numerical and Experimental Investigations of Wave-Induced Second Order Hydrodynamic Loads. *Ocean Eng.* **2017**, *131*, 197–212. [CrossRef]
42. El Moctar, O.; Sigmund, S.; Ley, J.; Schellin, T.E. Numerical and Experimental Analysis of Added Resistance of Ships in Waves. *J. Offshore Mech. Arct. Eng.* **2017**, *139*, 011301. [CrossRef]
43. Mei, T.-L.; Delefortrie, G.; Ruiz, M.T.; Lataire, E.; Vantorre, M.; Chen, C.; Zou, Z.-J. Numerical and Experimental Study on the Wave-Body Interaction Problem with the Effects of Forward Speed and Finite Water Depth in Regular Waves. *Ocean Eng.* **2019**, *192*, 106366. [CrossRef]
44. Van Zwijnsvoorde, T.; Ruiz, M.T.; Delefortrie, G. Sailing in Shallow Water Waves with the DTC Container Carrier: Open Model Test Data for Validation Purposes. In Proceedings of the 5th International Conference on Ship Manoeuvring in Shallow and Confined Water, Ostend, Belgium, 20–22 May 2019.
45. Rosetti, G.F.; Pinto, M.L.; de Mello, P.C.; Sampaio, C.M.P.; Simos, A.N.; Silva, D.F.C. CFD and Experimental Assessment of Green Water Events on an FPSO Hull Section in Beam Waves. *Mar. Struct.* **2019**, *65*, 154–180. [CrossRef]
46. Mayer, S.; Garapon, A.; Sørensen, L.S. A Fractional Step Method for Unsteady Free-Surface Flow with Applications to Non-Linear Wave Dynamics. *Int. J. Numer. Methods Fluids* **1998**, *28*, 293–315. [CrossRef]
47. Zhao, X.; Hu, C. Numerical and Experimental Study on a 2-D Floating Body under Extreme Wave Conditions. *Appl. Ocean Res.* **2012**, *35*, 1–13. [CrossRef]
48. Mori, N.; Liu, P.C.; Yasuda, T. Analysis of Freak Wave Measurements in the Sea of Japan. *Ocean Eng.* **2002**, *29*, 1399–1414. [CrossRef]
49. Cui, C.; Zhang, N.; Kang, H.; Yu, Y. An Experimental and Numerical Study of the Freak Wave Speed. *Acta Oceanol. Sin.* **2013**, *32*, 51–56. [CrossRef]
50. Klinting, P.; Sand, S.E. *Analysis of Prototype Freak Waves*; The National Academies of Sciences, Engineering, and Medicine: Washington, DC, USA, 1987.

51. Gómez-Gesteira, M.; Cerqueiro, D.; Crespo, C.; Dalrymple, R.A. Green Water Overtopping Analyzed with a SPH Model. *Ocean Eng.* **2005**, *32*, 223–238. [CrossRef]
52. Hu, C.; Kashiwagi, M.; Kitadai, A. Numerical Simulation of Strongly Nonlinear Wave-Body Interactions With Experimental Validation. In Proceedings of the Sixteenth International Offshore and Polar Engineering Conference, San Francisco, CA, USA, 28 May–2 June 2006.
53. Det Norske Veritas. *Rules for Ships*; Det Norske Veritas: Hovik, Norway, 2004.
54. Bureau Veritas. *Rules for the Classifications of Steel Ships, Part D—Container Ships*; Bureau Veritas: Neuilly-sur-Seine, France, 2009.
55. Lloyd Germanischer. *Rules for the Classifications and Construction of Seagoing Ships*; Lloyd Germanischer: Hamburg, Germany, 2009.
56. Korean Register. *Rules for the Classifications of Steel Ships*; Korean Register: Seoul, Republic of Korea, 2009.
57. Yang, B.; Wang, D. Numerical Study on the Dynamic Response of the Large Containership's Bow Structure under Slamming Pressures. *Mar. Struct.* **2018**, *61*, 524–539. [CrossRef]
58. Yang, S.-H.; Lee, Y.-J.; Chien, H.-L.; Chou, C.-M.; Tseng, K.-C. Dynamic Buckling Strength Assessment of Containership's Bow Structures Subjected to Bow Flare Slamming. In Proceedings of the ASME 2012 International Mechanical Engineering Congress and Exposition, Houston, TX, USA, 9–15 November 2012; pp. 727–733.
59. Xie, H.; Liu, F.; Tang, H.; Liu, X. Numerical Study on the Dynamic Response of a Truncated Ship-Hull Structure under Asymmetrical Slamming. *Mar. Struct.* **2020**, *72*, 102767. [CrossRef]
60. Cowper, G.R.; Symonds, P.S. *Strain-Hardening and Strain-Rate Effects in the Impact Loading of Cantilever Beams*; Brown University: Providence, RI, USA, 1957.
61. Paik, J.K. Practical Techniques for Finite Element Modeling to Simulate Structural Crashworthiness in Ship Collisions and Grounding (Part I: Theory). *Ships Offshore Struct.* **2007**, *2*, 69–80. [CrossRef]
62. China Classification Society. *Rules for Classification of Sea-Going Steel Ships*; China Classification Society: Beijing, China, 2022.
63. Wang, Q.; Yu, P.; Chang, X.; Fan, G.; He, G. Research on the Bow-Flared Slamming Load Identification Method of a Large Container Ship. *Ocean Eng.* **2022**, *266*, 113142. [CrossRef]

applied sciences

Article

Wind-Wave-Current Coupled Modeling of the Effect of Artificial Island on the Coastal Environment

Guowei Fu [1,2], Jian Li [2,*], Kun Yuan [1,*], Yanwei Song [1], Miao Fu [1], Hongbing Wang [1] and Xiaoming Wan [1]

1 Haikou Research Center of Marine Geology, China Geological Survey, Haikou 571127, China; fuguowei@mail.cgs.gov.cn (G.F.); songyanwei@mail.cgs.gov.cn (Y.S.); fumiao@mail.cgs.gov.cn (M.F.); wanghongbing@mail.cgs.gov.cn (H.W.); wanxiaoming@mail.cgs.gov.cn (X.W.)
2 College of Marine Science and Technology, China University of Geosciences, Wuhan 430074, China
* Correspondence: lijian_cky@hotmail.com (J.L.); yuankun01@mail.cgs.gov.cn (K.Y.)

Abstract: The effect of artificial island on the geomorphologic processes in the coastal area under the coupled hydrodynamics, wave, and sediment transport system is a complicated and multi-scale problem. Studying these dynamic processes will suggest how coastal ecological restoration should be conducted. In this study, a unified, unstructured, gridded coupled hydrodynamics, wave, and sediment transport model and a topographic evolution model were adopted. Based on the field observations of water depth, velocity, suspended sediment concentration, bed sand, and quaternary thickness, a high-spatiotemporal-resolution numerical simulation of the offshore dynamic environment under the disturbance of artificial island was performed, and the accuracy of the calculation was verified. The research showed that the coupling system with an unstructured mesh was able to reproduce the flow and sediment transport processes with acceptable accuracy. The contracted flow zone between the artificial island and the coastline, the runoff and alongshore current from the river, as well as the tidal flow from the ocean, worked together to mold the local complex morphology around the artificial island. The coupled modeling system, supported with parallel computation, can be used to study coastal environments with small-scale wading structures.

Keywords: coastal engineering; artificial island; coupled modeling system; parallel computation

Citation: Fu, G.; Li, J.; Yuan, K.; Song, Y.; Fu, M.; Wang, H.; Wan, X. Wind-Wave-Current Coupled Modeling of the Effect of Artificial Island on the Coastal Environment. *Appl. Sci.* **2023**, *13*, 7171. https://doi.org/10.3390/app13127171

Academic Editor: Marta Pérez Arlucea

Received: 26 April 2023
Revised: 9 June 2023
Accepted: 12 June 2023
Published: 15 June 2023

1. Introduction

With the rapid development of society and economies in coastal areas of China, the contradiction between the shortage of land resources and development is becoming more and more prominent. Artificial islands became an effective method for human beings to expand production and living space into the sea and alleviate the human–land contradiction in coastal areas [1]. Hainan is a province seeing the rapid development of artificial island. According to statistics, from 2000 to 2014, Hainan approved the construction of 11 artificial islands, totaling 1845.34 hm^2. However, unreasonable land reclamation also brings about corresponding negative environmental and ecological effects, such as the decline in the quality of the marine environment, habitat degradation, destruction of the ecological balance, changes in the hydrodynamic environment, erosion and siltation of the shoreline, etc., which is particularly significant for sandy coasts and coastal areas [2,3].

Under current national policy and because of the demand for the construct of an ecological civilization and shoal restoration, research on the coastal erosion and siltation caused by artificial island is also being carried out by various scientific and technological workers. The traditional method is to use multi-stage remote sensing interpretation analysis to check the scale of the impact of the artificial island's construction on the adjacent coastline. For example, Yang Yanxiong et al. [4] analyzed the changes in the coastline before and after the construction of an artificial island by using remote sensing image interpretation. Li Songzhe [5] applied a one-dimensional coastal shoreline evolution model based on the principle of the conservation of mass to evaluate the development of changes to the erosion and siltation of the nearby shoreline before and after the construction of man-made islands. However, using remote sensing image analysis and one-dimensional simulation, it is difficult to understand the dynamic development process of artificial island' impacts on the nearby sea area at high spatial and temporal resolutions.

Most studies on the stability of the shorelines of artificial island were carried out by establishing relevant numerical models and conducting a comprehensive analysis and evaluation of the influence of artificial islands on the stability of the shoreline through the simulation of regional wave, tide, and sediment change characteristics before and after the island's construction. Jetro [6] studied the feasibility of constructing an artificial island at the Pacific entrance of the Panama Canal in 2002. The artificial island was filled using earth excavated for the expansion of the Panama Canal. Bayyinch [7] studied the impact of Palm Tree Island in Dubai on the marine environment in the Persian Gulf in 2006. They analyzed the impact of artificial island construction on marine organisms and marine sediment movement and evaluated the compliance of artificial island construction with local marine environmental protection laws and regulations. In China, Gong Wenping et al. [8] used the DELFT3D model based on a structural grid to discuss the influence of building sun and moon islands under different schemes on the hydrodynamic conditions and sediment transport in coastal areas. However, a structural grid is unable to reflect the complex coastline and local geometric features of artificial islands. Tan and Gao [9] applied the FVCOM model of ocean dynamics based on an unstructured grid to evaluate the influence of the proposed construction of the new Sanya Airport artificial island in Hainan Province on the surrounding hydrodynamics. The results showed that the unstructured grid model can accurately reflect the boundaries of local buildings and is suitable for simulating problems across physical scales, which can significantly improve the simulation accuracy. However, there is a nonlinear response relationship between artificial island and natural driving forces (such as typhoons and waves, etc.). For example, Kuang Cuiping et al. [10] used the MIKE21 model based on an unstructured grid to study the influence of the nonlinear superposition effect on tidal currents and waves when artificial island and coastal remediation projects co-exist. Yang Fan et al. [11] used a mathematical model of typhoon winds and waves to analyze wave elements under extreme weather in the presence of the artificial island of the Hong Kong–Zhuhai–Macao Bridge.

According to previous studies, it is necessary to use an unstructured grid ocean dynamics model when considering the dynamic nonlinear processes of hydrodynamics, waves, sediment transport, and topographic evolution affecting the nearshore area under the coupling of artificial island and natural driving forces. When a complex coupled mathematical model of wind and wave flow is used to study offshore dynamic processes under the disturbance of artificial island, a large number of basic data-driven models are needed to obtain reliable initial and boundary conditions of the model, which was lacking in previous numerical simulation studies on the influence of artificial island on the offshore environment [12,13]. Meanwhile, when mathematical models with different spatial-temporal evolution scales are coupled, such as the hydrodynamic, wave, and sediment transport models mentioned above, couplers are usually used for spatial interpolation [8], and interpolation errors will be introduced no matter what interpolation algorithm is used.

In this study, the marine area around the artificial island of Qionghai City, Hainan Province, was taken as the research object, and a unified unstructured grid marine dynamics SCHISM simulation system [14,15] was adopted to complete the simulation of the wind, wave, current, and sediment transport systems. Combined with a large number of field-measured data including sounding, velocity, suspended sediment, and bed sand and quaternary bed sand thickness in the surrounding marine area collected in 2020 and 2021, a simulation study on the influence of artificial island on the coastal environment under a coupled model of the wind, waves, current [16], and sediment transport [17] was carried out. The proposed approach improved the simulation accuracy of inshore environments across multiple scales (transition from small-scale artificial island to large-scale open ocean boundary).

2. Materials and Methods

2.1. Study Area

The study area was the waters around an artificial island—in, Qionghai City, Hainan Island. The artificial island is the only artificially reclaimed island in Boao Town, Qionghai City, Hainan Province, about 6.8 km north of the mouth of the Wanquan River (Figure 1a). The study area is a tropical monsoon climate, with an average annual temperature of about 25.0 °C and an average annual precipitation of 2103.0 mm. According to the wind data from Qionghai-Wanning Meteorological Station, from 1981 to 2010, the average annual wind speed in Qionghai was 2.2 m/s, and the prevailing wind direction was mainly southerly in summer and northerly in winter. Spring and autumn represent the transition seasons between the two wind directions (Figure 1b). The tides were irregular semi-diurnal tides. The normal wave direction throughout the year was southeast, which was the main wave direction. At the same time, the coast will be affected by the Taiwan wave and the near sea wave; the sub-normal wave direction was SSE; the strong wave direction was ESE, SE, and SSE; and the SE and SSE wave directions account for 68% of the total frequency (Figure 1c).

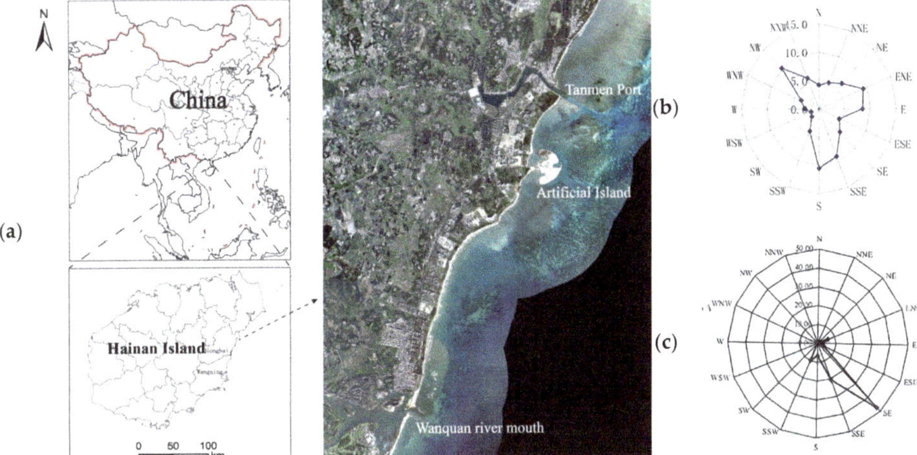

Figure 1. (a) Study area. (b) Rose chart of annual mean wind power in the study area. (c) Annual mean wave rose map of sea area around the study area.

The construction of an artificial island began in 2011, with a circumference of 4040 m and an area of 4.61×10^5 m^2. However, due to environmental problems along the shoreline, construction was stopped, and the shoreline restoration project was launched in 2021. The whole artificial island section presents the phenomenon of siltation in the middle channel and erosion on the north and south sides of the channel. Since the construction of the artificial island in 2011, the phenomenon of erosion and siltation was significantly intensified. The maximum annual average erosion and siltation distance were about 55.6 m/a and 10.2 m/a, respectively, which increased by about 22 times and 5 times compared to before construction. The monitoring data of the bank profile in 2020 showed that erosion and sedimentation occurred in both normal seasonal weather and typhoon weather, and the variation in erosion and sedimentation in the north and south coast sections increased by 2 times and 6 times, respectively, after the passage of typhoons. The reclamation of the artificial island changed the regional hydrodynamics, which were superimposed with the influence of extreme weather such as typhoons. Waves scour the loose shoals of the north and south coasts, which intensified erosion and increased the concentration of coastal suspended sediment. This sediment accumulated in the wave shadow area with weak hydrodynamics on the west side of the artificial island to form an intermediate silting salient [18]. In addition, the hydrodynamic environment around the artificial island was also affected by the inflow of inland rivers in Wanquan River and Tanmen Town, and the hydrodynamic process is very complicated. Therefore, it is necessary to establish a local 3D mathematical model of the artificial island and the Wanquan River.

2.2. Data Source and Model Setup

The data in this research came from the field survey data and part of the collected data were collected from the waters around Qionghai's artificial island in 2019, 2020, and 2021. The locations of the survey stations are shown in Figure 2. The pre-processing of the data and basic settings of the mathematical model are discussed below.

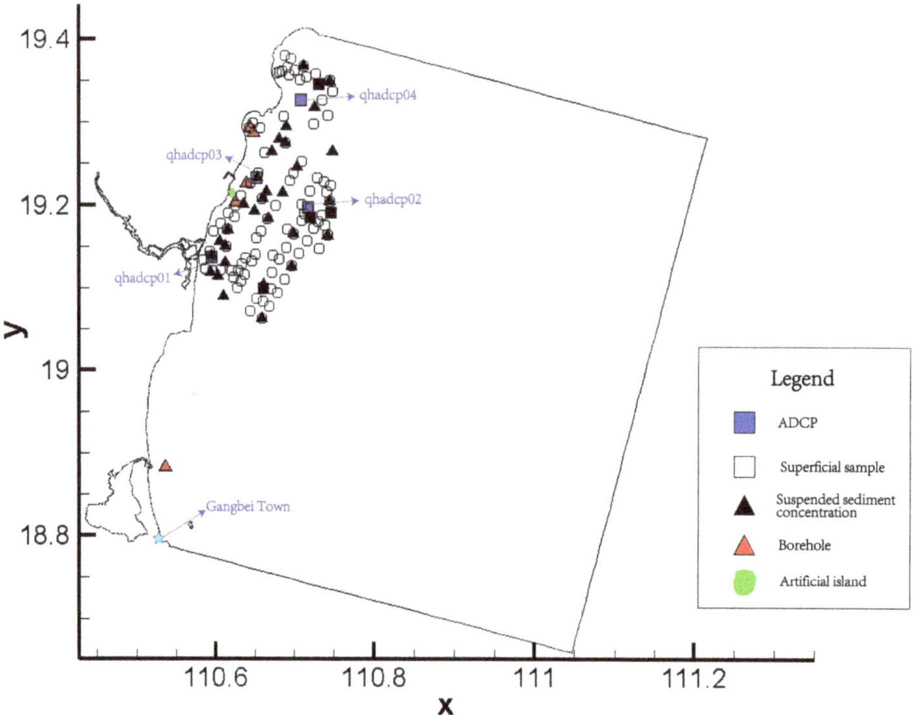

Figure 2. Field survey stations (The x-axis is the east longitude axis and the y-axis is the north latitude axis).

2.2.1. Initial Depth Condition

Based on the coastline data of Hainan Island collected by Bi Jingpeng et al. [19], we defined the calculation domain and generated a triangular unstructured grid in the study area, which was divided into 489,776 units and 249,782 nodes. The triangular grid used different lengths at the river channel, coastline, artificial island boundary line, and offshore boundary. After mesh quality optimization, the triangular mesh with uneven density was able to meet the needs of high-precision water and sediment simulation.

The topographic survey adopted a single-beam sounder to carry out an intensive survey of the marine area near the study area. The tracked line of the navigation survey is shown in Figure 3a (the x and y axes in the figure represents longitude and latitude, respectively, in decimal form of WGS1984 system, and the full text is unified). The intensive topographic survey reflects local topographic features under the influence of the artificial island, as shown in Figure 3b.

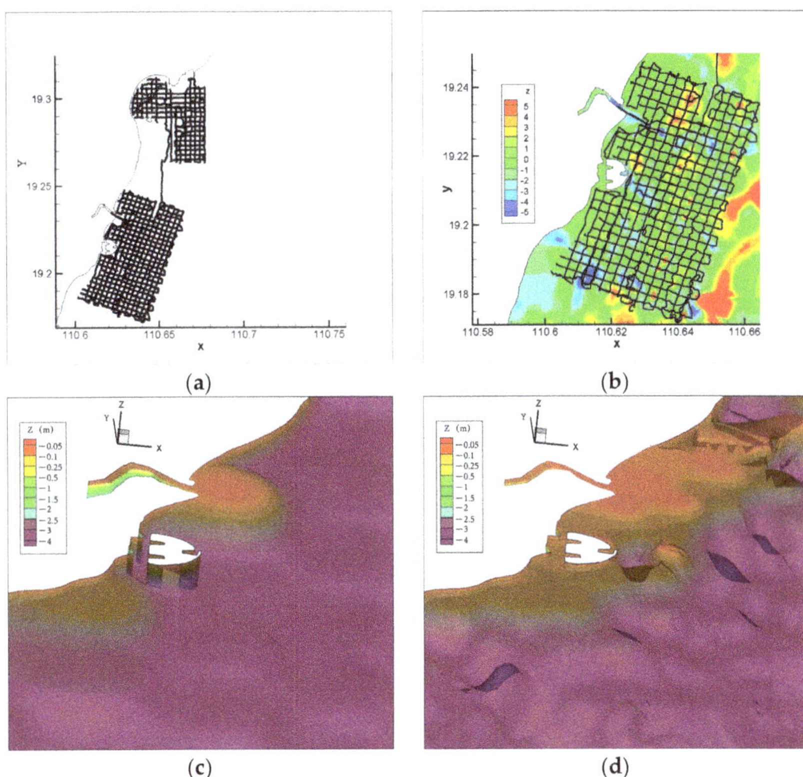

Figure 3. (**a**) Bathymetric survey line in the study area; (**b**) distribution of water depth around the artificial island; (**c**) bathymetric maps for ETOPO1 data collection; (**d**) three-dimensional topographic and geomorphic features of the study area for fusion of infill bathymetric data [The x-axis is the east longitude axis and the y-axis is the north latitude axis].

At the same time, we collected the spatial resolution of 2 km of global water depth ETOPO1 figures released by the American NOAA (https://www.ngdc.noaa.gov/mgg/global/, accessed on 15 July 2021). Due to the low spatial resolution of ETOPO1 data, which cannot accurately reflect the local topography around the artificial island (Figure 3c), the measured and collected data needed to be fused to form more accurate bathymetric data (Figure 3d) that can reflect the scour pits and silts around artificial island more clearly, in addition to more detailed siltation patterns at the mouth of Tanmen Port. Then, we interpolated the fused water depth data into the grid by using the inverse range interpolation method, which was used as the initial water depth condition for this study and laid a foundation for the numerical simulation of water and sediment in the complex near-shore environment.

2.2.2. Sediment Transport and Tidal Boundary Conditions

The open boundary included the water level boundary at the entrance of Wanquan River and the tidal driving boundary of the ocean. The 2020 daily recorded water level and suspended sediment concentration data from Jiaji Hydrographic Station at the entrance of Wanquan River were used, as shown in Figure 4. Except for flood season (concentrated from August to November), the flow and sediment transport rates of Wanquan River were very low.

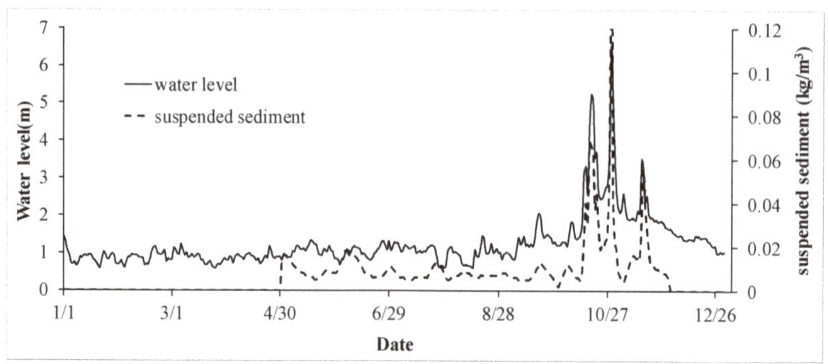

Figure 4. Changes in water level and suspended sediment concentration at the Wanquan River Jiaji Station (2020).

The tide boundary drive data from the Gangbei Tidal Station near the study area were adopted, as shown in Figure 5a. The T_TIDE program in MATLAB language was used to conduct harmonic analysis on the recorded tide data [20], and the tide value calculated by harmonic analysis was in good agreement with the measured value when applied to the ocean tidal boundary (Figure 5b).

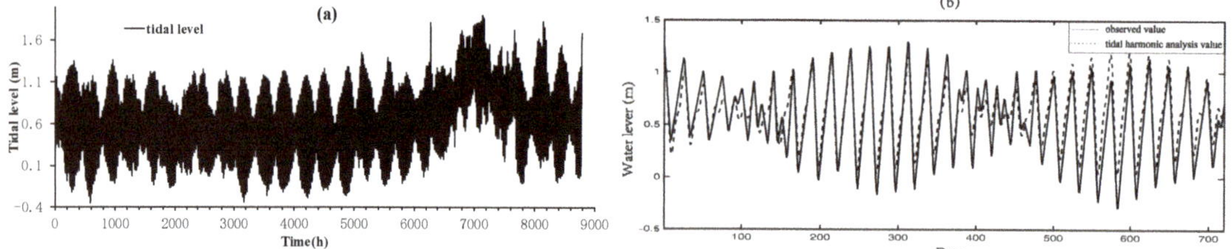

Figure 5. Tidal level and harmonic analysis recorded at Gangbei Town tidal level Station ((**a**) tidal level; (**b**) harmonic analysis).

2.2.3. Wind and Wave Conditions

The hydrodynamics and wave boundary conditions were determined using meteorological data of wind speed, wind direction, air temperature, and air pressure in 2020, which were collected from the meteorological station in Gangbei Town. In addition, for Typhoon 19 in July 2017, Gangbei Town recorded the wind field (including wind speed and direction) and water level process of the typhoon, as shown in Figure 6. The typhoon caused obvious water increases, and water increase and wind speed showed a positive correlation. At the same time, the wind direction during the typhoon was mainly northeasterly, which had a more obvious water increase effect on the Wanquan estuary and the area near Tanmen Port. Therefore, this study included both general meteorological conditions and extreme typhoon meteorological conditions, and the data were highly representative.

2.2.4. Suspended Sediment and Bed Sediment Conditions

The suspended sediment concentration affected the regional sediment movement trend and had a great influence on regional erosion and sedimentation. The data used in this study were based on the suspended sediment concentration in the middle and lower surface waters measured around the artificial island in 2020 and 2021. In addition, different types of seabed sediment also affected regional flow, suspended sand, and other indices. In this simulation, bed sand conditions were also taken into account. The data were based on bed sand and quaternary bed sand thickness surveys conducted using three shallow holes (depth < 60 cm) and 1 deep hole (about 20 m deep). Then, the suspended sediment was divided into 5 groups according to particle size and interpolated into the calculation grid by inverse distance to form the initial distribution conditions of the suspended sediment concentration, bed sand gradation, and quaternary bed sand thickness [21].

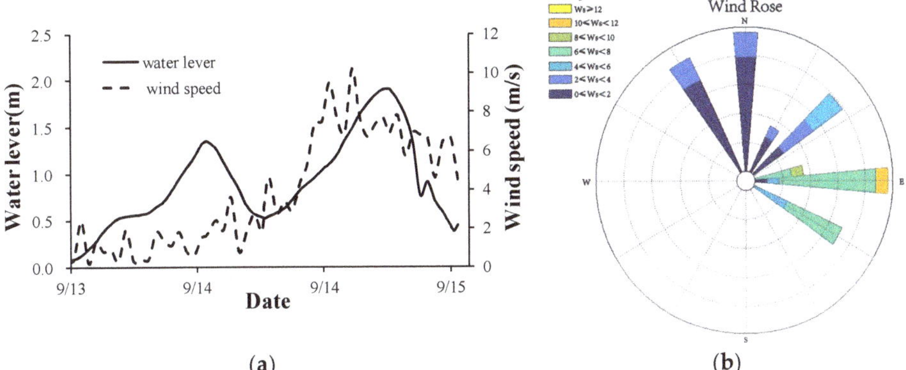

Figure 6. Water level and wind field recorded during Typhoon 19, 2017, in Gangbei Town. (**a**) Curve of typhoon water level and wind speed. (**b**) Typhoon transit wind speed rose chart.

2.3. Mathematical Model Building

In this study, SCHISM and WWM were used to simulate the water and sediment dynamics under the influence of the artificial island in the study area. Below, the mathematical model of the basic equations and modules used are introduced briefly, and the SCHISM of the specific physical principle and the numerical method can be seen at the following web site (https://geomodeling.njnu.edu.cn/modelItem, accessed on 10 August 2021).

2.3.1. Hydrodynamic Model

The hydrodynamic model adopts the SCHISM system, the full name of which is the Semi-implicit Cross-scale Hydroscience Integrated System Model, developed by the Virginia Oceanographic Institute [13]. SCHISM was widely applied in the research of many ocean engineering problems [12]. To be specific, SCHISM is not a model, but a simulation system. The core of the SCHISM simulation system is a three-dimensional hydrodynamic model based on the assumption of hydrostatic pressure, which drives other coupled modules, including a sediment model, water ecological model, water quality model, wave model, oil spill model, sea ice model, hydraulic building model, etc. The three-dimensional hydrodynamic calculation module was developed in unstructured grid mode, which was suitable for computationally complex geometric boundaries and complex terrain. The governing equations of the hydrodynamic model included a continuity equation and a momentum equation:

$$\frac{\partial u}{\partial x} + \frac{\partial v}{\partial y} + \frac{\partial w}{\partial z} = 0 \tag{1}$$

$$\frac{du}{dt} = fv - g\frac{\partial \eta}{\partial x} + K_{mh}(\frac{\partial^2 u}{\partial x^2} + \frac{\partial^2 u}{\partial y^2}) + \frac{\partial}{\partial z}(K_{mv}\frac{\partial u}{\partial z}) \tag{2}$$

$$\frac{dv}{dt} = -fu - g\frac{\partial \eta}{\partial y} + K_{mh}(\frac{\partial^2 v}{\partial x^2} + \frac{\partial^2 v}{\partial y^2}) + \frac{\partial}{\partial z}(K_{mv}\frac{\partial v}{\partial z}) \tag{3}$$

where (x, y) is the Cartesian plane coordinate (m); Z is the vertical coordinate, where upward is positive; t is the time (s); Z_{ini} is used to calculate the initial water level (m); η is the fluctuation of the free water level (m); h is the water depth (m); u, v are normal and tangential velocity (m/s); g is the acceleration of gravity (m/s^2) and is 9.81; f is the Coriolis force coefficient (1.0487974 × 10^{-4}); ρ is the density of water; and K_{mh}, K_{mv} are the horizontal and vertical turbulent viscosity coefficients (m^2/s) in the flow momentum equation.

2.3.2. Wave Model

The wind-generated wave model adopts the third-generation WWM (WWM-III) model, the full name of which is the Wind Wave Model. WWM-III solves the 2D wave spectrum equation of a plane [22]. The wave spectrum equation is:

$$\frac{\partial}{\partial t}N + \nabla_X(\dot{X}N) + \frac{\partial}{\partial \sigma}(\dot{\theta}N) + \frac{\partial}{\partial \theta}(\dot{\sigma}N) = S_{tot} \tag{4}$$

where the wave motion index N is defined as:

$$N_{(t,\mathbf{X},\sigma,\theta)} = \frac{E_{(t,\mathbf{X},\sigma,\theta)}}{\sigma} \tag{5}$$

where E represents the variance density of sea level elevation; σ is the relative fluctuation frequency; and θ is the direction of fluctuation.

Wave propagation from deep water to shallow water is accompanied by complex nonlinear processes, such as bed bottom friction, wave breaking and tumbling, and wave energy input imposed by the boundary, which are considered in the source terms of Equation (4). WWM-III is transformed by Roland into a triangular unstructured grid model to achieve seamless coupling with SCHISM.

2.3.3. Sediment Model

The sediment model adopts the MORSELFE model [17], which is based on a triangular unstructured mesh and can be applied to water simulation with complex geometric boundaries and complex topography. The MORSELFE model considers the effects of waves on sediment transport and riverbed evolution. In the suspended sediment module, the transport equation was solved based on SCHISM, the Soulsby formula was used to obtain the settling rate of suspended sediment, and the bed sediment exchange model used the active bed sediment adjustment model. The bedload sediment is calculated using the Meyer–Peter–Muller formula and the Van Rijn formula.

2.3.4. Hydrodynamic Wave–Sediment Coupling Model

SCHISM, WWM-III, and the MORSELFE model all adopt an MPI parallel communication mechanism and use the same subregions and calculation grid, and so, there is no need to perform interpolation calculations. Compared to the use of couplers to perform interpolation conversion between grids of different types and resolutions, the hard-coded coupling mode in this study avoided the problem of interpolation error. Because SCHISM, WWM-III, and the MORSELFE model used different discrete time formats, different computing time steps were adopted, and the information exchange of the models was performed at a certain frequency set by the user. As shown in Figure 7, during information exchange, water level (Zl), seabed topography (Zb), wet and dry terrain markers and velocity (U, V) of SCHISM and some meteorological driving data, including wind speed (Uwind, Vwind) and air pressure, are transferred to WWM-III. The wave direction (Dw), wave height (Hw), wave length (Lw), and wave period (Pw) of WWM-III were transferred to SCHISM. Radiation stress (Rs), energy dissipation (Es), and orbital wave velocity (So) calculated by WWM-III were unidirectionally transferred to the MORSELFE model, while the suspended sediment concentration (SS) and velocity (U, V) calculated by SCHISM were unidirectionally transferred to the MORSELFE model. The above variable exchange was completed in memory, and the efficient communication mechanism based on the MPI library realized the variable exchange on different processes.

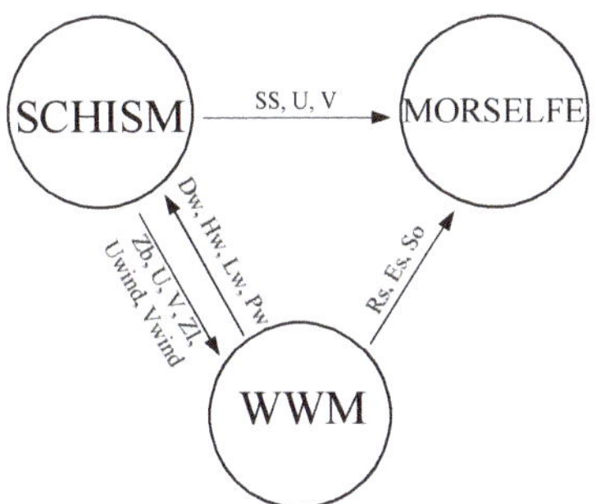

Figure 7. Framework of the hydrodynamic wave–sediment coupling simulation system.

3. Results

Model verification was required to complete a model's construction. The verified data measured were the water velocity, wave parameters (including significant wave height and wave period), and suspended sediment concentration at four stations. The four measuring points are shown in Figure 2, among which the two stations in the nearshore area, qwadcp01 and qwadcp03, were close to the artificial island. In addition, two sites, qwadcp02 and qwadcp04, were located in deep water. Based on the settings of the mathematical model above, the simulation period of 1 year was run, but only about 2 days of continuous synchronous observation data were used to verify the numerical simulation results of wind and wave flow. The accuracy of the model was verified using short-term measured data, and the influence of the artificial island on the topographic evolution of the coastal region was determined via long-term numerical simulation.

3.1. Velocity Verification

As shown in Figure 8a,b, the vertical velocity basically conformed to the decreasing law of the velocity of water facing the seabed. The velocity of water near the sea surface generally reached about 0.30 m/s, while the velocity of qwadcp01 and qwadcp03 stations in the offshore area decreased significantly and decreased to 0.02 m/s at the depth of 20 m. However, the vertical velocity distribution pattern of qwadcp02 and qwadcp04 sites in deep water was more complex. From 0 to 10 m, the velocity varied in the range of 0.2 to 0.3 m/s, while from 10 m depth below, the velocity gradually decreased to about 0.1 m/s, and the velocity was more uniform in the range of 20 to 40 m depth. Keep it at about 0.1 m/s. In general, the calculated flow velocity was in good agreement with the measured flow velocity. The accurate simulation of hydrodynamic conditions laid a foundation for the simulation of sediment transport and topographic evolution.

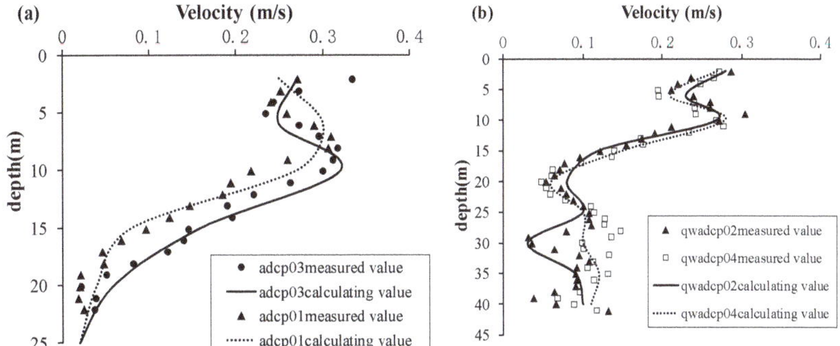

Figure 8. ADCP velocity verification ((**a**) qwadcp 01 and qwadcp 03 measurement points; (**b**) qwadcp 02 and qwadcp 04 measuring points).

3.2. Wave Simulation Verification

The significant wave height Hs (m) and wave period Tp (s) measured at the four measuring points in Figure 2 were used to verify the calculation accuracy of WWM-III. As shown in Figure 9a, the significant wave height of the qwadcp02 measuring point far from the coast reached 2.5 m, and the unsteady process was more obvious. At the qwadcp01 measuring point near the mouth of the Wanquan River in Figure 9b, the significant wave height of wind waves reached the maximum value of 1.2 m at 12 h, because it was affected by the runoff of the Wanquan River. The runoff of Wanquan River was opposite to the higher tidal current of open sea and cancelled each other out, making the wave height smaller than that of open sea area, which reflected the spatial difference of wind waves. The wave periods in Figure 9c,d show that the difference in the wave periods between these two measuring points was not significant, and both varied between 6 and 18 s. In general, the calculated results of WWM-III were in good agreement with the measured data and, thus, reflected the wave characteristics around the artificial island.

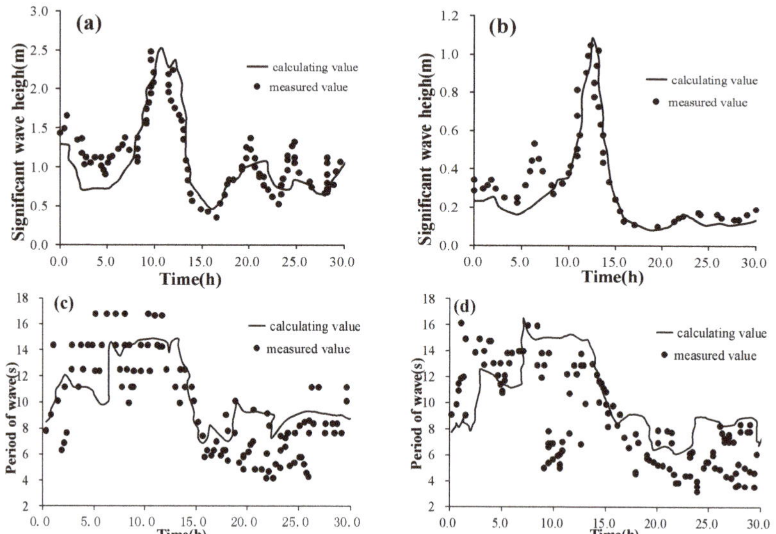

Figure 9. Verification of significant wave height and wave period ((**a**) significant wave height at qwadcp02; (**b**) significant wave height at qwadcp01; (**c**) wave period at qwadcp02; (**d**) wave period at qwadcp01).

3.3. Verification of Suspended Sediment Concentration

Due to the limitations of offshore real-time observation operation conditions, the effective time series values of the suspended sediment concentration were only obtained at qwadcp02, but the instantaneous values of suspended sediment concentration were obtained from 13 points around the artificial island and the waters near the mouth of the Wanquan River. These values were used as the initial distribution conditions of the suspended sediment concentration via spatial interpolation into the calculation grid, as shown in Figure 10a. The surface suspended sediment concentration over Tanmen Harbor formed two high concentration zones (about 0.15 kg/m^3) and one low concentration zone (sediment concentration <0.01 kg/m^3); the concentration at the entrance of Tanmen Port was very low, and a high sediment content area was formed in the offshore area (sediment concentration >0.35 kg/m^3). The formation of areas with high sediment concentrations was related to human activities near the artificial island and fishing ports. The measured suspended sediment concentration at qwadcp02 showed no significant rule, but it was roughly consistent with the variation trend of the calculated value, as shown in Figure 10b, which showed a decreasing trend during the simulation period.

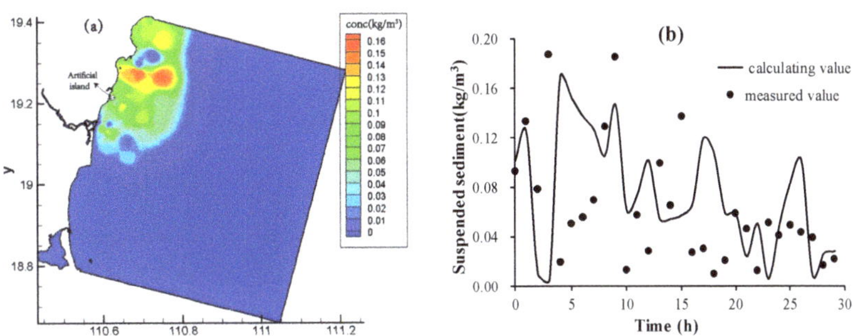

Figure 10. (**a**) Distribution of suspended sediment concentration [the x-axis is the east longitude axis and the y-axis is the north latitude axis]; (**b**) suspension sediment simulation validation.

4. Discussion

Based on the verified coupled model driven by measured water and sediment data presented above, the evolution of the near-shore topography under the disturbance of the artificial island and its causes were analyzed, and the trend of the topographic evolution after the removal of the artificial island is discussed.

The existence of the artificial island will interfere with the local flow field, causing the formation and shedding of vortices. Different hydrodynamic characteristics were shown during the different advances and retreats of the tides relative to the coastline, as shown in Figure 11a. In the tidal surge stage, an obvious beam flow area was formed between the artificial island and the coastline, and a vortex area was formed behind the island. In addition, there was a dry terrain area near Tanmen Port, and the dry and wet terrains changed with the change in the tidal current, which also shows the superiority of SCHISM's dry and wet terrain processing algorithm. In the stage of an abrupt tidal current (Figure 11b), the channel runoff from Tanmen Harbor was large, forming a significant coastal current from bottom to top (from west to east), constantly transporting the sediment from the Wanquan River. Meanwhile, the seabed around the artificial island was mainly composed of sand and silt, which were non-cohesive and easy to stir up, contributing to the sediment accumulation between the artificial island and the coastline. With the passage of time, the tombolo bar was gradually formed. In general, multi-directional currents and sediment transport from the artificial island and shoreline-confined areas, channel runoff and coastal currents, and ocean-directed tidal currents contributed to the local seabed topography of the artificial island.

In order to further simulate the influence of the artificial island on regional changes in scour and siltation, we simulated an ideal scenario of the complete removal of the artificial island. The simulation compared the distribution of topographic scour and siltation thickness around the artificial island one year after its removal and combined the results with the actual situation of active beds of the seabed surveyed by drilling, that is, the thickness of the quaternary was generally less than 2 m. The results show that in the case of the artificial island, as shown in Figure 12a, about 0.1 m of siltation occurred between the island and shoreline and on both sides of the island, with a −0.1 m scour thickness behind it. After the removal of the artificial island, as shown in Figure 12b, the erosion and siltation on both sides of and behind the island decreased (the thickness was less than 0.05 m) and tended to return to the natural bed morphology. However, the siltation morphology remained at the local location of the shore at 0.1 m.

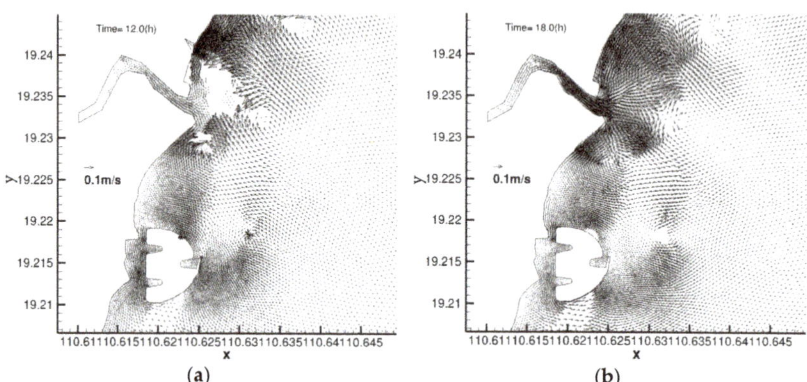

Figure 11. Vector diagram of velocity around the artificial island ((a) surge tidal current field; (b) plunging tidal current field) [the x-axis is the east longitude axis and the y-axis is the north latitude axis].

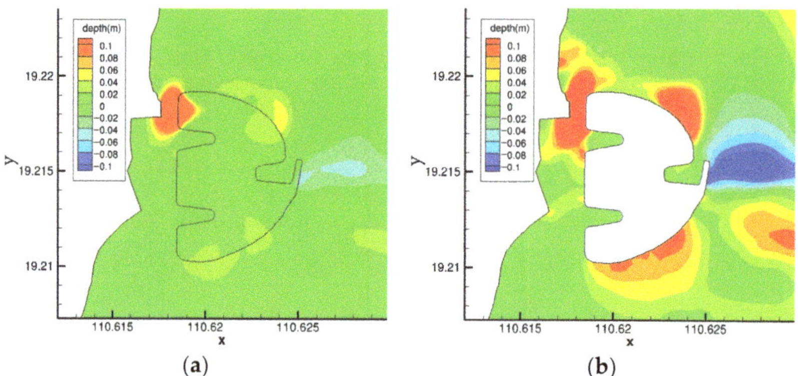

Figure 12. Impact assessment of the artificial island's demolition ((a) thickness distribution of scour and silt around the artificial island after demolition; (b) distribution of scour and silt thickness around the artificial island under standard conditions [the x-axis is the east longitude axis and the y-axis is the north latitude axis].

A comprehensive analysis of the sediment transport and topographic evolution mechanism around the artificial island concluded that when the artificial island exists, a beam will form between the artificial island and the coastline, which will meet the coastal current coming to Tanmen Port (see Figure 11), resulting in the siltation between the artificial island and the coast. An eddy wake was formed on the back surface of the artificial island, which interacted with the tidal current and formed scour pits on the back surface (see Figure 3b). After the complete removal of the artificial island, the sea bed developed towards a balance of erosion and silt. In conclusion, due to the synthetic effect of the beam and the water-retaining effect on the man-made island, the coastal current and the tidal current led to the artificial island having a complex influence on the water and sediment transport and topographic evolution.

5. Conclusions

In this study, a coupled simulation system based on an unstructured grid SCHISM, the WWM-III wave model, and the MORSELFE sediment transport and topographic evolution model was used to simulate the dynamic processes of local water and sediment transport and the topographic evolution under interference from an artificial island in the offshore area of Qionghai, Hainan Province. The field-measured data can drive complex coupled simulations of the wind, waves, and sediment transport. In particular, the local small-scale geometric characteristics of the artificial island significantly affect the distribution of hydrodynamics, suspended sediment, and topographic evolution in local waters. The coupled hydrodynamic wave–sediment simulation system based on a unified unstructured grid can reflect detailed information of the cross-scale, near-shore environmental evolution.

Local disturbance from the artificial island led to the formation of eddies in the surrounding waters during the ebb and flow of the tides, resulting in the formation of local siltation and erosion pits, especially near the angle between the island and the coastline. Over time, siltation will form into tombolo bars, and erosion will be caused behind local walls such as the sharp corners of the artificial island. The simulation results showed that after the removal of the artificial island, most of the thickness of the scouring and silting will be reduced, and the seabed can return to its natural equilibrium state; however, the silting caused by local coastline morphology will still exist.

There were also shortcomings in this study, such as the insufficient observational data on suspended sediment concentration and topographic evolution, which limited the demonstration and disclosure of more detailed information about near-shore sediment transport and topographic evolution under interference from the artificial island. Due to the lack of observational data on hydrodynamics, sediment transport, and topographic evolution during typhoons, the topographic evolution during typhoons was not analyzed in this paper. The influence of individual extreme typhoon events on the evolution of the coastal environment combined with the influence of the artificial island also needs to be studied in the future.

Author Contributions: Thanks to all the authors for their joint efforts, which were as follows. Investigation, K.Y., M.F., H.W. and G.F.; writing—original draft preparation, G.F.; writing—review and editing, J.L., K.Y., Y.S. and X.W.; project administration, J.L. All authors have read and agreed to the published version of the manuscript.

Funding: This research was funded by Science and Technology Innovation Fund of Command Center of Integrated Natural Resources Survey Center (KC20220009) and funded by the Comprehensive Survey of Natural Resources in HaiChengWen Coastal Zone, grant number DD20230414.

Institutional Review Board Statement: Not applicable.

Informed Consent Statement: Not applicable.

Data Availability Statement: The data are unavailable due to privacy.

Acknowledgments: Thanks to Li Jian for his guidance and help. I would like to thank all the authors for writing this article together. Finally, I would like to express my sincere thanks to the editors and reviewers for their comments and suggestions.

Conflicts of Interest: The authors declare no conflict of interest.

References

1. Yan, K. *Coastal Engineering*; Ocean Press: Beijing, China, 2002; pp. 120–130.
2. Peng, B.; Hong, H.; Chen, W.; Xue, X.; Cao, X.; Peng, J. Ecological Damage Appr aisal of Sea Reclamation: Theory8 Method and Application. *J. Nat. Resour.* **2005**, *5*, 714–726.
3. Zhu, G.; Xu, X. Research review on environmental effects of land reclamation from sea. *Ecol. Environ. Sci.* **2011**, *20*, 761–766.
4. Yang, Y.; Liu, X.; Qiu, R.; Yang, W. The analysis of shoreline changes respond to artificial island project of Dongjiao coco forest based on DSAS and SMC. *Period. Ocean. Univ. China* **2017**, *47*, 162–168.
5. Li, S. Study on the influence of artificial island on dynamic sediment environment and beach erosion and deposition evolution of sandy coast. *Ocean. Eng.* **2021**, *39*, 144–153.
6. JETRO. *Feasibility Study of the Construction of an Artificial Island at the Pacific Entrance to the Panama Canal*; Belgian Press: Brussels, Belgium, 2004.
7. Salahuddin, B. The Marine Environmental Impacts of Artificial Island Construction. *J. Environ. Manag.* **2006**, P15.
8. Gong, W.P.; Li, C.Y.; Lin, G.Y.; Mo, W.Y. Application of DELFT 3D model for plan design of an artificial island—A case study for the artificial island construction in The Riyue Bay, Wanning City, Hainan Island. *Ocean Eng.* **2012**, *30*, 35–44.

9. Tan, X.; Gao, J. Impact of the Sanya new airport artifial islands project on tidal dynamics of the Hongtang Bay. *Mar. Sci. Bull.* **2019**, *21*, 1–15.
10. Kuang, C.; Jiang, L.; Ma, Y.; Qiu, R. Wave-current coupled hydrodynamic responses to artifical island and beach nourishment projects. *J. Tongji Univ.* **2019**, *47*, 38–46.
11. Yang, F.; Pan, J.; Wang, H. Study on the design wave on the artificial island of Hong Kong-Zhuhai-Macao Bridge while considering the extreme weather I: Selection method of the typhoon track in numerical wave simulation. *Adv. Water Sci.* **2019**, *30*, 892–901.
12. Warner, J.C.; Sherwood, C.R.; Signell, R.P.; Harris, C.K.; Arango, H.G. Development of a three-dimensional, regional, coupled wave, current, and sediment-transport model. *Comput. Geosci.* **2008**, *34*, 1284–1306. [CrossRef]
13. Zhang, Y.; Baptista, A.M. SCHISM: A semi-implicit Eulerian-Lagrangian finite-element model for cross-scale ocean circulation. *Ocean Model.* **2008**, *21*, 71–96. [CrossRef]
14. Zhang, Y.; Ye, F.; Stanev, E.V.; Grashorn, S. Seamless cross-scale modeling with SCHISM. *Ocean Model.* **2016**, *102*, 64–81. [CrossRef]
15. Roland, A. Development of WWM II: Spectral Wave Modeling on Unstructured Meshes. Ph.D. Thesis, Technische Universität Darmstadt, Darmstadt, Germany, 2009; pp. 25–38.
16. Roland, A.; Zhang, Y.J.; Wang, H.V.; Meng, Y.; Teng, Y.-C.; Maderich, V.; Brovchenko, I.; Dutour-Sikiric, M.; Zanke, U. A fully coupled 3D wave-current interaction model on unstructured grids. *J. Geophys. Res. Oceans* **2012**, *117*, C00J33. [CrossRef]
17. Pinto, L.; Fortunato, A.B.; Zhang, Y.; Oliveira, A.; Sancho, F.E.P. Development and validation of a three-dimensional morphodynamic modelling system. *Ocean. Model.* **2012**, *57–58*, 1–14. [CrossRef]
18. Fu, G.; Song, Y.; Yuan, K. A study on the erosion and siltation evolution of the neighboring coast caused by the enclosure of Boao Coral island. *Mar. Environ. Sci.* **2022**, *41*, 174–179.
19. Bi, J.; Zhang, L.; Song, Q.; Sui, Y.; Wen, L. Coastline Dataset of Hainan Island during 1987–2017. [DB]. V1. Science Data Bank: 2019. Available online: https://www.scidb.cn/en/detail?dataSetId=633694461133586432 (accessed on 25 April 2023).
20. Pawlowicz, R.; Beardsley, B.; Lentz, S. Classical tidal harmonic analysis including error estimates in MATLAB using T_TIDE. *Comput. Geosci.* **2002**, *28*, 929–937. [CrossRef]
21. Song, Y.W.; Yuan, K.; Fu, G.W. *Report for the Geological and Coastal Surveying Project of Qionghai-Wanning in Hainan Province (No. DD20208012)*; Haikou Marine Geological Surveying Center affiliated with China Geological Survey Bureau: Beijing, China, 2021.
22. Clodman, S. A Generalized Parametric Wind Wave Model. *J. Great Lakes Res.* **1994**, *20*, 613–624. [CrossRef]

Article

Spatial Distribution and Main Controlling Factors of Nitrogen in the Soils and Sediments of a Coastal Lagoon Area (Shameineihai, Hainan)

Kun Yuan [1], Yanwei Song [1], Guowei Fu [1,*], Bigui Lin [2], Kaizhe Fu [1] and Zhaofan Wang [1]

1 Haikou Marine Geological Survey Center, China Geological Survey, Haikou 570100, China; yuankun01@mail.cgs.gov.cn (K.Y.)
2 Institute of Environment and Plant Protection, Chinese Academy of Tropical Agricultural Sciences, Haikou 571101, China
* Correspondence: fuguowei@mail.cgs.gov.cn

Abstract: As the relationship between the spatial distribution characteristics and physicochemical properties of nitrogen in lagoon soil and sediment is still unclear, we obtained and systematically analyzed soil and sediment samples from the surroundings of a lagoon in the Shamei Inland Sea, Qionghai City, Hainan Province. The spatial distribution of nitrogen forms was investigated, and the soil physicochemical properties that predominantly influenced the nitrogen distribution were characterized. The results were as follows: (1) There are differences in nitrogen content in the soil and sediment around the Shamei Inland Sea. The total nitrogen levels in the Shamei Inland Sea were low, and organic nitrogen was dominant. The soil samples showed higher organic nitrogen concentrations than the sediment samples. Thus, the characteristics of dispersion in offshore waters differ from those in lagoons. The average contents of nitrate, ammonium and nitrite in the soil around the lagoon were higher than those in the sediment, and they were especially high in the lagoon mouth area. The nitrate content was greater in the estuary area in the northwest of the lagoon, and the contents of ammonium and nitrite in the estuary area in the south of the lagoon were the highest. (2) The changes in the basic physical and chemical properties of the soils and sediments in the Shamei Inland Sea area have an important coupled effect on the enrichment, preservation and mineralization of nitrogen. Through redundancy analysis (RDA), the explanatory degree of the organic matter content in relation to the nitrogen content was determined to be approximately 17.2%, contributing 80.9% of all nitrogen. The organic matter content, cation exchange capacity and clay content were positively correlated with the nitrogen content, indicating that changes in the basic physical and chemical properties of the soil and sediment had an important impact on nitrogen enrichment, preservation and mineralization processes. The total nitrogen and the organic matter content, cation exchange capacity and clay content of the soil and sediment were positively correlated. The high proportion of nitrate in the soil, the high proportion of ammonium in the sediment, the heavy texture of the sediment, poor soil ventilation and weak nitrification were related.

Keywords: Shamei Inland Sea; soil–sediment; nitrogen form; spatial distribution; main controlling factor

Citation: Yuan, K.; Song, Y.; Fu, G.; Lin, B.; Fu, K.; Wang, Z. Spatial Distribution and Main Controlling Factors of Nitrogen in the Soils and Sediments of a Coastal Lagoon Area (Shameineihai, Hainan). *Appl. Sci.* **2023**, *13*, 7409. https://doi.org/10.3390/app13137409

Academic Editor: Fulvia Chiampo

Received: 29 April 2023
Revised: 11 June 2023
Accepted: 16 June 2023
Published: 22 June 2023

1. Introduction

Nitrogen is an important element for many organisms, and nitrogen cycling is often coupled with the carbon cycle and affects the operation of ecosystems. The relatively low nitrogen content in the marine environment is often an important limiting factor for the primary productivity of marine ecosystems [1]. A lagoon is a dynamic and complex ecosystem located at the intersection of land, sea, brackish water and freshwater. Lagoons are sensitive to weather and climate change, making them important areas for studying climate change [2]. Lagoons also face ecosystem degradation issues such as eutrophication, and nitrogen is the main driver of eutrophication in coastal ecosystems. Thus, it is important to clarify the spatial distribution and main controlling factors in lagoon soil and sediment to maintain the health and productivity of the associated ecosystems.

Research on lagoons started early, and the research mainly focused on the development and formation conditions of lagoons, salinity evolution, migration mechanisms, eutrophication and other factors. Kroon FJ et al. [3] reported that the average total nitrogen load has increased by a factor of 5.7, severely impacting coastal ecosystems. Hayn M et al. [4] observed a nitrogen and phosphorus exchange in the West Falmouth Harbor lagoon and coastal waters off the coast of the United States. Shallow water systems dominated by benthic producers have the potential to retain substantial terrestrial nitrogen loads when adequate phosphorus supplies are available through exchange with coastal waters. Nazneen et al. [5] studied the largest lagoon in Asia and found that primary production within the lagoon, terrestrial inputs from river discharge and anthropogenic activities near the lagoon controlled the distribution of total nitrogen in surface and core sediments.

Other scholars have also conducted research on the transformation mechanism of nitrogen in lagoons, and the research has typically involved saturated sediments from inland lakes and marine continental shelves. Liu Xiaotong et al. [6] found that the nitrogen content in the vertical direction changed greatly according to the nitrogen geochemical characteristics of the sedimentary column in the lagoon of the East Island of the Xisha Islands. Based on this, they speculated that the nitrogen content was greatly influenced by the ocean and made a useful attempt to infer the sedimentary environment. Ge Chendong et al. [7] conducted an isotopic analysis of total nitrogen and other indicators based on sediment column samples from the Shamei Inland Sea and concluded that the lagoon in the Shamei Inland Sea is gradually closing. Since the 19th century, the lagoon has been continuously filling with sediment and shallowing, with stable sedimentation at a rate of approximately 0.15 cm/a. Zhou Meiling et al. [8] collected surface sediment samples from the Qilihai Lagoon and analyzed the factors influencing different nitrogen forms and the external environment; they found that the nitrogen content was related to the median particle size and the organic matter content. Zhang et al. [9] analyzed the relationship between the particle size and total nitrogen in sediments from Shuang Lagoon, Lingshui County, Hainan Province, and found that total nitrogen was correlated with the average particle size of the sediments. Jiangjun Jia et al. [10] studied the sources and burial patterns of organic matter in Shamei Lake, Hainan. Ruihuan Li et al. [11] conducted a study on the nutrient dynamics of the Wanquan River estuary. However, these studies were limited to a single sediment or soil location, and there is a lack of integrated soil–sediment research.

Therefore, this study focuses on nitrogen in Hainan coastal lagoon sediments and surrounding soils, which have been rarely studied in China. The high-resolution spatial distribution characteristics of nitrogen in the inland sea and surrounding areas are considered, the relationships between the basic physical and chemical properties of the lagoon soil and sediment and the pH, organic matter content, cation exchange capacity, clay content and nitrogen forms are analyzed and the main factors affecting the spatial distribution of nitrogen forms are identified. Furthermore, this study provides theoretical support for eutrophication prevention, protection and restoration in lagoon ecosystems and comprehensive coastal zone management in the Shamei Inland Sea area. Moreover, the results of this study can guide ecological restoration, ecological protection and coastal agricultural management and can provide theoretical support for assessing the transformation mechanisms of typical lagoons in southern China from a new perspective.

2. Research Area and Research Method

2.1. Study Area

The soil–sediment samples used in this study came from the Shamei Inland Sea (Figure 1), which is a naturally formed lagoon to the southeast of Qionghai City, Hainan Province. It is located in the tropical monsoon climate zone. From north to south, the Wanquan River, Jiuqu Jiang River and Longgun River flow into the lagoon. The area spans latitudes from 19°05′~19°09′ N and longitudes from 110°32′~110°34′ E. The lagoon is surrounded by a north–south spit known as the Jade Belt Beach on the coast and has a slender shape, shallow water depth and gentle slope. The southern part is approximately 6 km long and 500–700 m wide, the northern part is approximately 2.5 km long and the narrowest part is only approximately 10 m, with a total area of approximately 25.87 km^2. Due to the barrier of the Jade Belt Beach, there is no direct water exchange between the inner sea and the outer sea, and the salinity of the inner sea is less than 1% [7,10]. No lagoon dredging work has been carried out in the past three years.

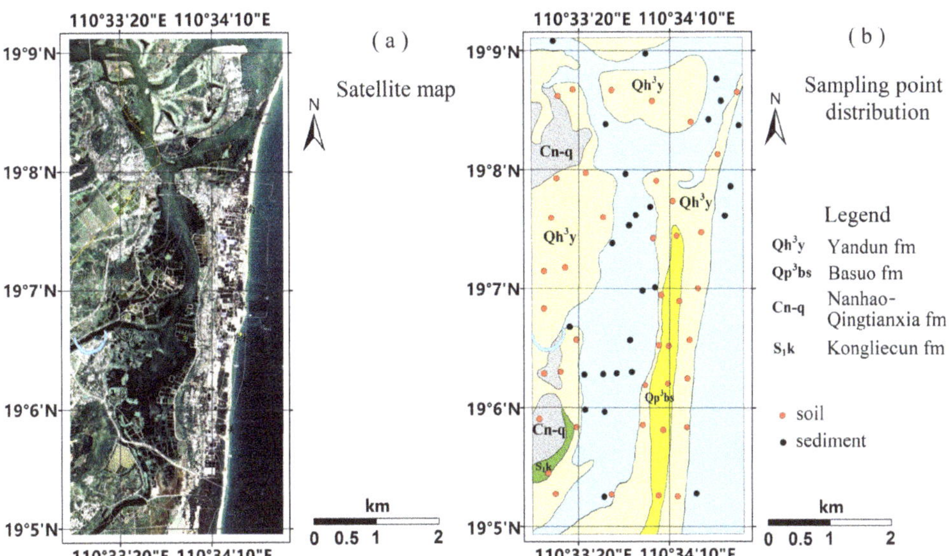

Figure 1. (**a**) Satellite map of the study area; (**b**) sampling point distribution map.

2.2. Research Materials

The sampling was performed in the summer of 2021, the working scale was 1:50,000 and the sampling resolution was 2 samples/km^2. A total of 69 samples were collected, including 41 soil (25 inland soil samples and 16 sand bar soil samples) and 28 sediment samples (19 lagoon deposits and 9 offshore deposits). The distribution of the sample points and the elevation of the study area are shown in Figures 2 and 3. The soil samples were collected at a depth of 20–40 cm and passed through a 2 mm sieve on site, and the sample weight was not less than 0.5 kg. A grab sampler was used to collect sediment samples, visible foreign objects were removed and the sediment was passed through a 2 mm nylon sieve on location to ensure that the sieved sediment was no less than 1.5 kg. The wet soil that had been sieved was wrapped with 60-mesh nylon mesh, and the water was drained to the extent possible.

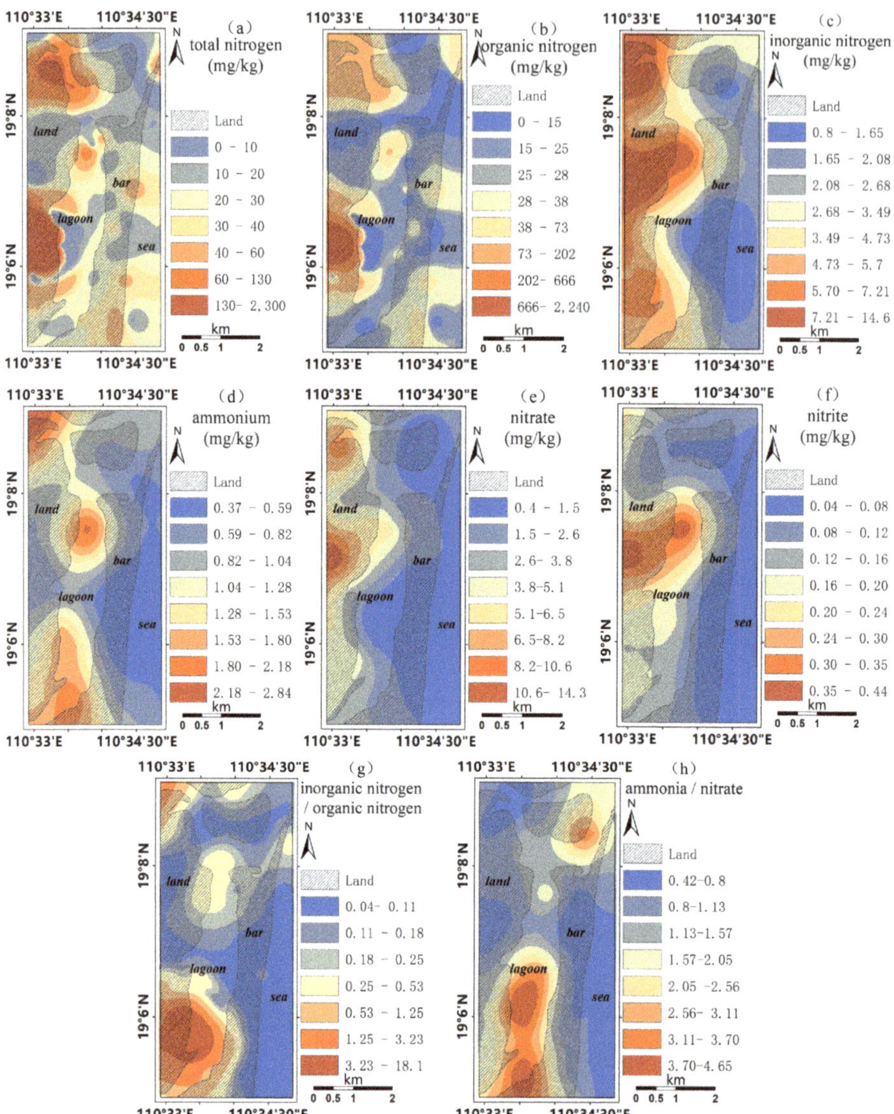

Figure 2. (**a**) Total nitrogen distribution, (**b**) organic nitrogen distribution, (**c**) inorganic nitrogen distribution, (**d**) ammonium distribution, (**e**) nitrate distribution, (**f**) nitrite distribution, (**g**) inorganic nitrogen/organic nitrogen distribution and (**h**) ammonium/nitrate distribution.

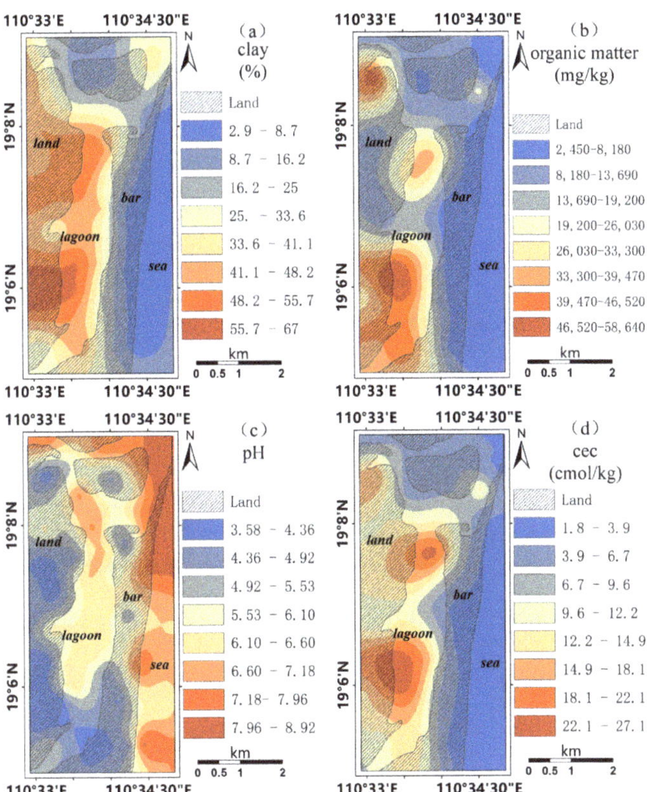

Figure 3. Distributions of (**a**) clay, (**b**) organic matter (OM), (**c**) pH and (**d**) cation exchange capacity.

2.3. Test Method

1. Total nitrogen and forms of nitrogen

Total nitrogen: To determine the total nitrogen content, soil samples passed through a 0.25 mm sieve were weighed, placed into deionized water and boiled in concentrated sulfuric acid to convert all nitrogen-containing compounds into ammonium nitrogen. Then, a standard hydrochloric acid titration solution, distillation and titration were used in conjunction with an automatic nitrogen analyzer.

Ammonia and nitrite were extracted by potassium chloride solution in alkaline and acidic environments, respectively. After the addition of the chromogenic agent, calibration curves were drawn at 630 nm and 543 nm by a Perkin Elmer Lambda 900 ultraviolet/visible/near-infrared spectrophotometer, and the absorbance was measured. The nitrate was reduced to nitrite, the total amount of nitrite was measured and the amount of nitrate was obtained by subtracting the amount of original nitrite obtained in the previous step [12].

The detection limits for ammonia, nitrite nitrogen and nitrate nitrogen in the sample were 0.2 mg/kg, 0.15 mg/kg and 0.25 mg/kg, respectively, with an error range of 5%.

2. Granularity

Air-dried soil samples were passed through a 2 mm sieve, and hydrogen peroxide and hydrochloric acid were added to remove organic matter and $CaCO_3$. Finally, the soil samples were rinsed with deionized water to remove excess HCl and other chlorides; then, the soil particle size was determined by a Mastersizer 2000 laser particle size analyzer [13].

3. Organic matter

While being heated, excess potassium dichromate-sulfuric acid solution was used to oxidize soil organic carbon. The excess potassium dichromate was titrated with a ferrous sulfate standard solution, and the organic carbon content was calculated from the amount of consumed potassium dichromate according to the oxidation correction coefficient amount multiplied by the constant 1.724 to obtain the organic matter content [14].

4. pH

Air-dried soil samples that had been passed through a 2 mm sieve were weighed, deionized water was added according to the soil–water ratio of 1:2.5 to prepare a solution and three standard buffer solutions of phthalate, phosphate and borate at room temperature were added. To measure the pH, a 3510 Portable Multiparameter Water Quality Analyzer for Determination of Solution pH was used [15].

5. CEC (cation exchange capacity)

Air-dried soil samples that had been passed through a 2 mm sieve were weighed, and ammonium acetate solution was added repeatedly to treat the soil. The excess ammonium acetate was rinsed off with 95% ethanol; then, deionized water was used to wash all the soil sample residues into the NKB-3200 automatic nitrogen analyzer for digestion. The tube was distilled, the absorption solution was titrated with a 0.05 mol/L hydrochloric acid standard solution and a blank experiment was performed at the same time [16].

3. Results

3.1. Spatial Distribution of Soil–Sediment Total Nitrogen

Overall, the soil–sediment total nitrogen content was 0–2360 mg/kg, the average content was 90 mg/kg, the variation coefficient was 3.88 and the fluctuation range was large. In other regions, the total nitrogen content in the Qilihai Lagoon in Hebei Province ranges from 90 to 1180 mg/kg [8], the total nitrogen content in the sediments of Dianchi Lake in Yunnan Province ranges from 1600 to 5560 mg/kg [17] and the total nitrogen content in the sediments of Puppet Lake in Jiangsu Province ranges from 670–2570 mg/kg [18]. A comparison indicated that the soil–sediment total nitrogen content in the lagoon of the Samet Inland Sea was low. Using ArcGIS software, the spatial distribution of nitrogen content in the study area was obtained by kriging interpolation analysis. In Figure 2a, it can be seen that the total nitrogen content of terrestrial soil is higher than that of aquatic sediments, and the highest concentration of total nitrogen appeared in the southern part of the inland soil, followed by the Wanquan River estuary and the adjacent land areas in the north. Overall, the distribution of total nitrogen showed characteristics of diffusion from inland to offshore waters. This indicated that a large amount of nitrogen-containing nutrients carried by the rivers that enter the lagoon may be input into the lagoon from the inland sea of Samet, and at the same time, under the influence of hydrodynamics, nitrogen becomes enriched near the estuary of the lagoon. The overall nitrogen deficiency in the offshore area may be related to the exchange of water between the lagoon and the ocean, and frequent coastal currents and tidal action accelerate the process of nitrogen volatilization. This result confirmed that the denitrification effect in the offshore area is strong [19]. The concentrated distribution area of total nitrogen is basically consistent with the distribution of organic matter, suggesting the connection between the two in terms of genesis and enrichment, which is consistent with previous research results [20].

3.2. Spatial Distribution of Soil–Sediment Organic Nitrogen

The soil–sediment organic nitrogen content ranged from 0 to 2354.29 mg/kg (Table 1), with an average content of 82.88 mg/kg and a coefficient of variation of 4.08, reflecting a large fluctuation range. Based on Figure 2b, the organic nitrogen content of terrestrial soil is higher than that of aquatic sediments. As shown in Figure 2b, organic nitrogen is enriched in terrestrial areas, and the enrichment of organic nitrogen in terrestrial areas may be related to the surface return of terrestrial ecosystems and is also greatly affected by human activities. The distribution of organic nitrogen is roughly consistent with the distribution of organic matter, which may be due to the abundant precipitation in the region, abundant river flow, strong erosion and transport capacity on the surface and the organic matter and nitrogen-containing nutrients carried by the three rivers in the north, west and south. Nutrient salts have been shown to be injected into the lagoon of the inland sea of Samet, which is hydrodynamically affected, and are not transported offshore over long distances [12].

Table 1. Statistical results of different forms of nitrogen in soil and sediment.

Type	Index	Total Nitrogen (mg/kg)	Organic Nitrogen (mg/kg)	Inorganic Nitrogen (mg/kg)	Ammonium (mg/kg)	Nitrate (mg/kg)	Nitrite (mg/kg)
Terrestrial soil	Max	2360	2354.29	37.71	2.28	35.87	1.49
	Min	0	0.00	0.23	0.00	0.12	0.00
	Average value	290	280.53	7.82	0.49	7.16	0.17
	Standard deviation	680	679.20	10.53	0.60	9.96	0.37
	Coefficient of variation	2.35	2.42	1.35	1.22	1.39	2.21
Lagoon sediment	Max	110	84.12	26.51	8.78	16.07	1.67
	Min	0	0.00	0.21	0.15	0.00	0.04
	Average value	30	20.63	4.46	2.47	1.64	0.34
	Standard deviation	20	19.81	5.88	2.44	3.65	0.36
	Coefficient of variation	0.97	0.96	1.32	0.99	2.22	1.04
Bar soil	Max	70	68.49	18.26	1.38	17.56	0.26
	Min	0	0.00	0.20	0.07	0.00	0.00
	Average value	30	26.09	2.39	0.23	2.12	0.04
	Standard deviation	20	19.28	3.51	0.25	3.38	0.06
	Coefficient of variation	0.63	0.74	1.47	1.10	1.59	1.70
Oceanic sediment	Max	140	136.23	8.82	4.07	4.66	0.23
	Min	0	0.00	0.76	0.34	0.00	0.00
	Average value	20	20.72	2.62	1.37	1.17	0.09
	Standard deviation	40	44.06	2.57	1.34	1.37	0.08
	Coefficient of variation	1.87	2.13	0.98	0.98	1.17	0.97
Region of interest	Max	2360	2354.29	37.71	8.78	35.87	1.67
	Min	0	0.00	0.20	0.00	0.00	0.00
	Average value	90	82.88	4.25	1.06	3.03	0.16
	Standard deviation	340	337.93	6.59	1.67	5.91	0.29
	Coefficient of variation	3.88	4.08	1.55	1.58	1.95	1.82

3.3. *Spatial Distribution of Soil–Sediment Inorganic Nitrogen*

The soil–sediment inorganic nitrogen content was 0–37.71 mg/kg (Table 1, Figure 2), with an average content of 4.25 mg/kg and a coefficient of variation of 1.55. The overall fluctuation was small. The average content of inorganic nitrogen in the soil around the lagoon was higher than that in the surface layer. The sediments exhibit the terrigenous enrichment characteristics of nitrogen. The lagoon sediments have the highest ammonium content, the ammonium content of the terrestrial soils is lower than that of the aquatic sediments and the nitrate content of the terrestrial soils is higher than that of the aquatic sediments. The content of nitrite was slightly higher in the aquatic sediments.

Ammonium and nitrate, as forms of soil inorganic nitrogen, can be absorbed and utilized by plants in large quantities, which is of great significance for crop growth. However, the action of running water may cause nitrogen to migrate into the water body, and an excessive amount may cause eutrophication in the water body, which is potentially harmful. From the distribution in Figure 2c, inorganic nitrogen is enriched in the terrestrial area, and the inorganic nitrogen in the young Yandun Formation is slightly higher, which may be related to the rapid nitrification process of organic nitrogen under oxidative conditions in this area. In addition, the saturation levels of the soil near the lagoon and the offshore water body fluctuate greatly due to the fluctuation in the tidal process, and a dry soil effect may occur, which enhances the mineralization rate, increases the release of ammonium and leads to inorganic nitrogen supplementation.

From the distribution point of view, the content of ammonium was higher in the estuary area of the lagoon because the river flowing into the lagoon flows through a rural area, and the concentrations of nitrogen pollutants carried in the runoff were high, potentially due to nitrogen from agricultural sources. Overall, ammonium, nitrate and nitrite were enriched in the lagoon mouth area, which may be related to a stable hydrodynamic environment and colloidal condensation and sedimentation in the area between the lake mouth and the sea (Figure 2d).

The content of nitrate in the estuary area of the northwestern part of the study area and in the lagoon was higher than that in other areas (Figure 2e). The upper reaches of the Wanquan River, which flows into the northwestern lagoon, flows through urban areas that are densely populated. Rapid urbanization has led to the increased discharge of nitrogen-containing wastewater, thereby providing more nitrogen from industrial sources.

The content of nitrous nitrogen in the estuary of the lagoon in the west and south of the study area was relatively high, the distribution in different geological layers was heterogeneous and the distribution was high in the relatively young Yandun Formation (Figure 2f). The Jiuqu River and Longgun River, which flow into the southwestern lagoon, flow through rural areas, and the runoff from agricultural production areas and livestock areas has high concentrations of nitrogen pollutants.

Based on the ratio of inorganic nitrogen/organic nitrogen, the northern, western and southern estuaries of the study area were favorable for the accumulation of inorganic nitrogen, while the land and other water areas were favorable for the accumulation of organic nitrogen (Figure 2g), which may be associated with higher mineralization rates. Figure 2h shows that from the ratio of ammonium/nitrate, the proportion of ammonium in river estuaries and lagoons was high, but the proportion of ammonium in land and sea areas was low. This may be related to the adsorption of ammonium ions during the flocculation and sedimentation of clay minerals in the estuary area, which is related to the rapid nitrification of nitrogen caused by the oxidative environment of terrestrial and offshore currents.

3.4. Spatial Distribution of the Soil–Sediment Particle Size

The average content of sand in the soil around the lagoon was higher than that in the sediment (Table 2), while the average contents of silt and clay were lower than those in the sediment, indicating that the grain size of the soil is coarser than that of sediment. This was related to the sorting effect that moving water has on soil particles. Regarding the distribution of single particle size components, as shown in Figure 3a, a high-value area was observed for clay in the terrestrial zone in the western part of the study area, and this area was found to extend to the lagoon water area. The content of sand in the sea area was the lowest; the content of sand in the study area showed a completely opposite distribution trend to that of clay and silt. Specifically, the sand content was highest in the lagoon bar and the sea area to the east and gradually decreased in the transition to the land area. This showed that the texture of the land area was the coarsest, the texture of the lagoon bar and the sea area was the finest and the sedimentary texture of the lagoon was in the middle. The distribution characteristics of median-size particles were basically consistent with the distribution of sand particles, reflecting the overall coarse characteristics of soil particles in the study area. In addition, in the western continental area, the soil–sedimentary material geography displayed obvious parent material differentiation; that is, the old continental strata were finer in texture, the younger continental strata were macroscopically granular and the lagoon sedimentary texture was in the middle. This may be due to the longer weathering and soil-forming process experienced by the old terrestrial bottom layer. By comparing additional particle size groups, it can be seen that the 0.25–1 mm component was consistent with sand; the content of this component was high in the soil, and the content of the component smaller than 0.25 mm was high in the sediment.

Table 2. Soil–sediment texture characteristics.

Type	Index	Clay (%)	Silt (%)	Sand (%)	<0.031 mm (%)	0.031~0.053 mm (%)	0.053~0.25 mm (%)	0.25~1 mm (%)	1~2 mm (%)
Terrestrial soil	max	448.22	85.2	65.2	83.4	90.1	9.8	51.5	40.3
	min	5.27	16.6	12.8	14.8	19.7	4.4	5.2	0.0
	average value	56.26	48.5	39.0	51.5	54.1	7.4	23.3	14.1
	standard deviation	108.58	17.0	14.3	17.0	17.7	1.7	9.7	10.1
	coefficient of variation	1.93	0.4	0.4	0.3	0.3	0.2	0.4	0.7
Lagoon sediment	max	467.50	73.6	64.2	96.7	83.1	19.4	59.8	79.4
	min	8.49	3.3	3.1	26.4	4.3	0.3	2.7	0.0
	average value	62.75	50.8	46.2	49.2	60.7	9.9	17.3	11.6
	standard deviation	118.89	20.0	18.1	20.0	23.3	5.3	14.1	20.9
	coefficient of variation	1.89	0.4	0.4	0.4	0.4	0.5	0.8	1.8
Bar soil	max	448.22	56.6	51.8	81.4	66.5	9.8	30.3	40.3
	min	15.91	18.6	15.1	43.4	21.6	4.4	15.2	7.7
	average value	127.11	43.2	36.2	56.8	50.1	8.3	20.6	16.4
	standard deviation	214.12	17.1	15.4	17.1	19.8	2.6	6.9	15.9
	coefficient of variation	1.68	0.4	0.4	0.3	0.4	0.3	0.3	1.0
Oceanic sediment	max	541.03	68.0	62.3	100.0	80.6	11.4	94.5	67.5
	min	12.31	0.0	0.0	32.0	0.0	0.4	6.9	1.1
	average value	244.43	15.7	14.2	84.3	18.3	3.2	43.2	29.8
	standard deviation	204.87	23.4	21.3	23.4	27.5	3.7	39.3	30.0
	coefficient of variation	0.84	1.5	1.5	0.3	1.5	1.2	0.9	1.0
Region of interest	max	706.85	85.2	68.7	100.0	90.1	12.0	51.5	85.6
	min	5.27	0.0	0.0	14.8	0.0	0.5	3.2	0.0
	average value	237.63	28.8	23.4	72.6	32.4	4.9	19.8	41.2
	standard deviation	205.46	26.4	21.7	25.0	29.0	3.3	10.8	29.1
	coefficient of variation	0.86	0.91	0.93	0.34	0.90	0.68	0.55	0.71

3.5. Spatial Distribution of Soil–Sediment Organic Matter

Soil–sediment organic matter is an important part of soil–sediment fertility and is an important indicator reflecting soil–sediment maturity and fertility level. The level of soil–sediment organic matter also affects other nutrients. In turn, nutrients affect crop growth and development, as well as yield and quality. The average content of soil organic matter around the lagoon was 13,080 mg/kg, and the content of organic matter in the surface sediments was 23,580 mg/kg. The former was significantly lower than the latter, which may be related to the favorable soil aeration conditions for the mineralization and decomposition of organic matter. Erosion of terrestrial soils and subsidence in lagoon water environments may also affect the organic matter content. Specifically, in terms of its distribution, as shown in Figure 3b, the most concentrated areas of organic matter were the land area and water area near the entrance to the river in the southwestern part of the study area, followed by the land area on the south bank of the entrance of the Wanquan River in the northwest. Additionally, a high-value area of organic matter appeared at the mouth of the lagoon. Enrichment of organic matter was observed in the land and waters near the mouth of the lagoon. This pattern reflects the erosion of terrigenous organic matter and the recharge process of the lagoon. The enrichment of organic matter at the mouth of the lagoon may be related to the condensation and deposition of organic matter in the context of high ion concentrations at the lake–sea interface.

3.6. Spatial Distribution of the Soil–Sediment pH

The average pH value of the soil in the study area was 5.1, which is acidic, reflecting the leaching of base ions, such as K, Na, Ca and Mg, and the enrichment of acid ions, such as H and Al. The average pH of the sediment was 7.3, which is neutral and is largely from the acceptance of terrestrial salt ions, reflecting the sediment's enrichment process.

From the distribution point of view, as shown in Figure 3c, the soil pH in the western land area was the lowest and slightly acidic, the pH value in the sea area east of the sand bar was the second highest and neutral and the pH value in the northeastern sea area was the highest. The values ranged from acidic to neutral and exhibited transitional characteristics. Overall, the pH difference in the study area reflected the enrichment and migration process of base ions.

3.7. Spatial Distribution of the Soil–Sediment Ion Exchange Capacity

The CEC is one of the most important features of sediments. The average CEC of the soil around the lagoon was 8.97 cmol/kg, and the average CEC of the surface sediment was 12.52 cmol/kg. This reflected the sorption phenomenon of base ions. As shown in Figure 3d, the distribution characteristics of the CEC and organic matter were basically consistent; that is, the highest values were found in the southwestern land area and its adjacent waters, as well as in the lagoon mouth area. This result indicated that the migration process of cations was related to their adsorption by organic matter. In addition, by comparing the two high-value areas, it was found that the diffusion of marine CEC values in the estuary was more obvious than that of the continental facies to the lagoon waters, which may also be related to the lower organic matter content of the open water that ensures CEC diffusion.

4. Discussion

4.1. Correlations among Different Forms of Nitrogen in the Soil–Sediment

1. Total nitrogen and organic nitrogen

Total nitrogen in soil and sediment can generally be divided into inorganic nitrogen and organic nitrogen. In different regions, the content of each form of nitrogen and the proportions of the total nitrogen have great differences. Figure 4a shows that, overall, the soil–sediment organic nitrogen and total nitrogen contents in the lagoon of the Samet Inland Sea were significantly and positively correlated, and the correlation between the two was more similar in different soil–sediment samples. Tables 3–5 support this result, and organic nitrogen was the main component of total nitrogen in the soil and sediment.

2. Total nitrogen and inorganic nitrogen

The relationship between total nitrogen and inorganic nitrogen is complex among the different soil–sediment samples. Table 6 shows that, overall, the contents of inorganic nitrogen and total nitrogen in the soil–sediment samples from the inland sea lagoon were positively correlated, but the correlation was not strong. This was probably because the inorganic nitrogen proportion of the total nitrogen was small, and the inorganic nitrogen content in aquatic sediments was lower than that in terrestrial soils. According to Figure 4c, the total nitrogen in the sediment was positively correlated with the ammonium content. Although the nitrogen and ammonium contents in the soil were also positively correlated, the correlation was not strong. The competition ability for ammonium ions in the sediment was weak, so it was challenging to desorb ammonium, and ammonium could be preserved in the sediment. Figure 4d shows that the total nitrogen in the soil was positively correlated with nitrate, and the total nitrogen in the sediment was also positively correlated with nitrate. Be-cause the grain size of the soil is coarser than that of sediment, the aeration was good, nitrification was easy to carry out and ammonium was easily oxidized to nitrate and nitrite. Figure 4e shows that the total nitrogen was positively correlated with the nitrite in the sediment, the total nitrogen was negatively correlated with the nitrite in the soil and the nitrite in the sediment was more abundant than that in the soil. This may be because the studied waters are mainly in a reducing environment, and in an anaerobic sedimentary environment, nitrate is easily reduced to nitrite.

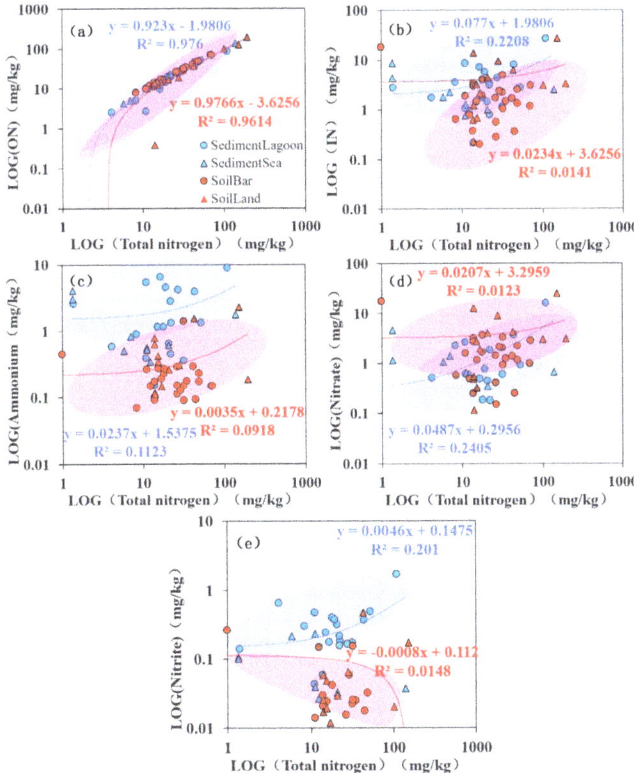

Figure 4. (**a**) Total nitrogen and organic nitrogen scatter plot, (**b**) total nitrogen and inorganic nitrogen scatter plot, (**c**) total nitrogen and ammonium scatter plot, (**d**) total nitrogen and nitrate scatter plot and (**e**) total nitrogen and nitrite scatter plot.

Table 3. Correlation analysis of the basic physical and chemical properties of inland soil and nitrogen forms (25 samples).

	pH	Organic Matter	CEC	Clay	Total Nitrogen	Organic Nitrogen	Inorganic Nitrogen	Ammonium	Nitrate	Nitrite
pH	1.00									
organic matter	−0.42	1.00								
CEC	−0.37	0.693 **	1.00							
clay	−0.28	0.06	0.28	1.00						
total nitrogen	−0.37	0.34	0.36	0.12	1.00					
organic nitrogen	−0.37	0.33	0.35	0.12	1.000 **	1.00				
inorganic nitrogen	−0.24	0.42	0.35	0.33	−0.04	−0.05	1.00			
ammonium	−0.32	0.832 **	0.517 *	0.01	−0.14	−0.15	0.47	1.00		
nitrate	−0.23	0.39	0.33	0.33	−0.03	−0.05	0.999 **	0.43	1.00	
nitrite	−0.03	0.05	0.05	0.31	−0.02	−0.04	0.795 **	0.14	0.794 **	1.00

* At the 0.05 level (two-tailed), the correlation is significant. ** At the 0.01 level (two-tailed), the correlation is significant.

Table 4. Correlation analysis between the basic physicochemical properties and nitrogen forms of lagoon sediments (19 samples).

	pH	Organic Matter	CEC	Clay	Total Nitrogen	Organic Nitrogen	Inorganic Nitrogen	Ammonium	Nitrate	Nitrite
pH	1.00									
organic matter	−0.485 *	1.00								
CEC	−0.31	0.878 **	1.00							
clay	−0.10	0.38	0.40	1.00						
total nitrogen	0.13	0.20	0.23	0.06	1.00					
organic nitrogen	0.13	0.17	0.23	0.02	0.985 **	1.00				
inorganic nitrogen	0.13	0.24	0.17	0.18	0.811 **	0.697 **	1.00			
ammonium	−0.17	0.518 *	0.34	0.20	0.556 *	0.43	0.862 **	1.00		
nitrate	0.30	0.03	0.04	0.13	0.857 **	0.768 **	0.950 **	0.662 **	1.00	
nitrite	0.19	0.04	0.11	0.22	0.797 **	0.718 **	0.872 **	0.602 **	0.905 **	1.00

* At the 0.05 level (two-tailed), the correlation is significant. ** At the 0.01 level (two-tailed), the correlation is significant.

Table 5. Correlation analysis between the basic physical and chemical soil properties and nitrogen forms on the sand bar (16 samples).

	pH	Organic Matter	CEC	Clay	Total Nitrogen	Organic Nitrogen	Inorganic Nitrogen	Ammonium	Nitrate	Nitrite
pH	1.00									
organic matter	−0.490 *	1.00								
CEC	−0.35	0.609 **	1.00							
clay	−0.508 **	0.854 **	0.861 **	1.00						
total nitrogen	0.04	−0.15	−0.35	−0.24	1.00					
organic nitrogen	0.06	−0.21	−0.484 *	−0.33	0.985 **	1.00				
inorganic nitrogen	−0.15	0.34	0.883 **	0.590 **	−0.27	−0.438 *	1.00			
ammonium	−0.37	0.944 **	0.580 **	0.801 **	−0.03	−0.10	0.37	1.00		
nitrate	−0.13	0.28	0.859 **	0.541 **	−0.28	−0.439 *	0.997 **	0.30	1.00	
nitrite	−0.08	0.495 *	0.827 **	0.635 **	−0.33	−0.454 *	0.777 **	0.524 **	0.750 **	1.00

* At the 0.05 level (two-tailed), the correlation is significant. ** At the 0.01 level (two-tailed), the correlation is significant.

Table 6. Correlation analysis between the basic physicochemical properties and nitrogen forms of offshore sediments (9 samples).

	pH	Organic Matter	CEC	Clay	Total Nitrogen	Organic Nitrogen	Inorganic Nitrogen	Ammonium	Nitrate	Nitrite
pH	1.00									
organic matter	−0.55	1.00								
CEC	−0.52	0.996 **	1.00							
clay	−0.15	0.34	0.36	1.00						
total nitrogen	−0.53	−0.12	−0.12	−0.22	1.00					
organic nitrogen	−0.52	−0.11	−0.10	−0.24	0.998 **	1.00				
inorganic nitrogen	−0.15	−0.28	−0.30	0.28	−0.12	−0.18	1.00			
ammonium	−0.20	−0.28	−0.28	0.49	0.02	−0.04	0.946 **	1.00		
nitrate	−0.05	−0.31	−0.33	0.02	−0.23	−0.28	0.948 **	0.795 *	1.00	
nitrite	−0.63	0.803 **	0.765 *	0.42	−0.26	−0.25	0.02	0.01	−0.03	1.00

* At the 0.05 level (two-tailed), the correlation is significant. ** At the 0.01 level (two-tailed), the correlation is significant.

3. Organic nitrogen and inorganic nitrogen

Organic nitrogen and inorganic nitrogen jointly constitute the total nitrogen in soil and sediment, but their correlation varies greatly among the different soil and sediment samples. As shown in Figure 5a, inorganic nitrogen in soil shows a negative correlation with organic nitrogen, while inorganic nitrogen in sediment shows a positive correlation with organic nitrogen. This indicates that in inland and sandbar soils, there is a competitive effect between organic and inorganic nitrogen, but in lagoons and nearshore areas, there is a synergistic evolution effect between organic and inorganic nitrogen. This may be due to the strong conversion between inorganic nitrogen and organic nitrogen in soil because of the large biomass in the soil. The amount of organic nitrogen increases through the transformation of inorganic nitrogen by organisms. In addition, inorganic nitrogen also forms organic nitrogen-containing compounds through chemical interactions with soil organic matter. Due to the low terrain of the lagoon, the nitrogen carried by the rivers entering the lake accumulates in the sediment, making the sediment a "nitrogen sink".

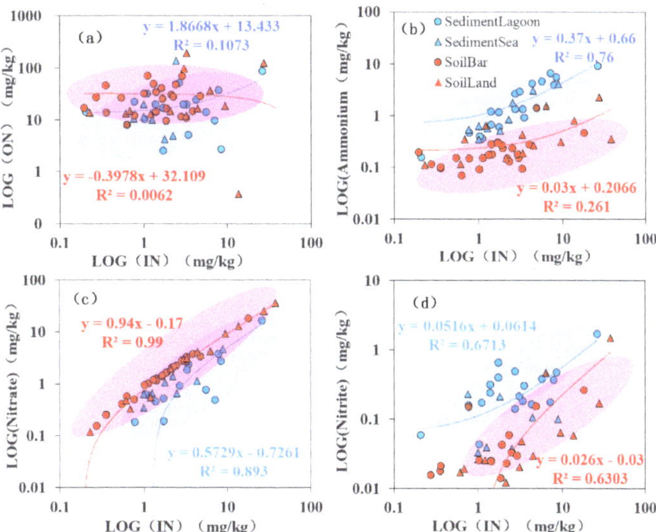

Figure 5. (**a**) Inorganic nitrogen and organic nitrogen scatter plot, (**b**) inorganic nitrogen and ammonium scatter plot, (**c**) inorganic nitrogen and nitrate scatter plot and (**d**) inorganic nitrogen and nitrite scatter plot.

4. Distribution of inorganic nitrogen

Ammonium includes ammonium and ammonia, and there is a dynamic equilibrium between ammonium ions and ammonia. In the core soil layer, fixed ammonium is the main form of inorganic nitrogen. From Figure 5b, overall, ammonium and inorganic nitrogen showed a significant positive correlation. In terms of classification, the correlation between ammonium and inorganic nitrogen in lagoons and offshore sediments was stronger, and the content was higher. This may be because the closer the soil water is to saturation, the weaker the nitrification, while the lagoon and offshore sediments feature a hypoxic environment with weak nitrification and the accumulation of ammonium. Figure 5c shows that, overall, nitrate and inorganic nitrogen had a significant positive correlation. In terms of classification, the correlation between nitrate and inorganic nitrogen in inland and sand bar soils was stronger, and the content was higher. This may be because nitrate mainly comes from the erosion and scouring of the surface by the three rivers entering the lake in the north, west and south. In addition, because of the oxidative environment both inland and in the sand bars, nitrification is strong, and nitrate can accumulate. In addition, denitrification easily occurs under the insufficient oxygen supply in the offshore area, and nitrate is easily reduced to nitrogen, nitrogen oxides and other gases, resulting in the loss of nitrate from the sediments. According to Figure 5d, overall, nitrite was significantly and positively correlated with inorganic nitrogen. In terms of classification, the correlation between nitrite and inorganic nitrogen in lagoons and offshore sediments was strong, and the content of nitrous nitrogen and inorganic nitrogen was high. This may be because in lagoons and offshore areas, ammonium may generate nitrite through the ammonia oxidation process, but there are insufficient oxidation conditions to further generate nitrate, although the originally accumulated nitrate is easily reduced to nitrite.

4.2. Correlation between the Soil–Sediment Particle Size and Nitrogen

It can be seen from the correlation analysis in Table 7 that, overall, the soil–sediment clay content in the study area was weakly, positively correlated with total nitrogen and organic nitrogen and showed a significant positive correlation with all inorganic nitrogen. In terms of classification (Tables 3–6), the clay content of inland soil and lagoon sediments was positively correlated with all nitrogen forms, but the correlation was not strong. The clay content in sand bar soil and offshore sediments was negatively correlated with total nitrogen and organic nitrogen and was significantly and positively correlated with all inorganic nitrogen. This is because clay minerals account for a large part of the clay particles, and the fixed ammonium generated by the fixation of nitrogen by clay minerals is the main form of soil inorganic nitrogen, especially in the core soil layer [21]. This also confirms that the more clay and organic matter in the soil, the lower the leaching rate of nitrate [22]. This shows that soil particle size is the main factor affecting soil–sediment inorganic nitrogen but not the main factor affecting total nitrogen and organic nitrogen, which is consistent with previous research results [23,24]. Previous studies have provided little discussion on the correlation between soil particle size and total nitrogen.

Table 7. Correlation analysis between the basic soil–sediment physical and chemical properties and nitrogen forms (69 samples).

	pH	Organic Matter	CEC	Clay	Total Nitrogen	Organic Nitrogen	Inorganic Nitrogen	Ammonium	Nitrate	Nitrite
pH	1.00									
organic matter	−0.23	1.00								
CEC	−0.23	0.828 **	1.00							
clay	−0.295 *	0.560 **	0.670 **	1.00						
total nitrogen	−0.271 *	0.240 *	0.18	0.14	1.00					
Organic nitrogen	−0.267 *	0.23	0.17	0.13	1.000 **	1.00				
inorganic nitrogen	−0.21	0.370 **	0.360 **	0.380 **	0.08	0.06	1.00			
ammonium	0.18	0.514 **	0.388 **	0.360 **	−0.07	−0.08	0.413 **	1.00		
nitrate	−0.284 *	0.253 *	0.274 *	0.301 *	0.11	0.09	0.962 **	0.15	1.00	
nitrite	0.06	0.293 *	0.340 **	0.420 **	0.01	−0.01	0.714 **	0.533 **	0.596 **	1.00

* At the 0.05 level (two-tailed), the correlation is significant. ** At the 0.01 level (two-tailed), the correlation is significant.

4.3. Correlation between Soil–Sediment Organic Matter and Nitrogen

It can be seen from the correlation analysis in Table 7 that, overall, organic matter was positively correlated with soil–sediment total nitrogen and nitrogen forms. In terms of classification (Tables 3–6), the organic matter contents of inland soil, lagoon sediments and sand bar soil were positively correlated with all nitrogen forms and significantly and positively correlated with ammonium, which is similar to the conclusion of He Jun et al. [25] that offshore sediment organic matter was negatively, but not significantly, correlated with all nitrogen species and was significantly and positively correlated with nitrous nitrogen. This is because most of the nitrogen in the soil–sediment surface exists in the organic matter, and the colloidal adsorption by the organic matter can also improve the nitrogen preservation ability in the soil–sediment. The ammonium obtained by the ammonification of organic matter is the substrate source of the nitrification process and affects the activity of nitrifying bacteria in soil. In general, soil organic matter helps to improve the adsorption and retention capacity of ammonium ions and ammonia and reduce the volatilization of soil nitrogen. This result shows that organic matter is the main factor influencing the soil–sediment total nitrogen and organic nitrogen, which is consistent with the research results of Li Hui et al. [26].

4.4. Correlation between the Soil–Sediment pH and Nitrogen

It can be seen from the correlation analysis table (Table 7) that, overall, pH was negatively correlated with soil–sediment total nitrogen, organic nitrogen, total inorganic nitrogen and nitrate, but the correlations were not strong. Ammonium and nitrite showed a weak positive correlation, and the effect of pH on nitrogen forms in sediments was not obvious. The specific pattern (Tables 3–7) showed that inland soil pH and all nitrogen forms were negatively correlated, but not significantly. Lagoon sediment pH was positively correlated with total nitrogen, organic nitrogen, inorganic nitrogen, nitrate and nitrite and negatively correlated with ammonium, but not significantly. The soil pH in the sand bar was weakly, positively correlated with total nitrogen and organic nitrogen and negatively correlated with inorganic nitrogen. Offshore sediment pH was negatively correlated with all nitrogen species, but not significantly. This may be because pH is related to whether nitrogen compounds can undergo redox changes, and in soil with a high pH, clay has a stronger ability to fix nitrogen [21]. The pH of the study area ranges from weakly acidic to weakly alkaline from west to east, the suitable pH range for denitrification is 6.50–7.50 and the suitable pH range for nitrification is 7.50–8.50 [26]. However, in the normal pH range, the change in pH will not have an important effect on the mineralization, nitrification and denitrification of organic nitrogen [27]. This indicates that pH is not the main factor influencing soil–sediment nitrogen.

4.5. Correlation between the Soil–Sediment Cation Exchange Capacity and Nitrogen

From the correlation analysis table (Table 7), it can be observed that the CEC was positively correlated with all nitrogen forms in the soil–sediment samples and had a significant positive correlation with all inorganic nitrogen. The specific performance of different types of environments was as follows (Tables 3–7): inland soil CEC was positively correlated with all nitrogen forms and significantly and positively correlated with ammonium; lagoon sediment CEC was negatively correlated with all nitrogen forms but not significantly correlated with nitrite; sand bar soil CEC was negatively correlated with total nitrogen and organic nitrogen and significantly and positively correlated with all inorganic nitrogen and offshore sediment CEC was negatively correlated with all nitrogen forms but not significantly correlated with nitrous nitrogen, similar to the findings of Zhao Ziwen [28] and Geng Ruonan [29]. Because the higher the soil CEC is, the easier it is to adsorb and fix ammonium ions in the soil, and it is not easy to nitrify; the above analysis showed that the cation exchange CEC exhibited the same pattern as the clay content in the soil–sediment samples. The main factors influencing inorganic nitrogen identified in this work are consistent with previous research results [30].

4.6. Analysis of the Main Factors That Influence the Spatial Variations in Soil–Sediment Nitrogen Forms

To further understand nitrogen in the soil and sediment, the relationship between morphospatial differentiation and environmental factors [31,32] must be studied. Thus, an RDA study on soil–sediment nitrogen content and environmental factors was carried out, and nitrogen content in the soil and sediment was selected as the response variable, and particle size Caly, organic matter OM, pH and cation exchange capacity CEC were selected as explanatory variables. The RDA sorting results showed that the gradient length of the first sorting axis was 1.2, indicating that linear sorting was suitable for further RDA. The results showed that the first 2 RDA axes explained 18.86% and 2.39% of the total variance in the soil–sediment nitrogen content, and 21.2% of the variation in the soil–sediment nitrogen content could be explained by environmental factors, which indicated that environmental factors had a certain influence on the soil–sediment nitrogen content. The explanatory degrees of environmental factors from large to small were 17.2% for organic matter, 2.1% for pH, 1.8% for cation exchange capacity and 0.2% for particle size. The cation exchange capacity was 8.5%, and the particle size was 0.8% (Table 8). This showed that among the four explanatory variables, particle size, organic matter, pH and cation exchange capacity, organic matter had the largest contribution to the nitrogen content in soil–sediment and was the main factor affecting the distribution of nitrogen in soil and sediment in the Shamei Inland Sea.

Table 8. Redundancy analysis results between the soil–sediment nitrogen content and environmental variables.

Environmental Variable	Explanation %	Contribution %	Pseudo-F	p
OM	17.2	80.9	13.9	0.002
pH	2.1	9.9	1.7	0.206
CEC	1.8	8.5	1.5	0.184
Caly	0.2	0.8	0.1	0.88

5. Conclusions

This study was carried out in the Shamei Inland Sea, Boao town, Qionghai City, Hainan Province. Based on samples of the soil and sediments around the Shamei Inland Sea, analyses of the samples and relevant data obtained from the study area, the following conclusions were drawn:

(1) Nitrogen in soils and sediments around the Shamei Inland Sea showed obvious variations among inland areas, lagoon areas, sand bar areas and offshore areas. There is an overall deficiency of total nitrogen in the inland sea area of Shamei; the nitrogen form is mainly organic nitrogen, the organic nitrogen level in the soil is greater than that in the sediment and the distribution of total nitrogen is characterized by inland diffusion to offshore waters. Consequently, the productivity of terrestrial ecosystems in this area is greater than the productivity of aquatic ecosystems. The average contents of nitrate, ammonium and nitrite in the soil around the lagoon were higher than those in the sediment, and they were significantly enriched in the lagoon mouth area, which may be related to the dynamic sedimentation and colloidal condensation in the lake–sea boundary area. The content of nitrate is highest in the estuary area to the northwest of the lagoon, while the contents of ammonium and nitrite are highest in the estuary area to the south of the lagoon, which may be related to the difference in upstream parent rocks and the different types of industrial and agricultural activities.

(2) There is a positive correlation between soil–sediment total nitrogen and the organic matter content, cation exchange capacity and clay content in the Shamei Inland Sea, indicating that changes in basic physical and chemical soil–sediment properties have important effects on nitrogen enrichment, preservation and mineralization. In the lagoon soil–sediment, the organic nitrogen content in the soil is greater than that in the sediment, indicating that the productivity of the terrestrial ecosystem is greater than that of the aquatic system. Additionally, the high proportion of ammonium in the sediment and the high proportion of nitrate in the soil may be related to heavy sedimentary material, poor soil ventilation and weak nitrification.

(3) Changes in the basic physical and chemical properties of soils and sediments in the Shamei Inland Sea have important coupled effects on nitrogen enrichment, preservation and mineralization. Through RDA, it was found that the organic matter content is closely related to the nitrogen content and is the main factor influencing (or controlling) the nitrogen content, with an explanatory degree of 17.2% and a contribution degree of 80.9% in the nitrogen content.

Author Contributions: Data curation, K.Y.; formal analysis, K.Y. and Y.S.; funding acquisition, Y.S., K.Y. and G.F.; investigation, K.F. and Z.W.; methodology, Y.S.; project administration, K.Y. and Y.S.; writing—original draft, K.Y.; writing—review and editing, K.Y., B.L., G.F. and Y.S.; visualization, K.Y.; supervision, G.F. All authors have read and agreed to the published version of the manuscript.

Funding: Comprehensive Survey of Natural Resources in HaiChengWen Coastal Zone, grant number DD20230414; Comprehensive Survey of Natural Resources in Huizhou-Shanwei Coastal Zone, grant number DD20230415.

Institutional Review Board Statement: Not applicable.

Informed Consent Statement: Not applicable.

Data Availability Statement: Not applicable.

Acknowledgments: HaiChengWen Coastal Zone, grant number DD20230414; Huizhou-Shanwei Coastal Zone, grant number DD20230415.

Conflicts of Interest: The authors declare no conflict of interest.

References

1. Paytan, A.; Mclaughlin, K. The Oceanic Phosphorus Cycle. *Chem. Rev.* **2007**, *107*, 563–576. [CrossRef]
2. Cederwall, H. Biological effects of eutrophication in the Baltic Sea, particularly the coastal zone. *Ambio* **1990**, *19*, 109–112.
3. Kroon, F.J.; Kuhnert, P.M.; Henderson, B.L.; Henderson, B.L.; Wilkinson, S.N.; Henderson, A.K.; Abbott, B.; Brodie, J.E.; Turner, R.D.R. River loads of suspended solids, nitrogen, phosphorus and herbicides delivered to the Great Barrier Reef lagoon. *Mar. Pollut. Bull.* **2012**, *65*, 167–181. [CrossRef]
4. Hayn, M.; Howarth, R.; Marino, R.; Ganju, N.; Berg, P.; Foreman, K.H.; Giblin, A.E. Exchange of Nitrogen and Phosphorus Between a Shallow Lagoon and Coastal Waters. *Estuaries Coasts* **2014**, *37* (Suppl. S1), 63–73. [CrossRef]
5. Nazneen, S.; Raju, N.J. Distribution and sources of carbon, nitrogen, phosphorus and biogenic silica in the sediments of Chilika lagoon. *J. Earth Syst. Sci.* **2017**, *126*, 13. [CrossRef]
6. Liu, X.; Ge, C.; Zou, X.; Huang, M. Carbon, Nitrogen geochemical characteristics and their implications on environmental nge in the lagoon sediments of the Dongdao Island of Xisha Islands in South China Sea. *Haiyang Xuebao* **2017**, *39*, 43–54.
7. Chendong, G.; Ying, W. Variability of organic carbon isotope, nitrogenisotope, and c/n in the Wanquan riverestuary, eastern Hainan island, China, and its environmental implications. *Quat. Sci.* **2007**, *5*, 845–852.
8. Meiling, Z. Distribution and Occurrence Characteristics of Nitrogen Forms in Sediments of the Qilihai Lagoon. Ph.D. Thesis, Normal University, Beijing, China, 2018; pp. 23–24.
9. Zhang, X.; Ge, C.D.; Dong, T.T.; Zong, X. Distribution patterns of sedimentary organic matter in Xincun and Li-an lagoons of lingshui county, Hainan island and their source implications. *Quat. Sci.* **2016**, *36*, 78–85.
10. Jia, J.; Gao, J.H.; Liu, Y.F.; Gao, S.; Yang, Y. Environmental changes in Shamei Lagoon, Hainan Island, China: Interactions between natural processes and human activities. *J. Asian Earth Sci.* **2012**, *52*, 158–168. [CrossRef]
11. Li, R.; Liu, S.; Zhang, G.; Ren, J.; Zhang, J. Biogeochemistry of nutrients in an estuary affected by human activities: The Wanquan River estuary, eastern Hainan Island, China. *Cont. Shelf Res.* **2013**, *57*, 18–31. [CrossRef]
12. *HJ634-2012*; Determination of Soil Ammonia Nitrogen, Nitrite Nitrogen, Nitrate Nitrogen. AbeBooks Seller Since: Victoria, BC, Canada, 2009.
13. Li, H.; Tang, Q.; Zhang, H.; Li, T.; Duan, L. Quantitative sampling for grain size analysis by MS2000 laser analyzer. *Mar. Geol. Quat. Geol.* **2020**, *40*, 200–207. [CrossRef]
14. *NY/T 1121.6-2006*; Method for Determination of Soil Organic Matter. National Agricultural Technology Extension and Service Center: Beijing, China, 2006.
15. *NY/T 1377-2007*; Determination of Soil pH. Jiangxi Lyujuren Ecological Environment Co.: Nanchang, China, 2009.
16. *GB 7863-87*; Determination of Cation Exchange Capacity in Forest Soil. Ministry of Forestry of the China: Beijing, China, 1988.
17. Meng, Y.Y.; Wang, S.R.; Jiao, L.X.; Liu, W.B.; Xiao, Y.B.; Zu, W.M. Characteristics of Nitrogen Pollution and the Potential Mineralization in Surface Sediments of Dianchi Lake. *Environ. Sci.* **2015**, *36*, 471–480.
18. Xue, J.; Jiang, X.; Yao, X.; Li, M.; Zhang, L. Dissimilatory nitrate reduction processes at the sediment-water interface in Lake Kuilei. *China Environ. Sci.* **2018**, *38*, 2289–2296.
19. Jinyu, Y.; Jinming, T.; Xianghui, G.; Shuhji, K. Nitrogen cycling processes and its budget in China marginal sea:case studies in the south China sea. *Oceanol. Limnol. Sin.* **2021**, *52*, 314–322.
20. Yu, H. Distribution and Fluxes of Methane and Nitrous Oxide in Different Coastal Water Systems of Eastern Hainan. Ph.D. Thesis, Ocean University of China, Qingdao, China, 2011; pp. 43–44.
21. Jingguo, W. *Biogeochemical Material Cycle and Soil Processes*; China Agricultural University Press: Beijing, China, 2017; pp. 181–255.
22. Bing, H.; Hao, S.; Zhao, W.; Sun, T. A review on nitrogen cycle in wetland soils. *Territ. Nat. Resour. Study* **2015**, *5*, 67–71.
23. Chen, A.; Lei, B.; Liu, H.; Zhai, L.; Wang, H.; Mao, Y.; Zhang, D. Adsorption and desorption of NH-N in the different soil genesis layers in the nearshore vegetable field of Erhai Lake. *J. Agro Environ. Sci.* **2017**, *36*, 345–352.
24. Xinzhong, W.; Guoshun, L.; Zhengyang, Z.; Qinghua, L.; Zhenhai, W. Spatial Distribution of Soil Particle Composition and its Relationship with Soil Nutrients in Tobacco Planting Soils. *Chin. Tob. Sci.* **2011**, *32*, 47–51.

25. Zhang, J.; Liu, Y.; Wen, M.; Zheng, C.; Chai, S.; Huang, L.; Liu, P. Distribution characteristics and ecological risk assessment of nitrogen, phosphorus and heavy metals in sediments of typical inland lakes: A case study of Wuhu Lake in Wuhan. *Geol. Surv. China* **2022**, *9*, 110–118.
26. Li, H.; Lei, P.; Li, X.; Liu, H.; Zhou, B.; Chen, C.; Xin, M.; Li, X.; Kong, W. Distribution characteristics and pollution assessment of nitrogen and phosphorus in sediments from Beidagang Wetland in Tianjin City. *Acta Sci. Circumstantiae* **2021**, *41*, 4086–4096.
27. Jiabing, L.; Dangyu, Z.; Chunshan, W.; Yinghong, W.; Ningyu, Z.; Mengyu, L.; Rongrong, X. Effects of pH on the Key Nitrogen Transformation Processes of the Wetland Sediment in the Min River Estuary. *J. Soil Water Conserv.* **2017**, *31*, 272–278.
28. Ziwen, Z. Responses of Meadow Soil Nitrogen Pool to Degradation and Simulated Warming in Wugong Mountain. Ph.D. Thesis, Jiangxi Agricultural University, Nanchang, China, 2016; pp. 22–23.
29. Ruonan, G. Nitrogen and Phosphorus Forms and Loss Risk in a Typical Artificial Woodland North of Huaibei Plain. Ph.D. Thesis, Hefei University of Technology, Hefei, China, 2017; pp. 43–44.
30. Kong, L.; Yin, S.; Liu, J.; Liang, C. Distribution characteristics of soil cation exchange capacity of saucer-shaped depressions to is-land forests in the Sanjiang Plain. *Sci. Technol. Eng.* **2021**, *21*, 8828–8833.
31. Wang, Z.; Zhao, Q.; Yang, J.; Cui, H.; Cao, W.; Min, W. Factors Influencing Provenance Authentication of Meihe Rice Based on RDA. *J. Jilin Agric. Univ.* **2021**, *43*, 12.
32. Ren, C.; Zhang, W.; Zhong, Z.; Han, X.; Yang, G.; Feng, Y.; Ren, G. Differential responses of soil microbial biomass, diversity, and compositions to altitudinal gradients depend on plant and soil characteristics. *Sci. Total Environ.* **2017**, *610–611*, 750–758. [CrossRef] [PubMed]

MDPI
St. Alban-Anlage 66
4052 Basel
Switzerland
Tel. +41 61 683 77 34
Fax +41 61 302 89 18
www.mdpi.com

Applied Sciences Editorial Office
E-mail: applsci@mdpi.com
www.mdpi.com/journal/applsci